Aquatic Functional Biodiversity

Aquatic Functional Biodiversity
An Ecological and Evolutionary Perspective

Edited by

Andrea Belgrano
Department of Aquatic Resources, Institute of Marine Research,
Swedish University of Agricultural Sciences, Lysekil;
Swedish Institute for the Marine Environment (SIME), Gothenburg,
Sweden

Guy Woodward
Department of Life Sciences, Imperial College London,
Ascot, Berkshire, United Kingdom

Ute Jacob
Institute for Hydrobiology and Fisheries Science,
University of Hamburg, Hamburg, Germany

AMSTERDAM • BOSTON • HEIDELBERG • LONDON
NEW YORK • OXFORD • PARIS • SAN DIEGO
SAN FRANCISCO • SINGAPORE • SYDNEY • TOKYO

Academic Press is an imprint of Elsevier

Academic Press is an imprint of Elsevier
125 London Wall, London EC2Y 5AS, UK
525 B Street, Suite 1800, San Diego, CA 92101-4495, USA
225 Wyman Street, Waltham, MA 02451, USA
The Boulevard, Langford Lane, Kidlington, Oxford OX5 1GB, UK

Copyright © 2015 Elsevier Inc. All rights reserved.

Cover Image, Lower © 2010: created using the software Foodweb3D, which was provided by Rich Williams, J.A. Dunne and N.D. Martinez (Williams, R.J. Network3D Software. Microsoft Research, Cambridge, UK)

No part of this publication may be reproduced or transmitted in any form or by any means, electronic or mechanical, including photocopying, recording, or any information storage and retrieval system, without permission in writing from the publisher. Details on how to seek permission, further information about the Publisher's permissions policies and our arrangements with organizations such as the Copyright Clearance Center and the Copyright Licensing Agency, can be found at our website: www.elsevier.com/permissions.

This book and the individual contributions contained in it are protected under copyright by the Publisher (other than as may be noted herein).

Notices
Knowledge and best practice in this field are constantly changing. As new research and experience broaden our understanding, changes in research methods, professional practices, or medical treatment may become necessary.

Practitioners and researchers must always rely on their own experience and knowledge in evaluating and using any information, methods, compounds, or experiments described herein. In using such information or methods they should be mindful of their own safety and the safety of others, including parties for whom they have a professional responsibility.

To the fullest extent of the law, neither the Publisher nor the authors, contributors, or editors, assume any liability for any injury and/or damage to persons or property as a matter of products liability, negligence or otherwise, or from any use or operation of any methods, products, instructions, or ideas contained in the material herein.

ISBN: 978-0-12-417015-5

British Library Cataloguing-in-Publication Data
A catalogue record for this book is available from the British Library

Library of Congress Cataloging-in-Publication Data
A catalog record for this book is available from the Library of Congress

For information on all Academic Press publications
visit our website at http://store.elsevier.com/

Publisher: Janice Audet
Senior Acquisitions Editor: Kristi A.S. Gomez
Senior Editorial Project Manager: Pat Gonzalez
Production Project Manager: Lucía Pérez
Designer: Matthew Limbert

Typeset by TNQ Books and Journals
www.tnq.co.in

Contents

Contributors   xi
Perspective: Functional Biodiversity during the Anthropocene   xv

Section I
Theoretical Background

1. **From Metabolic Constraints on Individuals to the Dynamics of Ecosystems**
   Samraat Pawar, Anthony I. Dell and Van M. Savage

Introduction	3
Individual Metabolic Rate, Biomechanics, and Fitness	6
The Size-and-Temperature Dependence of Metabolic Rate	6
From Metabolic Rate to Fitness	8
Evolution of Metabolic Rates and Thermal Physiology	10
From Individual Metabolism and Biomechanics to Interactions	11
A Metabolic Theory for Species Interactions	12
Empirical Support	16
From Interactions to Consumer–Resource Dynamics	18
Ecological Consumer–Resource Dynamics	19
Eco-Evolutionary Consumer–Resource Dynamics	22
From Consumer–Resource Pairs to Community and Ecosystem Dynamics	24
Conclusions	26
Abbreviations and Mathematical Symbols	28
Acknowledgments	29
References	29

2. **Ecological Effects of Intraspecific Consumer Biodiversity for Aquatic Communities and Ecosystems**
   Eric P. Palkovacs, David C. Fryxell, Nash E. Turley and David M. Post

Introduction	37
Case Studies	38
Migration and Foraging Trait Divergence in Alewife	41

v

Life History Divergence in the Trinidadian Guppy	42
Divergence due to Predators and Toxic Prey in *Daphnia*	43
Foraging Habitat Divergence in Threespine Stickleback	44
Within-Population Variation in Feeding Behavior in Pale Chub	45
Meta-Analysis	45
Conclusions	48
Acknowledgments	48
References	48

3. **How Does Evolutionary History Alter the Relationship between Biodiversity and Ecosystem Function?**
   David A. Vasseur and Susanna M. Messinger

Introduction	53
Methods	56
Resource Competition Models	57
Model 1: Partially Substitutable Resources	57
Case 2: Essential Resources	60
Model Analysis	62
Reanalysis of Empirical Data	63
Results	64
Discussion	70
Abbreviation	71
Acknowledgments	71
References	71

4. **Effects of Metacommunity Networks on Local Community Structures: From Theoretical Predictions to Empirical Evaluations**
   Ana Inés Borthagaray, Verónica Pinelli, Mauro Berazategui, Lucía Rodríguez-Tricot and Matías Arim

Introduction	75
Four Paradigms	77
Patch Dynamics and Mass Effect	79
Species Sorting	84
Neutral Mechanisms	85
Theory Data	88
Metacommunity Networks	89
Maximum Entropy	99
Acknowledgments	104
References	104

Section II
Across Aquatic Ecosystems

5. **Limited Functional Redundancy and Lack of Resilience in Coral Reefs to Human Stressors**
   Camilo Mora

Introduction	115
Data Quality	116
Pattern of Change	117
Drivers of Change	118
Are Coral Reefs Functionally Redundant?	119
Solutions to Ensure Resilience	121
Are there other Solutions Available?	121
Concluding Remarks	122
References	122

6. **Biodiversity, Ecosystem Functioning, and Services in Fresh Waters: Ecological and Evolutionary Implications of Climate Change**
   Guy Woodward and Daniel M. Perkins

Introduction	127
Climate Change: An Environmental Stressor That Is More Than Just the Sum of Its Parts?	129
Temperature and Metabolism: The Master Variables in Biological Responses to Global Warming	130
Theoretical Frameworks: The Metabolic Theory of Ecology and Beyond	133
Biodiversity—Ecosystem Functioning Relationships: How Many Species Do We Need to Maintain Functioning and Services in a Changing Climate?	136
Are We Measuring the Relevant Drivers and Responses in Biodiversity—Ecosystem Functioning Studies?	139
Traits and Functional Diversity in a Changing Climate: Beyond Body Size	141
Structure and Functioning of Freshwater Food Webs	141
From Averages to Individuals: The Common Currency of Freshwater Ecology	143
Scaling Up: Cross-System Subsidies and Source—Sink Dynamics in Fresh waters	144
Eco-evolutionary Dynamics: Reciprocal Feedbacks between Ecology and Evolution, and Interactions between Biotic and Abiotic Drivers in Multispecies Systems	147
Synergies between Multiple Stressors and the Modulation of Ecological and Evolutionary Responses in a Changing Climate	147
Future Directions and Concluding Remarks	149
References	150

7. **Global Aquatic Ecosystem Services Provided and Impacted by Fisheries: A Macroecological Perspective**
   Jonathan A.D. Fisher, Kenneth T. Frank and Andrea Belgrano

Introduction	157
Macroecological Variables and their Interactions within Aquatic Ecosystems	161
Species Richness	162
Abundance	164
Geographical Distribution	165
Body Size	167
A Central Challenge: Identifying Processes Underlying Macroecological Patterns	168
Physical and Biological Associations with Macroecological Patterns	170
Structural Relationships among Key Variables and Predictions at Multiple Scales	170
Dynamic Macroecological Patterns Driven by Anthropogenic and Natural Forces	171
A Traits-Based Focus on Aquatic Functional Diversity	173
Ecological and Evolutionary Effects of Selective Fisheries on Aquatic Ecosystem Functioning	177
Acknowledgments	180
References	180

8. **Valuing Biodiversity and Ecosystem Services in a Complex Marine Ecosystem**
   Ute Jacob, Tomas Jonsson, Sofia Berg, Thomas Brey, Anna Eklöf, Katja Mintenbeck, Christian Möllmann, Lyne Morissette, Andrea Rau and Owen Petchey

Introduction	189
Materials and Methods	192
Trophic Niche Dimensions and Parameters	192
Consumer Trophic Niche Position	193
Consumer Trophic Uniqueness	193
Consumer Trophic Flexibility (= Trophic Niche Width)	193
Assignment of Ecosystem Services on the Species Level	194
Lough Hyne Data Set	194
Statistical Analysis	195
Results	195
Trophic Flexibility and Trophic Uniqueness of Lough Hyne Consumers	195
Ecosystem Service Provisioning by Species of Lough Hyne	196
Discussion	196
Trophic Niche Dimensions and Parameters	201
Trophic Flexibility and Trophic Uniqueness of Lough Hyne Consumers	202
Distribution of Trophic Flexibility and Trophic Uniqueness	202
Ecosystem Service Provisioning by Species of Lough Hyne	203
Conclusions	203
References	204

Section III
In the Wild: Biodiversity and Ecosystem Service Conservation

9. The Role of Marine Protected Areas in Providing Ecosystem Services
Pierre Leenhardt, Natalie Low, Nicolas Pascal, Fiorenza Micheli and Joachim Claudet

Introduction	211
Introduction to Marine Protected Areas	212
Introduction to Ecosystem Services and the Link to Human Well-Being	213
Marine Protected Area Effects on Individual Ecosystem Services	215
Marine Protected Area Effects on Provisioning Services: The Example of Fisheries	215
Marine Protected Area Effects on Cultural Service: The Example of Recreational Activities	220
Marine Protected Area Effects on Supporting Services: The Example of Coastal Protection	221
Marine Protected Area Effects on Long-Term Ecosystem Function and the Provision of Multiple Services	221
The Role of Biodiversity: Expectations from Functional Diversity and Redundancy	222
Quantifying and Protecting Functional Diversity and Redundancy in Marine Protected Areas	224
Effects of Marine Protected Areas on Functional Diversity	226
Key Directions and Open Questions	229
References	230

10. Freshwater Conservation and Biomonitoring of Structure and Function: Genes to Ecosystems
Clare Gray, Iliana Bista, Simon Creer, Benoit O.L. Demars, Francesco Falciani, Don T. Monteith, Xiaoliang Sun and Guy Woodward

Introduction	241
Current Focus of Aquatic Biomonitoring and Conservation	241
State of the Art in the Science of Biomonitoring: From Species Traits to Community Structure and Ecosystem Functioning	245
Future Advances and New Perspectives—Genes to Ecosystems	251
Novel Molecular and Microbial Approaches	253

The Functional Analysis of Microbes, Metazoans,
   and Macrofaunal Communities   258
Concluding Remarks   261
Acknowledgments   261
References   262

Epilogue: The Robustness of Aquatic Biodiversity Functioning under
   Environmental Change: The Ythan Estuary, Scotland; by David Raffaelli   273
Index   283

Contributors

Matías Arim Departamento de Ecología y Evolución, Facultad de Ciencias and Centro Universitario Regional Este (CURE), Universidad de la República, Montevideo, Uruguay

Andrea Belgrano Department of Aquatic Resources, Institute of Marine Research, Swedish University of Agricultural Sciences, Lysekil, Sweden; Swedish Institute for the Marine Environment (SIME), Göteborg, Sweden

Mauro Berazategui Departamento de Ecología y Evolución, Facultad de Ciencias and Centro Universitario Regional Este (CURE), Universidad de la República, Montevideo, Uruguay

Sofia Berg EnviroPlanning AB, Göteborg, Sweden

Iliana Bista Molecular Ecology and Fisheries Genetics Laboratory, School of Biological Sciences, Environment Centre Wales, Bangor University, Gwynedd, UK

Ana Inés Borthagaray Departamento de Ecología y Evolución, Facultad de Ciencias and Centro Universitario Regional Este (CURE), Universidad de la República, Montevideo, Uruguay

Thomas Brey Alfred Wegener Institute for Polar and Marine Research, Bremerhaven, Germany

Joachim Claudet National Centre for Scientific Research, CRIOBE, CNRS-EPHE, Perpignan, France

Simon Creer Molecular Ecology and Fisheries Genetics Laboratory, School of Biological Sciences, Environment Centre Wales, Bangor University, Gwynedd, UK

Anthony I. Dell National Great Rivers Research and Education Center, Alton, IL, USA

Benoit O.L. Demars The James Hutton Institute, Aberdeen, Scotland, UK

Anna Eklöf Department of Physics, Chemistry and Biology (IFM), Linköping University, Linköping, Sweden

Francesco Falciani Institute of Integrative Biology, University of Liverpool, Liverpool, UK

Jonathan A.D. Fisher Centre for Fisheries Ecosystems Research, Fisheries and Marine Institute of Memorial University of Newfoundland, St. John's, NL, Canada

Charles W. Fowler Biology Department, Seattle University, Seattle, WA, USA

Kenneth T. Frank Department of Fisheries and Oceans, Bedford Institute of Oceanography, Dartmouth, NS, Canada

David C. Fryxell Department of Ecology and Evolutionary Biology, University of California, Santa Cruz, CA, USA

Clare Gray School of Biological and Chemical Sciences, Queen Mary University of London, London, UK; Department of Life Sciences, Imperial College London, Ascot, Berkshire, UK

Ute Jacob Institute for Hydrobiology and Fisheries Science, University of Hamburg, Hamburg, Germany

Tomas Jonsson Population Ecology Unit, Institute for Ecology, Uppsala, Sweden

Pierre Leenhardt CRIOBE, CNRS-EPHE, Perpignan, France

Natalie Low Hopkins Marine Station, Stanford University, Pacific Grove, CA, USA

Susanna M. Messinger Department of Ecology and Evolutionary Biology, Yale University, New Haven, CT, USA

Fiorenza Micheli Hopkins Marine Station, Stanford University, Pacific Grove, CA, USA

Katja Mintenbeck Alfred Wegener Institute for Polar and Marine Research, Bremerhaven, Germany

Christian Möllmann Institute for Hydrobiology and Fisheries Science, University of Hamburg, Hamburg, Germany

Don T. Monteith Centre for Ecology & Hydrology, Lancaster Environment Centre, Lancaster, UK

Camilo Mora Department of Geography, University of Hawaii, Honolulu, HI, USA

Lyne Morissette M-Expertise Marine, Sainte-Luce, Canada

Eric P. Palkovacs Department of Ecology and Evolutionary Biology, University of California, Santa Cruz, CA, USA

Nicolas Pascal Ecole Pratique des Hautes Etudes, CRIOBE, CNRS-EPHE, Perpignan, France

Samraat Pawar Department of Life Sciences, Imperial College London, Ascot, Berkshire, UK

Daniel M. Perkins Department of Life Sciences, Imperial College London, Ascot, Berkshire, UK

Owen Petchey Institute of Evolutionary Biology and Environmental Studies, University of Zurich, Zurich

Verónica Pinelli Departamento de Ecología y Evolución, Facultad de Ciencias and Centro Universitario Regional Este (CURE), Universidad de la República, Montevideo, Uruguay

David M. Post Department of Ecology and Evolutionary Biology, Yale University, New Haven, CT, USA

David Raffaelli Environment Department, University of York, York, UK

Andrea Rau Johann Heinrich von Thünen Institute for Baltic Sea Fisheries, Rostock, Germany

Lucía Rodríguez-Tricot Departamento de Ecología y Evolución, Facultad de Ciencias and Centro Universitario Regional Este (CURE), Universidad de la República, Montevideo, Uruguay

Van M. Savage Department of Biomathematics, David Geffen School of Medicine, University of California, Los Angeles, CA, USA; Department of Ecology and Evolutionary Biology, University of California, Los Angeles, CA, USA; Santa Fe Institute, Santa Fe, NM, USA

Xiaoliang Sun Department of Molecular Systems Biology, University of Vienna, Vienna, Austria

Nash E. Turley Department of Biology, University of Toronto at Mississauga, Mississauga, ON, Canada

David A. Vasseur Department of Ecology and Evolutionary Biology, Yale University, New Haven, CT, USA

Guy Woodward Department of Life Sciences, Imperial College London, Ascot, Berkshire, UK

Perspective: Functional Biodiversity during the Anthropocene

Andrea Belgrano, Ute Jacob, Charles Fowler, and Guy Woodward

We are living through a new temporal epoch, recently described as the Anthropocene (Latour, 2014), in which human activities are increasingly shaping the biota within and among Earth's ecosystems. These anthropogenic forces bring a variety of major consequences, including the significant loss of biodiversity in all its forms, and not just the more familiar measure of species richness (Mora et al., 2011; Cardinale, 2013). Biodiversity, in the broader sense of the term that we use here, includes patterns in the links between species assemblages and their functional organization within a web of interactions under environmental constraints. This shift of focus from the traditional emphasis on the "nodes" to the "links" represents an important philosophical change: it forces us to recognize that multispecies systems are not simply passively mapped onto an environmental template, but that their own internal dynamics influence higher level phenomena. Biodiversity is linked to ecosystem functioning and, by extension, to the socioeconomically valuable services they provide (Millennium Ecosystem Assessment (MA), 2005), yet the strength and form of these relationships are still surprisingly poorly understood. This emphasizes the urgent need to develop strategies that promote sustainable use of ecosystems at local scales, and of the biosphere at a global scale, in order to preserve a "safe operating space for humanity" (Naeem et al., 2012; Perrings et al., 2010; Dobson, 2009; Mace et al., 2015).

If we are to do this, first we need to understand how structure and functioning are connected. Functional diversity (FD) is a key component of biodiversity (Perrings et al., 2010; Mouillot et al., 2014) and provides a direct link between biodiversity and ecosystem processes (Naeem, 2006), yet comparisons of FD across systems has largely ignored perspectives that combine ecological and evolutionary principles and understanding (Perrings et al., 2010). It is imperative that a more comprehensive approach is developed for effectively conserving species and their functional roles that considers both ecological and evolutionary principles (Stouffer et al., 2012).

Macroecological patterns in the FD can be manifested across a wide range of taxa and systems, as well as across spatial and temporal scales, and organizational levels. This is often achieved by focusing on the explicit links between biodiversity and ecosystem functioning in experimental manipulations, a field that has gained huge traction in the past couple of decades (Reiss et al., 2009). The (usually positive) relationship between species richness and FD has profound implications for conservation and management (especially of our use of natural resources) and is also increasingly being linked to food web structure and ecosystem services. Combining macroecological studies of system properties with detailed analyses of community functional diversity patterns and food web structure (Petchey and Gaston, 2007) could provide one means of understanding which ecosystem services (Dobson, 2009) are particularly important for sustaining overall healthy environmental status, and also which might be most vulnerable to perturbations. This is only one of the many challenges that we need to resolve and understand (Figure 1), if we are

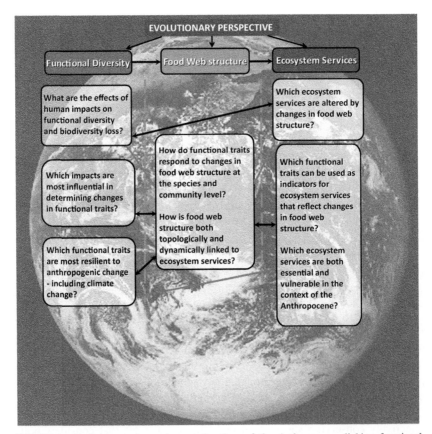

FIGURE 1 Challenges and questions at the time of the Anthropocene: linking functional diversity, food web structure, and ecosystem services.

to move toward a functionally based framework that can include the ecological and evolutionary roles that species (including humans) play within the local ecosystems on a global scale.

The biodiversity of ecosystems worldwide is undergoing major change, via human-induced extinctions set against gains via intentional and accidental introductions, with the balance increasingly on the red side of the ledger. These combined effects are altering the structure and function of ecosystems on a global scale, and we are now experiencing what has been widely described as the 6th Great Extinction. An especially challenging task is to understand how individual species combine to deliver ecosystem services in a changing world, yet a rigorous, systematic methodology for doing so still eludes us. This is, at least partially, because the dynamics of ecosystem services remain poorly characterized at local-to-regional scales, and the role of species in contributing to services is not fully understood. A priori, we know that functional traits, food web structure, and ecosystem services are interlinked and need to be considered in conjunction with the recognition that ecosystems change and require adaptive governance to ensure their long-term sustainability (Folke et al., 2005; Folke, 2007; Dietz et al., 2003).

Recently there has been an explosion of interest in employing network theory to disentangle and explore the relationships between biodiversity and ecosystem functioning (Reiss et al., 2009; Kéfi et al., 2015).

This volume *Aquatic Functional Biodiversity: An Ecological and Evolutionary Approach* is an attempt by some of the most prominent investigators in the field to provide a more general conceptual framework that can include new ecological and/or evolutionary approaches to the understanding of functional diversity. This is accomplished in a way that holds the promise of leading to effective and realistic conservation, especially in doing our best at ensuring the sustainability of ecosystem services in aquatic systems. The collection of chapters in this book represents a substantive contribution to: (1) defining common ground in terms of terminology and conceptual issues, (2) connecting conceptual frameworks from the ecological and/or evolutionary sciences with those from classical biodiversity theory, to make progress toward better practical application, and (3) providing examples of how biodiversity and ecosystem services might be conserved more effectively in the real world.

TERMINOLOGY AND CONCEPTUAL ISSUES IN ECOLOGICAL AND EVOLUTIONARY PERSPECTIVES

Understanding and predicting interactions between ecological and evolutionary processes are extremely challenging tasks. Pawar and colleagues (Chapter 1) provide a summary of recent theoretical and empirical advances for developing a mechanistic understanding of trophic interactions, and identify key methods and challenges for understanding and predicting the eco-evolutionary dynamics of aquatic ecosystems. These authors introduce a theoretical approach to understanding how the metabolic and biomechanical bases of trophic interactions can enhance general predictions regarding the eco-evolutionary dynamics and functioning of aquatic ecosystems. Their ideas will be applicable to other types of ecological interactions that involve metabolism and biomechanics (e.g., pollination, parasitism, and competitive interactions). Numerous relationships in which biodiversity contributes positively to ecosystem function have been identified across a wide variety of ecological communities in aquatic and terrestrial ecosystems. Even though these relationships are acknowledged in multiple ways, the underlying mechanisms are poorly understood and are often questioned. In their chapter, Vasseur and Messinger (Chapter 3) introduce and use two models that depict autotrophic and heterotrophic competitors. Their work includes a case study involving a series of experiments using in silico biodiversity and ecosystem function. They show that the "ancestry" of species, which is defined in terms of whether or not they have coevolved in the presence of competitors or in monocultures, is an important determinant in the extent to which transgressive overyielding can occur. Their research has important implications for the interpretation of previous meta-analyses, and may provide insight of importance to the management of human influence on aquatic communities and their recovery from abnormal human influence.

Borthagaray and coworkers (Chapter 4) present the state of the art of effects of metacommunity networks on local community structures; their chapter covers theoretical predictions and empirical evaluations. By using metapopulation models they point out the wide range of patterns predicted by different mechanisms. They stress that relative effects of dispersal on dominant and subordinate species determine the weakening or strengthening of patch dynamics important to fully understanding community structure.

CONCEPTUAL FRAMEWORKS IN ECOLOGICAL AND EVOLUTIONARY SCIENCES

In Chapter 6, Woodward and Perkins focus on the impacts and consequences of different drivers of change on freshwater ecosystems, which are particularly vulnerable to environmental stressors in general, and climate change in particular. A review of the current state of these relationships is presented—as a "jigsaw puzzle of our understanding of ecological and evolutionary responses to climate change." Freshwater ecology has a long history and is arguably much better understood than that of many marine systems: many of the biotic and abiotic constraints (i.e., chemical or physical properties) on species coexistence and food webs are now relatively well established. Although knowledge about how individuals and ecosystem processes are likely to respond to temperature change has improved over the years, the understanding of the community-level responses to warming, however, is still in its infancy. Woodward and Perkins conclude that obvious current gaps in freshwater ecology are increasingly being filled by a wide range of established and novel techniques to enable a better understanding of how freshwater ecosystems function, how they are valued, and how best to manage them more sustainably in the future. The monitoring of freshwater ecosystems is crucial for assessing their overall health and particularly for maintaining the supply of ecosystem services. However, the biomonitoring and conservation of fresh waters have failed historically to incorporate a fully ecological and evolutionary perspective. Clearly, the predictive capacity and outcomes of current biomonitoring in freshwater ecosystems will therefore be limited in their ability to adapt in the face of rapid and global habitat modification and change, especially as the reference conditions we used to gauge the strength of impacts shift away from their increasingly "obsolete" baselines as we move deeper into the Anthropocene. Gray and coworkers (Chapter 10) outline a list of limitations in the current state of biomonitoring, and suggest how these problems might be overcome to develop a more holistic ecological and evolutionary approach that could underpin new and more effective operational frameworks.

BIODIVERSITY AND ECOSYSTEM SERVICE CONSERVATION

Palkovacs and coworkers (Chapter 2) emphasize consumer diversity and the related intra- and interspecific impacts on aquatic communities and ecosystems. They provide a synthesis of the state of the art of work focused on the effects of biodiversity among consumer species in regard to the effects of removing species, compared to replacing species (biodiversity loss versus gain). The results of their meta-analysis reveal that, while the effects species have in their influence on communities are stronger on average, the more indirect elements of intraspecific effects can be important, suggesting that biodiversity within consumer species has important consequences for the ecology of aquatic systems. Biodiversity loss poses a global threat to ecosystem structure and function and subsequently a threat to the goods and services they provide (Cardinale et al., 2012); such losses were recently ranked among the dominant drivers of ecosystem change and the loss of ecosystem services (Hooper et al., 2012; Reich et al., 2012). We are now faced with the identification of the components of biodiversity, and the specification of which species in a system are most responsible for providing ecosystem services. Then we need to determine how vulnerable those species are. Jacob and coworkers (Chapter 8) introduce a new approach to the study of functional diversity and trophic structure within communities in relation to providing ecosystem services. This approach complements previous topological and dynamical analyses of community robustness and may be used to predict the response of food webs to the loss of species and resulting degradation of ecosystem services.

The chapter by Fisher and coworkers (Chapter 7) focuses on Integrated Ecosystem Assessments (IEAs) and the need to include full consideration of the ecological and evolutionary effects of selective fisheries. They propose that only by integration of a macroecological and evolutionary perspective IEAs can be informative for Management Strategy Evaluations that try to develop sustainable management solutions that link ecological outcomes, ecosystem services, and socioeconomic considerations and ultimately lead to the achievement of a healthy ecosystem-based management of the world oceans.

Leenhardt et al. (Chapter 9) provide an overview on how human impacts can affect the ecosystem functioning of marine ecosystems and reduce the associated provisioning of ecosystem services and subsequently the important role marine protected areas have in the conservation and/or restoration of marine biodiversity and its derived ecosystem goods and service. In their chapter they identify the relationships between the effects of MPAs on ecosystem functioning and service provision and they identify knowledge gaps on which future research efforts could or should focus. They conclude in proposing the importance of the quantification and monitoring of species functional traits distributions as promising approaches for assessing the effects of MPAs on ecosystem functioning and services.

Mora (Chapter 5) provide an overview on the world's coral reef in relation to functional redundancy, human stressor, and include solutions and conservation actions that promotes resilience.

There is an urgent need to conserve (or restore) functional traits across systems, as highlighted by the growing influence of overfishing (Bascompte et al., 2005; Worm et al., 2009), habitat loss, chemical pollution, and climate change in aquatic systems. The various challenges before us are extreme and complex and include the need to generate trust in ecological knowledge and means for transferring it to policy- and decision-makers complete with processes, tools, and information of use to managers and stakeholders. Their need is extreme as they find themselves engaged in the conservation of biodiversity on planet Earth at the time of the Anthropocene (Steffen et al., 2015; Latour, 2014, 2013; Thomas, 2013; Cardinale, 2013).

REFERENCES

Bascompte, J., Melian, C.J., Sala, E., 2005. Interaction strength combinations and the overfishing of a marine food web. PNAS 102, 5443–5447.

Cardinale, B.J., Duffy, J.E., Gonzalez, A., Hooper, D.U., Perrings, C., Venail, P., et al., 2012. Biodiversity loss and its impact on humanity. Nature 486, 59–67.

Cardinale, B.J., 2013. Towards a general theory of biodiversity for the Anthropocene. ELEMENTA Sci. Anthropocene 1–5.

Dietz, T., Ostrom, E., Stern, P.C., 2003. The struggle to govern the commons. Science 302, 1907–1912.

Dobson, A., 2009. Food-web structure and ecosystem services: insights from the Serengeti. Philos. Trans. R. Soc. B 364, 1665–1682.

Folke, C., et al., 2005. Adaptive governance of social-ecological systems. Annu. Rev. Environ. Resour. 30, 441–473.

Folke, C., 2007. Social-ecological systems and adaptive governance of the commons. Ecol. Econ. 22, 14–15.

Hooper, D.U., Adair, E.C., Cardinale, B.J., Byrnes, J.E.K., Hungate, B.A., Matulich, K.L., 2012. A global synthesis reveals biodiversity loss as a major driver of ecosystem change. Nature 486, 105–108.

Kéfi, S., Berlow, E.L., Wieters, E.A., Joppa, L.N., Wood, S.A., Brose, U., Navarrete, S.A., 2015. Network structure beyond food webs: mapping non-trophic and trophic interactions on Chilean rocky shores. Ecology 96 (1), 291–303.

Latour, B., 2014. Agency at the time of the Anthropocene. New Lit. Hist. 45, 1–18.

Latour, B., 2013. An Inquiry into Modes of Existence. Harvard University Press, Cambridge, Mass., USA, 520 pp.

Mace, G.M., Halls, R.S., Cryle, P., Harlow, J., Clarke, S.J., 2015. Towards a risk register for natural capital. Journal of Applied Ecology 52, 641–653.

Millenieum Ecosystem Assessment, 2005. Ecosystems and Human Well-being: General Synthesis. Island Press, Washington, DC.

Mora, C., Tittensor, D.P., Adl, S., Simpson, A.G.B., Worm, B., 2011. How many species are there on Earth and in the Ocean? PLoS Biol. 9 (8), 1–8.

Mouillot, D., Villéger, S., Parravicini, V., Kulbicki, M., Arias-González, J.E., Bender, M., et al., 2014. Functional over-redundancy and high functional vulnerability in global fish faunas on tropical reefs. PNAS 111, 13757–13762.

Naeem, S., et al., 2012. The functions of biological diversity in an age of extinction. Science 336, 1401–1406.

Naeem, S., 2006. Expanding scales in biodiversity-based research: challenges and solutions for marine systems. Mar. Ecol. Prog. Ser. 311, 273–283.

Perrings, C., et al., 2010. Ecosystem services for 2020. Science 330, 323–324.

Petchey, O.L., Gaston, K.J., 2007. Dendrograms and measuring functional diversity. Oikos 116 (8), 1422–1426.

Reich, P.B., Tilman, D., Isbell, F., Mueller, K., Hobbie, S.E., Flynn, D., Eisenhauer, N., 2012. Impacts of biodiversity loss escalate through time as redundancy fades. Science 336, 589–592.

Reiss, J., Bridle, J.R., Montoya, J.M., Woodward, G., 2009. Emerging horizons in biodiversity and ecosystem functioning research. Trends Ecol. Evol. 24, 505–514.

Steffen, W., et al., 2015. Planetary boundaries: guiding human development on a changing planet. Science 347 (6223).

Stouffer, D., et al., 2012. Evolutionary conservation of species roles in food webs. Science 335, 1489–1492.

Thomas, C.D., 2013. The Anthropocene could raise biological diversity. Nature 502 (7469), 7.

Worm, B., et al., 2009. Rebuilding global fisheries. Science 325, 578–585.

Section I

Theoretical Background

Chapter 1

From Metabolic Constraints on Individuals to the Dynamics of Ecosystems

Samraat Pawar[1], Anthony I. Dell[2] and Van M. Savage[3,4,5]
[1]*Department of Life Sciences, Imperial College London, Ascot, Berkshire, UK;*
[2]*National Great Rivers Research and Education Center, Alton, IL, USA;* [3]*Department of Biomathematics, David Geffen School of Medicine, University of California, Los Angeles, CA, USA;* [4]*Department of Ecology and Evolutionary Biology, University of California, Los Angeles, CA, USA;* [5]*Santa Fe Institute, Santa Fe, NM, USA*

INTRODUCTION

Abiotic factors, such as temperature or the dimensionality of space within which organisms live, move, and search for food, directly impact ecological systems at the level of metabolic rate (rate of energy use) of individual organisms. Individual metabolic rate sets the "pace of life" for populations through generation time and maximal growth rate, r_{max} (Brown et al., 2004; Savage et al., 2004), which also scales up to influence coupled ecological and evolutionary dynamics. Therefore, understanding how environmental factors constrain individual metabolic rate, and how these individual-level constraints influence population dynamics of the whole interacting community, is key for understanding ecosystems (Figure 1). Indeed, there is now increasing consensus that individual physiology is fundamental for predicting how global climate change affects the eco-evolutionary dynamics of ecosystems (Manila et al., 1990; Allen et al., 2005; Lavergne et al., 2010; Yvon-Durocher et al., 2011; Dell et al., 2011; Thuiller et al., 2013). Furthermore, understanding how these dynamics differ between aquatic and terrestrial ecosystems is an interesting and important problem (Cohen and Fenchel, 1994; Shurin et al., 2006).

The last two decades have heralded a golden age for research on physiological ecology. This new surge of research has led to the publication of key conceptual syntheses that on one front have advanced thermal biology and adaptation (Huey and Berrigan, 2001; Angilletta, 2009; Kingsolver, 2009; Dell et al., 2011; Schulte et al., 2011; Pörtner et al., 2012), and on another have

4 SECTION | I Theoretical Background

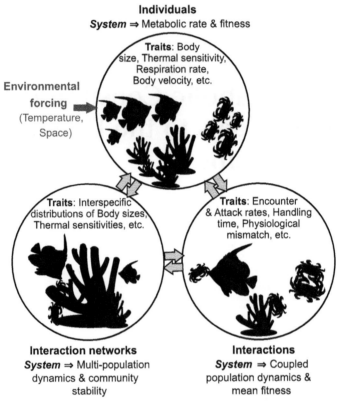

FIGURE 1 The study of coupled ecological and evolutionary dynamics is inherently hierarchical, each level (individuals, interactions, and interaction networks) comprising a system with distinct measurable dynamics. Note that although we deal mostly with consumer-resource (or trophic) interactions in this paper, we use the more general term "interaction" because many of the principles we discuss also apply to, or involve, other types of interactions such as intraspecific interactions. The bidirectional arrows connecting levels indicate that ecological (e.g., changes in abundance) and evolutionary (e.g., changes in trait distributions) feedback from one level can influence the system structure and dynamics in another. We include population dynamics and mean fitness as systemic properties of interactions, because no population grows or evolves in isolation from other populations in nature. Also note that community stability embodies the coupled dynamics of multiple populations, and that community-level traits, under our definition, consist of distributions of individual-level or interaction-level traits across species.

advanced the connections between metabolic traits and body size (West et al., 1997; Kooijman, 2000; Gillooly et al., 2001, 2002; Belgrano et al., 2002; Brown et al., 2004; Savage et al., 2004; West and Brown, 2005; Amarasekare and Savage, 2012; Pawar et al., 2012). All these research has been built up on seminal studies on thermal biology and metabolic scaling that culminated in several books (Kleiber, 1961; Johnson et al., 1974; Damuth, 1981; Schmidt-Nielsen, 1984; Peters, 1986). These were followed by more than a decade with little work or progress toward understanding the metabolic basis of species interactions. One exception to this period of relative dormancy was a paper by Yodzis and Innes (1992) that used allometric relationships between body size and life history, as well as interaction parameters, to develop a theory for metabolically driven consumer–resource (trophic) dynamics. This paper was ahead of its time, and little subsequent work was done until about 10 years ago, when a rapidly growing number of studies focused on the metabolic and biomechanical basis of consumer–resource interactions (Vasseur and McCann, 2005; McGill and Mittelbach, 2006; Weitz and Levin, 2006; Vucic-Pestic et al., 2010; O'Connor et al., 2011; DeLong and Vasseur, 2012; Pawar et al., 2012; Rall et al., 2012; Kalinkat et al., 2013; Dell et al., 2014) as well as community-level (food-web) dynamics (Jonsson and Ebenman, 1998; Emmerson and Raffaelli, 2004; Loeuille and Loreau, 2005; Brose et al., 2006b; Otto et al., 2007; Cohen, 2008; O'Connor et al., 2009; Petchey et al., 2010; Stegen et al., 2012a; Tang et al., 2014).

A logical and important outcome has been the beginning of the development of a unified "metabolic theory of ecology" (MTE) that combines both size and temperature effects (Brown et al., 2004). Such a theoretical unification of two fundamental aspects of individual-level metabolic constraints is necessary for understanding the effects of thermal physiology on size evolution, and in turn the effect of body size on thermal adaptation in eco-evolutionary dynamics. In this chapter, we draw from this new and exciting body of research to highlight recent theoretical and empirical advances, and describe important challenges to the inherently systems-oriented goal of scaling up individual physiology to whole communities (Figure 1), with a particular focus on aquatic ecosystems. We will consider each of the three levels—individuals, trophic interactions, and community/ecosystem—paying close attention to the integration between these levels and the coupling of ecological and evolutionary dynamics within them. Most of the interactions we consider will be of the consumer–resource (or trophic) variety. This focus is partly because trophic interactions dominate energy flows in ecosystems, and partly because theory and data on the metabolic basis of other interactions such as competition and mutualism are as yet rare.

Throughout the chapter, we refer to any empirical or theoretical approach that explains higher levels by going at least one level beneath as "mechanistic"—for instance, explaining ecosystems through models of interaction rates, interactions through models of individual rates, and individual rates through models of within-individual thermal and vascular biology as well as cellular metabolism (West and Brown, 2005) (Figure 1). Also, mechanistic approaches in ecology are often inherently "trait-based," because traits such as body size and rate of response to temperature change set limits both on the baseline physiological rates of individuals (such as respiration or photosynthesis rate), and on how these rates respond to a changing physical environment (Gibert et al., 2015; Enquist et al., 2015). A consideration of traits is important in any eco-evolutionary study, because traits that govern metabolic rate also affect fitness and are thus prime targets of evolution by selection. For example, because body size is a major determinant of differences in r_{max} (Brown et al., 2004; Savage et al., 2004) across species, the distribution of body sizes within a species' population may change due to natural selection to achieve higher or lower population growth rates (fitness) (Brown et al., 1993; Chown and Gaston, 1997; Allen et al., 2006). Finally, note that we use the terms "physiology" and "metabolism" interchangeably throughout this chapter.

INDIVIDUAL METABOLIC RATE, BIOMECHANICS, AND FITNESS

The Size-and-Temperature Dependence of Metabolic Rate

A model capturing the dominant factors that determine whole-individual metabolic rate P (J·s^{-1}) is

$$P = P_0 m^b e^{-\frac{E}{kT}} l(T) \tag{1}$$

where P_0 is a taxon-and-metabolic-state-dependent normalization constant (J·(s·kg^b)$^{-1}$), m is body mass (kg), b is a scaling exponent (dimensionless), E is average activation energy (eV) of rate-limiting steps in underlying biochemical reactions (1 eV = 96.49 kJ·mol^{-1}), k is the Boltzmann constant (8.62 × 10^{-5} eV·K^{-1}), T is temperature (in kelvin), and $l(T)$ is a function that captures the decrease in metabolic rates at higher-than-optimal temperatures (beyond the physiological temperature range (PTR); see Figure 2). The Boltzmann–Arrhenius factor exp($-E/kT$) in Eqn (1) can also be expressed in terms of the familiar Q$_{10}$ coefficient favored by many experimentalists (Gillooly et al., 2001).

The size-scaling component of metabolic rate in Eqn (1), when measured across species, is allometric with $b \sim 0.75$ for multicellular eukaryotes (Peters, 1986; Gillooly et al., 2001; Brown et al., 2004; Nagy 2005), but has been shown to vary across the major domains of life. In particular, according to a meta-analysis of DeLong et al. (2010), it may be steeper ($b > 0.75$ or even $b > 1$ (superlinear)) in unicellular protists and prokaryotes. This is important to consider because protists and prokaryotes form an important part of most aquatic food webs. This steeper scaling of metabolic rate in protists may also be linked to the superlinear scaling of consumption rate recently reported by Pawar et al. (2012) in three-dimensional environments such as pelagic zones (also see Giacomini et al., 2013; Pawar et al., 2013; and section 'From individual metabolism and biomechanics to interactions' in this chapter). Assuming that the decrease in metabolic rate at higher temperatures is mainly due to changes in the kinetics of rate-limiting enzymes, we can use the Johnson and Lewin (1946) model for the $l(T)$ component of Eqn (1):

$$l(T) = \frac{1}{1 + e^{-\frac{1}{kT}\left(E_D - \left(\frac{E_D}{T_{pk}} + k \ln\left(\frac{E}{E_D - E}\right)\right)T\right)}} \tag{2}$$

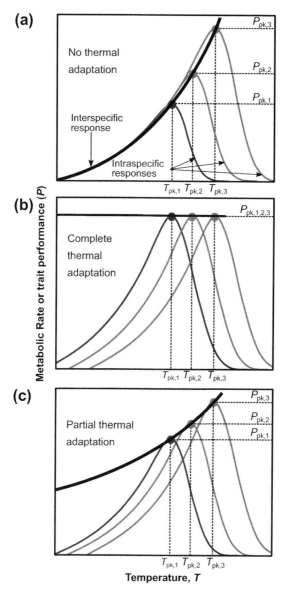

FIGURE 2 The effect of temperature on individual physiology is unimodal and strongly constrained by the thermodynamics of biochemical reactions (here modeled by the Johnson–Lewin model; Eqn (2)). The three panels show different feasible scenarios of adaptation or acclimation of thermal responses: (a) In a "hotter is better" scenario, biochemical constraints dominate, and adaptation (or acclimation) to a different environmental temperature can only occur by moving peak performance of an *intraspecific* thermal response along a universal, across-species (*interspecific*) exponential curve representing Boltzmann–Arrhenius reaction kinetics (numerator of Eqn (2)). The result is that the peak rates (P_{pk}s) of different species adapted to their respective thermal environment themselves fall on a single "interspecific" curve; (b) In a "hotter is not better" scenario, complete adaptation can take place by equalization of peak rates across environments; and (c) An intermediate scenario of partial biochemical adaptation. Note that whole-organism metabolic rate also scales with body mass (Eqn (1))—therefore, thermal adaptation can in principle be achieved by a change in body size (which would shift the intercept of these unimodal curves). We define the thermal zone below T_{pk} to be the "physiological temperature range" (PTR) within which organisms typically operate.

where T_{pk} is the temperature at which metabolic rate reaches its peak value, and E_D can be interpreted as the average energy constant at which proteins denature, therefore controlling the steepness of decline beyond T_{pk} (Johnson et al., 1974; Hochachka and Somero, 1984; Ratkowsky et al., 2005; Corkrey et al., 2012). As such, Eqn (2) is the simplest among a class of thermodynamic models, and can be replaced by more elaborate equations that account for deviations from standard enzyme kinetics at physiologically stressful low and high temperatures, at the cost of additional parameters (Sharpe and DeMichele, 1977; Farquhar et al., 1980; Schoolfield et al., 1981; Ratkowsky et al., 2005; Corkrey et al., 2014). Figure 2 illustrates the thermal response component (Eqn (2)) of the full metabolic rate model (Eqn (1)).

Decreases in metabolic rates at temperatures beyond T_{pk} may be attributable to mechanisms and factors that are additional to protein denaturation, including behavioral thermoregulation—and in the case of aquatic environments, the nonlinear decrease in solubility of gases (O_2 for respiration or CO_2 for photosynthesis) in water that occurs with increasing temperature within biologically relevant temperature ranges (Pörtner and Knust, 2007; Schulte et al., 2011). In such cases, the $l(T)$ model in Eqn (2) may be replaced by a phenomenological one that affords more flexibility to the shape of the decline (Martin and Huey, 2008; Angilletta, 2009; Amarasekare and Savage, 2012).

In all this, it is important to note that most organisms live within their PTR—a range of ambient temperatures with a maximum that lies somewhat below the temperature for peak performance (T_{pk}) (Savage et al., 2004; Deutsch et al., 2008; Martin and Huey, 2008; Huey and Kingsolver, 2011). Thus, understanding thermal responses within the PTR is particularly important, because the increase in environmental temperatures of up to $\sim 4\,°C$ projected to take place by the end of this century (IPCC, 2007) should impact most organisms by shifting the operational temperatures up within the PTR. Therefore, in the rest of this chapter we will mainly focus on the PTR, which can be modeled by

$$P_{PTR} = P_0 m^b e^{-\frac{E}{kT}} \tag{3}$$

where we have ignored the $l(T)$ component of Eqn (1).

From Metabolic Rate to Fitness

The dependence of individual metabolic rate on size and temperature means that a fundamental measure for fitness—i.e., the population's average intrinsic growth rate or the Malthusian parameter (r_{max})—is also dependent upon body size and temperature (Savage et al., 2004; Deutsch et al., 2008; Amarasekare and Savage, 2012). There are multiple ways to link metabolic rate to r_{max}

(Kooijman, 2000; Savage et al., 2004). All of them, however, must include the principle that r_{max} is the result of how, on average, individuals in a population allocate their metabolic rate between the mass-specific power needed for individual growth and maintenance and that needed to produce new individuals. In aquatic ecosystems, most primary producers and a large proportion of primary and secondary consumers have simple life cycles—this includes most unicellular protists and all prokaryotes. In such organisms, the metabolic dependence of r_{max} is simply proportional to the difference between the population-averaged rate of mass-specific energy sequestration and energy loss to respiration, without any strong dependence upon life-stage specific allocation. In particular, for phytoplankton (unicellular algae and cyanobacteria), which account for the majority of primary production in most aquatic ecosystems, r_{max} can be expressed as

$r_{max} \propto$ Net production = Gross photosynthesis rate − Respiration rate

where the rates are mass-specific (Yvon-Durocher et al., 2012; Yvon-Durocher and Allen, 2012). For organisms with overlapping generations, Savage et al. (2004) have derived the prediction that the general relationship between size, temperature, and r_{max} should be

$$r_{max} = r_0 m^{b-1} e^{-\frac{E}{kT}} l(T) \qquad (4)$$

where r_0 is a normalization constant (which includes P_0), and the original model derived by Savage et al. (2004) has been modified to include unimodality in thermal response through $l(T)$ (e.g., in the form of Eqn (2)). Savage et al. (2004) also show that this model is well supported across a wide range of species including protists and algae. In particular, when corrected for body mass r_{max} measured across species close to the respective T_{pk} (Figure 2), it is well fitted by a Boltzmann–Arrhenius response, implicitly supporting a "hotter is better" scenario (see below and Figure 2). However, within-species (see Figure 2) tests for the temperature dependence of Eqn (4) that include data across a wide range of species are still lacking.

As such, the simplicity of Savage et al.'s (2004) model (Eqn (4)) partly stems from the assumption that the thermal dependencies of energy sequestration, use, and allocation are the same—that is, they follow the Boltzmann–Arrhenius equation with identical activation energies. However, while scaling up individual metabolic rates to r_{max} in autotrophs, it is important to note that the temperature dependences of photosynthesis and respiration are different due to the fundamental differences in their biochemical pathways and rate-limiting enzymes (Allen et al., 2005; López-Urrutia et al., 2006; Yvon-Durocher et al., 2012; Yvon-Durocher and Allen, 2012). Specifically, at light-saturating conditions, the temperature dependence of photosynthetic rate is determined mainly by the net difference between the Rubisco enzyme-catalyzed carboxylation (CO_2 fixation) and photorespiration (O_2 fixation). It has been argued that because photorespiration tends to increase faster with temperature than carboxylation, it results in a relatively low overall activation energy of light-saturated photosynthesis in the region of $E \sim 0.35$ eV (Coleman and Colman, 1980; Bernacchi et al., 2001; Allen et al., 2005; Walker et al., 2013). However, clear evidence to support this is still missing. In contrast, the activation energy of respiration due to the temperature dependence of ATP synthesis in respiratory complexes is high (between 0.6 and 1.0 eV) (Allen et al., 2005; Beke-Somfai et al., 2010). As a result, over a temperature range of 0–30 °C, respiration may increase by up to 67-fold, whereas photosynthesis may increase by only fourfold. Hence, warming may change the balance of photosynthetic carbon sequestration and heterotrophic carbon production globally (Allen et al., 2005; López-Urrutia et al., 2006; Yvon-Durocher et al., 2012; Yvon-Durocher and Allen, 2012), accentuated by the increase in heterotrophy in mixotrophic organisms faced with metabolic deficits (Chen et al., 2012; Wilken et al., 2013).

Along with temperature, the effect of body size in aquatic ecosystems is also important. For example, at the trophic level of primary producers, the small size of phytoplankton means that they have higher mass-specific metabolic rates (Eqn (4)) and therefore higher biomass turnover rates than higher plants (Allen et al., 2005; Schramski et al., 2015). Indeed this is the primary reason why oceanic phytoplankton contribute $\sim 50\%$ of global net primary productivity, although they comprise only $\sim 0.2\%$ of global plant biomass (López-Urrutia et al., 2006).

Evolution of Metabolic Rates and Thermal Physiology

Using r_{max} as a measure of fitness, a few studies have begun to investigate whether biochemical constraints on thermal responses of individual metabolism can be overcome by adaptation (Frazier et al., 2006; Knies et al., 2009; Angilletta et al., 2010), or whether hotter is typically better (Figure 2). These preliminary studies suggest that evolution cannot fully overcome biochemical constraints. Indeed, at least over short timescales, thermal acclimation is expected to follow biochemical constraints, leading to a hotter-is-better pattern. However, evidence for a universal hotter-is-better pattern is far from conclusive. In particular, our recent work (Pawar et al., in press) shows that compared with terrestrial organisms, aquatic organisms may in fact show a stronger pattern of thermal adaptation (Figure 2(b) and (c)), likely because of differences in thermal environment between terrestrial and aquatic environments. For example, aquatic habitats typically provide fewer opportunities for thermal refuges and have smaller temporal fluctuations in temperature than terrestrial habitats, making it less possible for individuals to control their temperature by moving to different regions in space or modifying their period of activity. This emphasizes the importance of the physical environment in the physiological responses of individuals. The question of whether organisms can acclimate and adapt to new thermal environments is particularly important due to global climate change, especially because ongoing shifts in species' ranges and phenologies are resulting in physiologically mismatched species coming into contact (Parmesan and Yohe, 2003; Dell et al., 2014). This issue will be considered in more detail in the following section.

FROM INDIVIDUAL METABOLISM AND BIOMECHANICS TO INTERACTIONS

At trophic levels above primary producers (autotrophs), consumption rate and fitness are determined by metabolic and biomechanical constraints on traits that govern the interactions between consumer and resource individuals (Figure 3) (Domenici, 2001; McGill and Mittelbach, 2006; Pawar et al., 2012). Indeed, even in the case of the autotrophs discussed above, "top-down" control from primary consumers often plays a major role in determining fitness. Because metabolic rate creates energy demands and also provides the power for consumers to search, attack, and ingest resources to satisfy these energy demands (Figure 3), understanding the metabolic basis of components of consumer–resource interactions is crucial (Peters, 1986; Blake and Domenici, 2000; Domenici, 2001; Pawar et al., 2012; Dell et al., 2014), and is the first step toward a truly general theory for organismal fitness (r_{max}). Decades after Holling's influential work on the "components of predation" (Holling, 1959a,b, 1966), a mechanistic, empirically grounded

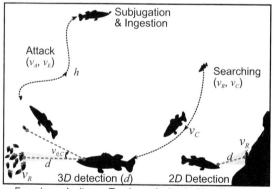

v: Foraging velocity; v_θ: Turning velocity; h: Handling time;
d: Detection distance; v_A: Attack velocity; v_E: Escape velocity

FIGURE 3 The components of species interactions, most of which are driven by individual level metabolic rate. Note that foraging velocity refers to the velocity before detection both of the consumer and of the resources moving in the landscape (active foraging), which together determine relative velocity. For fish eating the cladoceran, the interaction is 3D because the consumer searches and detects the resource in a volume, while for fish eating shrimp the interaction is 2D because search and detection occurs on a surface. Also, note that the velocity of the shrimp is much smaller than the fish's ($v_R \ll v_C$), effectively making this a grazing interaction.

theory for the metabolic and biomechanical bases of consumer–resource interactions is beginning to emerge (McGill and Mittelbach, 2006; Brose et al., 2008; Pawar et al., 2012; Dell et al., 2014). We will outline this theory here, with due emphasis on issues that need further empirical and theoretical work.

A Metabolic Theory for Species Interactions

In general, encounter and consumption rates depend upon various metabolic and biomechanical traits that constrain the sequence of search, detection, attack, pursuit, capture, subjugation, and ingestion (Figure 3). A general formulation of per-capita biomass consumption rate c (mass·time^{-1}) that includes these components is (Pawar et al., 2012; Dell et al., 2014)

$$c = aAf(x_R) \qquad (5)$$

Here, x_R is resource biomass density (mass·area^{-1} or volume^{-1}), a is search rate (area or volume·time^{-1}), A is probability of attack success (conditional on attack), and $f(x_R)$ is the prey risk function that determines shape of the consumer's functional response. The search rate (a) of a consumer, throughout the landscape, governs the number of potential attacks a consumer can make and is partly determined by relative velocity v_r (units of distance/time), which is the population average for how fast a consumer and a resource, with velocities v_C and v_R, respectively, move across the physical landscape toward each other (Figure 3). When both consumer and resource individuals move randomly prior to detection ("active-capture" interaction), relative velocity can be shown to be proportional to the root-mean-square of the average velocities of the consumer (v_C) and resource (v_R)—i.e., $v_r = \sqrt{v_C^2 + v_R^2}$ (Okubo, 1980). The assumption of random movement might appear an oversimplification, but animal foraging in nature often follows diffusion-like movement (distance scaling with square root of time) more than directional (distance scaling linearly with time) (Skellam, 1958; Okubo and Levin, 2001; Viswanathan et al., 2011). Furthermore, many of our conclusions are not sensitive to movement patterns that include a directional component. By assuming that the power devoted to locomotion is a constant proportion of whole-body metabolic rate and that consumer–resource interactions typically take place within the PTR (Eqn (3)), we obtain

$$v_r = v_0 \sqrt{m_C^{2p_v} e^{-\frac{2E_C}{kT_C}} + m_R^{2p_v} e^{-\frac{2E_R}{kT_R}}} \qquad (6)$$

where v_0 is a constant that depends on locomotory mode and physical medium (Schmidt-Nielsen, 1972, 1984; Hein et al., 2012; Pawar et al., 2012), with maximum speeds, resistance, and energy expended being notably different in aquatic than in terrestrial habitats. The metabolic scaling constant P_0 (Eqn (1)) and the exponent p_v also depend upon details of the locomotory mode and physical medium. Note that Eqn (6) has a common v_0 and p_v for both consumer and resource velocities, because we assume pairs of interacting species have a similar locomotory mode and interaction medium. Equation (6) predicts that for a given resource density, warmer temperatures (within the PTR) and larger-sized consumer–resource pairs will result in faster convergence in space. Furthermore, when the consumer moves with a velocity much faster than the resource (Figure 3), or the resource is sessile, relative velocity is well approximated by the velocity of the consumer and we have a "grazing" interaction where $v_r \sim v_C = v_0 m_C^{p_v} e^{-\frac{E_C}{kT}}$. In contrast, when the resource moves much faster than the consumer, or the consumer is sessile, relative velocity is well approximated by resource velocity and we have a "sit-and-wait" foraging interaction where $v_r \sim v_R = v_0 m_R^{p_v} e^{-\frac{E_R}{kT}}$.

Upon convergence in space, an encounter occurs when the consumer detects the resource or vice versa (Figure 3), which depends on the maximum distance (reaction distance, d) at which the consumer and resource can sense and react to each other as well as the shape of their detection region. For a homogeneous habitat, consumers will search some portion of a D-dimensional volume independent of detection modality. Assuming that the detection region is a D-dimensional sphere $S_D = (\pi^{(D-1)/2}/\Gamma(D+1)/2)d^{D-1}$, where $\Gamma(\cdot)$ is the gamma function (Abramowitz and Stegun, 1964); that is,

$$S_D = 2d \text{ when } D = 2 (\text{i.e., 2D}) \text{ and } SD = \pi d^2 \text{ when } D = 3 (\text{i.e., 3D}) \quad (7)$$

Because a 1D search space is rare in both aquatic and terrestrial environments, we will consider only 2D and 3D. Thus, the detection region is classified as 2D when both consumer and resource move in 2D (e.g., both are benthic) or if a consumer moves in 3D and a resource in 2D (e.g., pelagic consumer on benthic resource). The detection region is classified as 3D when both consumer and resource move in 3D (e.g., both pelagic) or if the consumer moves in 2D and resource in 3D (e.g., benthic consumer and pelagic resource) (Figure 3). That is, the movement space of the resource defines the search space of the consumer. For consumers that search for resources visually, d is expected to depend on properties of the eye, height of the eye above the foraging surface, and the size of the prey (Kirschfeld, 1976; Kiltie, 2000; Pawar et al., 2012). Because none of these are known to depend on temperature directly (Dell et al., 2014), d is expected to vary with size only and can be shown to scale as (Pawar et al., 2012)

$$d = d_0 (m_R m_C)^{p_d} \quad (8)$$

where d_0 depends upon the detection medium (e.g., water vs air) and the exponent p_d. Equation (8) predicts that d should increase with both consumer and resource size because larger individuals have a higher vantage point and a longer line of sight, and are easier to resolve at greater distances. For sensory modalities such as hearing, smell, or touch, the temperature dependence of d is also expected to be weak. For example, smell and hearing may vary as a square root due to temperature influences diffusion (for smell) and the density of the environmental medium through which sound waves travel (for hearing). Because of the mathematical form of this dependence, these effects are expected to be much weaker than the exponential (Boltzmann–Arrhenius-like; Eqn (1)) initial thermal dependencies of other traits. Thus, after substituting Eqn (8) into Eqn (7), we get

$$S_D = 2d_0(m_R m_C)^{p_d} \text{ in 2D and } S_D = \pi d_0^2 (m_R m_C)^{2p_d} \text{ in 3D} \quad (9)$$

Multiplying v_r by S_D gives search rate a—that is, $a = 2v_r d$ (units of area/time) when the consumer searches for resources in 2D, and $a = \pi v_r d^2$ (units of volume/time) in 3D (Figure 3). Then, upon substituting the biomechanical and metabolic dependencies (Eqns (6) and (9)) and simplifying we get (Pawar et al., 2012; Dell et al., 2014)

$$a = a_0 m_C^{p_v + 2p_d} I^{p_d} e^{-\frac{E_C}{kT_C}} \sqrt{1 + I^{2p_v} \Delta^2} \text{ in 2D and}$$
$$a = a_0 m_C^{p_v + 4p_d} I^{2p_d} e^{-\frac{E_C}{kT_C}} \sqrt{1 + I^{2p_v} \Delta^2} \text{ in 3D} \quad (10)$$

where the constant $a_0 = 2v_0 d_0$ in 2D and $\pi v_0 d_0^2$ in 3D,

$$I \equiv m_R/m_C \text{ (size ratio, or size mismatch)}, \quad (11)$$

and

$$\Delta \equiv e^{-\frac{1}{k}\left(\frac{E_R}{T_R} - \frac{E_C}{T_C}\right)} \text{ (Boltzmann–Arrhenius factor ratio, or thermal sensitivity mismatch)} \quad (12)$$

Equation (10) is for active foraging (Figure 3), but it is straightforward to derive similar equations for grazing and sit-and-wait foraging interactions (Pawar et al., 2012; Dell et al., 2014). Equation (10) for a is a generalization of Holling's (1959a,b) "attack coefficient" to include effects of foraging strategy, interaction dimensionality, body size, and thermal physiology. It predicts that search rate increases with consumer size, temperature, and mismatches both of size and of thermal sensitivity. Substituting Eqns (10) into (5), we get

$$c = \alpha_0 m_C^{p_v + 2p_d} I^{p_d} e^{-\frac{E_C}{kT_C}} \sqrt{1 + I^{2p_v} \Delta^2} Af(R) \text{ in 2D and}$$
$$c = \alpha_0 m_C^{p_v + 4p_d} I^{2p_d} e^{-\frac{E_C}{kT_C}} \sqrt{1 + I^{2p_v} \Delta^2} Af(R) \text{ in 3D} \quad (13)$$

This means that consumption rate can *potentially* scale with size and vary with temperature like search rate does. This is a *potential* relationship because we have not yet accounted for the biomechanical and metabolic basis of attack success A, or the functional response $f(R)$, as illustrated in Figure 3. That is, upon detection, if the consumer decides to attack, attack success (A) depends upon consumer and resource attack and escape velocities (v_A and v_E, respectively) and turning velocities ($v_{\theta,C}$ and $v_{\theta,R}$) in pursuit–escape maneuvers. Along with attack success, handling time (h)—the total time taken from detection up to subjugation and ingestion of the resource (conclusion of a successful attack)—is also important. This delay between encounter and completion of ingestion prevents consumers from exploiting resources in direct proportion to their availability, resulting in the commonly observed saturating Type II functional response (Holling, 1959a; Jeschke et al., 2004)

$$f(x_R) = \frac{x_R}{1 + ahx_R} \quad (14)$$

Deriving a general mechanistic model for A and h is difficult, because the relative importance of attack velocity, acceleration, and body maneuverability during attack or escape, all of which can determine A and h, varies greatly with organism and foraging strategy (Alexander and Goldspink, 1977; Blake and Domenici, 2000; Domenici, 2001; Higham, 2007). For example, the success of a dragonfly nymph preying on a tadpole depends mainly on its burst speed and maneuverability, with endurance playing almost no role, whereas the success of orcas hunting a sperm whale depends on endurance during long chases as well as their burst speed and maneuverability during attack. Indeed, this remains a poorly developed area in metabolic theory for species interactions. However, there is accumulating empirical evidence that A declines and h increases at extreme size ratios (as resources get very large relative to consumer size, or $m_R \gg m_C$), both as power laws (McArdle and Lawton, 1979; Persson et al., 1998; Aljetlawi et al., 2004; Weitz and Levin, 2006; Vucic-Pestic et al., 2010; Pawar et al., 2012; Rall et al., 2012). One general model for h, which accounts for uncertainties in the scaling exponents and effects of size ratio, was proposed by Rall et al. (2012):

$$h = h_0 e^{\frac{E_C}{kT}} m_C^{-(0.66 \text{ to } 1)} m_R^{(0 \text{ to } 1)} \quad (15)$$

where h_0 is a metabolic-state dependent constant, and as above we have considered only the thermal response within the PTR. A more specific version of this was used by Pawar et al. (2012) and Dell et al. (2014):

$$h = h_0 e^{\frac{E_C}{kT}} m_C^{-0.75} m_R \quad (16)$$

Both Eqns (15) and (16) predict that resource mass-specific handling time declines with consumer mass to some exponent (of resting metabolic rate scaling 0.75 in the case of Eqn (16)) as well as consumer body temperature. Clearly, these models (McArdle and Lawton, 1979; Persson et al., 1998), and indeed most such models (see Aljetlawi et al., 2004; Weitz and Levin, 2006; Vucic-Pestic et al., 2010; Rall et al., 2012), do not decline at high size ratios and are therefore relevant to only certain size ratios. Nor do they capture the potential effects of differences in

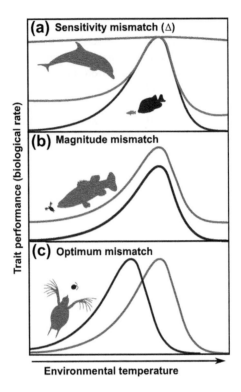

FIGURE 4 Mismatches between species in thermal responses of traits (physiological mismatches) affect the components of consumer–resource interactions (Figure 3), and can therefore influence ecological and evolutionary dynamics of ecosystems: (a) Mismatch in rates of response (either within the PTR or during the fall after T_{pk}), with an extreme case being when one species responds to environmental temperature and the other practically does not (i.e., endotherm, such as a dolphin; red line (gray in print versions)); (b) Mismatch due to difference in baseline performance, which can arise from the scaling effect of body size on metabolic rate (Eqn (1)) or differences in thermal adaptation and acclimation (Dell et al., 2011, 2014; Pawar et al., 2012); and (c) Mismatch in temperature for peak performance (T_{pk}). Different mismatches are here illustrated in different trophic levels because they can be observed in all types of consumer–resource interactions.

the thermal responses of traits underlying h, which can lead to physiological mismatches at particular temperatures (Figure 4).

Empirical Support

But how useful is this mechanistic theoretical framework? That is, can it actually predict observed consumption rates across a wide suite of organisms and interactions by scaling up individual metabolism to species interactions? Recent work suggests that the answer is yes, with both the size and the temperature components of consumption rate matching predictions well (Dell et al., 2011, 2014; Pawar et al., 2012; Rall et al., 2012). In particular, by focusing on the size component of the scaling of consumption rate—that is, by correcting for the effects of temperature—Pawar et al. (2012) have shown that consumption rates across a wide range of interactions and organisms, and across both terrestrial and aquatic environments, do indeed scale with size and size ratio as predicted by Eqn (13) (also see DeLong and Vasseur, 2012) (Figure 5). In particular, Pawar et al. (2012) found that the scaling exponent of consumption and search rates, as expected, was much steeper in 3D (due to the amplification of the detection region) than in 2D, and that both 2D and 3D scaling exponents are steeper than the canonical value for the scaling exponent of metabolic rate (0.75) (Peters, 1986; Brown et al., 2004). Also, Pawar et al. (2012) have shown that not only is the scaling of consumption rate steeper in 3D, but the baseline consumption rate itself is much higher. For example, a 1 kg organism is expected to have a 10 times higher consumption rate in a 3D environment such as a pelagic zone than in a 2D environment such as on land (5.00 ± 3.01 vs 0.50 ± 0.24 mg·s^{-1}) (Pawar et al., 2012). Subsequent work has shown that these results hold across ontogenic life stages as well (Giacomini et al., 2013; Pawar et al., 2013).

Similarly, recent work has shown that after controlling for size effects, the temperature dependence of consumption rate within the PTR is well fitted by the Boltzmann–Arrhenius model (Eqn (3)) (Dell et al., 2011, 2014;

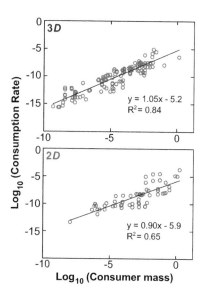

FIGURE 5 Scaling of per-capita consumption rate (kg·s^{-1}) with consumer body mass (kg) in aquatic organisms. Solid black lines were fitted using OLS regression (Pawar et al., 2012). The 3D scaling exponent (1.05 ± 0.06, $N = 129$) is steeper than the 2D exponent (0.90 ± 0.09, $N = 59$), and both are steeper than the canonical value for the scaling exponent of metabolic rate (0.75). Note that as explained in Pawar et al. (2012), these are directly measured consumption rates and not those obtained by fitting functional response models (Vucic-Pestic et al., 2010; Rall et al., 2012).

Rall et al., 2012). However, the potentially important issue of physiological mismatches (see next section) that emerges from the above theory (Vasseur et al., 2014; Dell et al., 2014) still needs to be quantified empirically. In particular, it would be enlightening to calculate "standing" levels of physiological mismatches between coexisting species within local communities that interact with each other.

FROM INTERACTIONS TO CONSUMER–RESOURCE DYNAMICS

We now consider what implications this new, mechanistic (and increasingly, empirically grounded) theoretical framework has for eco-evolutionary dynamics of consumer–resource pairs. To this end, we will use a general model for a community of S species of consumers and resources (including inorganic substrate), biomass density of the ith species (x_i units of mass·area^{-1} or volume^{-1}) given by

$$\frac{dx_i}{dt} = g_i(\cdot) + x_i \left[\sum_{k \in \text{res}(i)} e_{ki} c_{ki} - \sum_{j \in \text{con}(i)} c_{ij} - a_{ii} x_i - z_i \right], \quad i = 1, 2, \ldots, S$$

That is, from Eqn (5),

$$\frac{dx_i}{dt} = g_i(\cdot) + x_i \left[\sum_{k \in \text{res}(i)} e_{ki} a_{ki} A_{ki} f_{ki}(x_k) - \sum_{j \in \text{con}(i)} a_{ij} A_{ij} f_{ij}(x_j) - a_{ii} x_i - z_i \right],$$
$$i = 1, 2, \ldots, S$$

(17)

Here, for the ith population, the function $g_i(\cdot)$ is the intrinsic biomass production rate (mass·time^{-1}·area^{-1} or volume^{-1}), res(i) and con(i) are sets of its resources and consumers, respectively, a_{ii} is a coefficient (area or volume·mass^{-1}·time^{-1}) for biomass loss rate due to intraspecific interference, a_{ij} is mass-specific search rate (Eqn (10)) of consumer j for resource i (area or volume·mass^{-1}·time^{-1}), z_i (time^{-1}) is intrinsic biomass loss rate due to respiration and mortality, e_{ij} (a proportion) is conversion efficiency of resource to consumer biomass, and $f(\cdot)$ is the prey risk function (e.g., Eqn (14)) that determines the functional response. We propose Eqn (17) as a general model because different specifications of its parameters yield particular models, including Lotka–Volterra (for $g(\cdot) = r_{\max} x_i$ and $f(\cdot) = 1$) (May 1974; Pawar, 2009; Tang et al., 2014), Rosenzweig–MacArthur ($g(\cdot) = r_{\max} x_i$, $a_{ii} = 0$ and $f(\cdot) = $ Type II functional response) (Yodzis and Innes, 1992; Weitz and Levin, 2006; Pawar et al., 2012), the recent family of "bio-energetic" models ($f(\cdot) = $ multispecies functional response) (Brose et al., 2006b;

Otto et al., 2007) or Monod-like ($g(\cdot)$ = dilution rate-dependent substrate flux, $a_{ii} = 0$ and $f(\cdot)$ = saturating uptake function) (Tilman, 1977).

As the first step toward studying community-level dynamics, one can study the effects of biomechanical and metabolic constraints on dynamics of consumer–resource pairs (Yodzis and Innes, 1992; Persson et al., 1998; Aljetlawi et al., 2004; Vasseur and McCann, 2005; Weitz and Levin, 2006; Vucic-Pestic et al., 2010; Pawar et al., 2012; Dell et al., 2014). To this end, we can specify Eqn (17) such that $g_1(\cdot) = rx_1(1 - x_1/K)$ for the resource (say a producer, which we will call "sp. 1"), $g_2(\cdot) = 0$ for the consumer (say a primary consumer, "sp. 2"), $a_{22} = 0$ and $f(\cdot) = x_1/(1 + ahx_1)$ Type II functional response (Eqn (14)), which gives the following pair of Rosenzweig–MacArthur type equations:

$$\frac{dx_1}{dt} = rx_1\left(1 - \frac{x_1}{K}\right) - \frac{aAx_1x_2}{1 + ahx_1} \text{ and } \frac{dx_2}{dt} = \frac{eaAx_1x_2}{1 + ahx_1} - z_1x_2 \quad (18)$$

where r is sp. 1's intrinsic (primary) biomass production rate (r_{max}, time^{-1}), K is the resource carrying capacity (mass, area, or volume), and h is handling time. Then, by substituting the size and temperature dependencies of the parameters of Eqn (18), we can study ecological and evolutionary dynamics of the consumer–resource pair. This includes the size and temperature dependences of a (Eqn (10)) and h (Eqns (15) and (16)), as well as the life–history parameters r, z, and K (Savage et al., 2004). In addition, we assume that attack success probability A declines at high size ratios as a power law (Pawar et al., in preparation):

$$A = \frac{1}{1 + I^{\gamma}} \quad (19)$$

where γ governs decrease in attack success as resources get very large relative to consumer size ($m_R \gg m_C$).

Ecological Consumer–Resource Dynamics

Assuming a fixed temperature, we can obtain temperature-normalized versions of Eqns (10), (15), and (16) to study the effects of body size and size ratio alone. Using this approach, Pawar et al. (2012) have shown that due to their steeper scaling as well as higher baseline consumption rates (Eqn (10)), 3D interactions (such as in pelagic zones) are more likely to: (a) allow coexistence at more extreme size ratios (wider range of consumer–resource size combinations feasible), (b) show persistent and larger cycles (boom-bust dynamics) (Figure 6), and (c) show steeper scaling of numerical abundance with body size across species (Pawar et al., 2012). All of these suggest fundamental differences in ecosystem functioning and eco-evolutionary dynamics between aquatic and terrestrial habitats. Pelagic habitats may be inherently unstable relative to more 2D benthic or terrestrial habitats, which may be offset by the

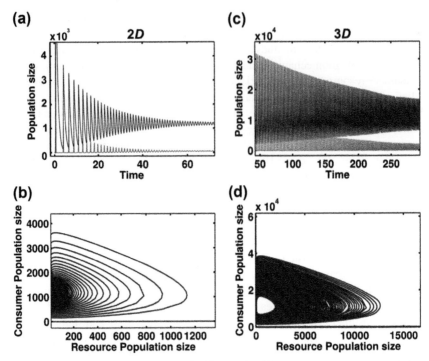

FIGURE 6 Effect of interaction dimensionality on size-driven consumer–resource population dynamics. Note that 3D interactions (c,d) show persistent cycles and larger amplitude booms and busts, while 2D interactions converge on a stable fixed point (a,b). All of these results are for a fixed temperature (15 °C). Varying the temperature realistically (e.g., periodically) would typically result in more complex dynamics.

wider range of feasible size ratios. This result is consistent with empirical observations that pelagic communities appear less stable than terrestrial communities (Rip and McCann, 2011), yet show a wider range of and more extreme size ratios (Pawar et al., in preparation; Brose et al., 2006a).

The stronger 3D consumption rates for a wide range of body sizes also suggest that pelagic ecosystems may experience more frequent top-down regulation than 2D terrestrial or benthic ones (Chase, 2000; Shurin et al., 2006). That the scaling of numerical abundance at equilibrium may be steeper in pelagic environments relative to benthic or terrestrial ones (Pawar et al., 2012) means that only 2D abundance scaling exponents are predicted to be close to Damuth's $-3/4$ rule, which was indeed derived from data on terrestrial mammals (2D consumers) (Damuth, 1981; Savage et al., 2004). The 3D ecosystems may show much steeper, superlinear (<-1) exponents (Pawar et al., 2012). This may help explain deviations from energetic equivalence ($-3/4$ power mass-abundance scaling) in local communities (Damuth, 1981; Cyr et al., 1997a,b; Leaper et al., 1999; Reuman et al., 2009).

What about the effects of temperature? By assuming a fixed consumer size and resource–consumer size ratio, we can study the effect of changes in temperature and the effect of variation in parameters of the thermal response (E as well as the normalization constant). By using this approach, recent work (Vasseur and McCann, 2005; Dell et al., 2014) suggests that mismatches between consumer and resource activation energies E (the factor Δ; Eqn (12), also see Figure 4) can have nontrivial effects on consumer–resource dynamics and equilibrium abundances. For example, in active–capture interactions, equilibrium biomass of the resource is predicted to be $\hat{x}_1 = 1/\sqrt{1+a\Delta^2}$, and for consumers it is predicted to be $\hat{x}_2 = 1/\sqrt{1+a\Delta^2} = \Delta/\sqrt{1+a\Delta^2}$. Then, for a mismatch wherein the resource and consumer activation energies (E_1 and E_2) can range between ($E \sim 0.2–1.2$ eV), resource equilibrium biomass $\hat{x}_1$ can be anywhere between 1.24×10^{-17} (for $E_1 = 0.2$, $E_2 = 1.2$ eV) and 1 (for $E_1 = 1.2$, $E_2 = 0.2$ eV), and consumer equilibrium biomass $\hat{x}_2$ can be anywhere between 1.6×10^{-34} and 8.0×10^{16}, an astounding 50 orders of magnitude difference. This highlights not just the potential importance of physiological mismatches themselves, but also the need for further research on the typical levels of mismatches seen between species that are likely to interact with each.

Along with commonly observed differences in activation energies between consumers and resources (Δ) (Dell et al., 2011, 2014), differences in thermy (e.g., ectotherm vs endotherm), or thermoregulation (e.g., two ectotherms with different body temperatures) are common in food webs. For example, across most temperatures ectotherms will likely be relatively slower at low temperatures, and endothermic consumers feeding on these ectothermic resources will therefore have higher success rates for capture and attack at these lower temperatures (Christian and Tracy, 1981). Also when both consumer and resource are ectothermic, escapes and failed attacks may be more common at low temperatures because escape body velocity typically remains close to peak levels and is thus higher than attack body velocity (Dell et al., 2011). Indeed, each of the mismatch scenarios in Figure 4 will become more likely as species change their geographic and temporal niches in response to climate change. It is now well established that warm-adapted species are moving into regions that were previously too cold, and that climate change is altering the phenology of many plants (including phytoplankton) and animals (Kareiva et al., 1993; Parmesan and Yohe, 2003; Reid et al., 2007; Hallegraeff, 2010; Chen et al., 2011; Saikkonen et al., 2012). Climate change could elicit such shifts when warming cues occur earlier in the year, while other cues, such as seasonal light conditions, remain constant. These differences in environmental drivers could potentially cause matched species interactions to become uncoordinated (Pörtner and Farrell, 2008) and new combinations of interacting species to arise.

Finally, the effects of physiological mismatches on consumer–resource dynamics may be amplified because aquatic environments in general, and pelagic zones in particular, show stronger size and size-difference scaling than terrestrial ones (Pawar et al., 2012) (Figure 5). Thus overall, the issue of physiological mismatches introduces an important new perspective on the effects of temperature and climate change on consumer resources and therefore community food web and ecosystem dynamics (see section on community and ecosystem dynamics below). This perspective is important because conditions promoting mismatches in the response of traits relevant to trophic interactions will exist in virtually all ecosystems including aquatic ones, and mismatches are likely to be a major factor driving invasion and community assembly dynamics.

Eco-Evolutionary Consumer–Resource Dynamics

Clearly, ecological and evolutionary mechanisms do not operate in isolation, and integration of these processes is necessary for a general theory for the metabolic and biomechanical bases of consumer–resource, community food web, and ecosystem dynamics (Matthews et al., 2011; Thuiller et al., 2013). This is particularly important considering the fact that unlike terrestrial ecosystems, most aquatic ecosystems are built upon a small-sized primary producer base (the phytoplankton), which means that they have short generation times and high mass-specific metabolic rates and therefore potentially rapid acclimation and adaptation rates. As a first step toward studying coupled eco-evolutionary dynamics, we can ask, given a species pair that can stably coexist above threshold abundances, what size or thermal performance curve is evolutionarily optimal for the consumer and the resource? The optimum size ratio will likely differ depending on whether the optimization is from the perspective of the consumer or the resource. To answer these questions, one can calculate conditions under which a stable consumer–resource pair cannot be invaded by a novel (mutant or immigrant) consumer (or resource) with a body size or thermal performance (e.g., activation energy E) different than that of the resident (Marrow et al., 1996; Geritz et al., 1997; Weitz and Levin, 2006; Reuman et al., 2014)—the first step toward an evolutionary change in the resident consumer (resource) population.

As an illustration, consider again the consumer–resource system governed by Eqn (18), in which we assume temperature is fixed and into which a mutant consumer arises that differs from the resident consumer in size. A size-mutant consumer's population dynamics are governed by (cf. dx_2/dt part of Eqn (18) above)

$$\frac{dx_{2,u}}{dt} = \frac{ea_{2,u}x_{2,u}\hat{x}_1}{1 + a_{2,u}h_u\hat{x}_1} - z_u x_{2,u} \tag{20}$$

where subscript u indicates a parameter linked to the mutant. We assume that the resident resource population is at biomass equilibrium ($\hat{x}_1$) and the mutant's and resident's conversion efficiencies are the same (e). Then, by solving for positive growth of the invader population, we find that the evolutionarily stable strategy (ESS) is the one that maximizes the quantity $(a/m_C)(e - zhm_C)/z$ (Pawar et al., in preparation). That is, the consumer mutant with size that best exploits resources though its search rate is the one that cannot be invaded by any other mutant (has an ESS). Then, upon substituting the size scaling of a (Eqn (10)), h (Eqn (16)), and z (Savage et al., 2004), we can determine what optimal consumer sizes and consumer–resource size ratios this ESS translates to. Similarly, the fate of a rare resource mutant's population ($x_{1,u}$) is governed by the equation (cf. dx_1/dt part of Eqn (18) above)

$$\frac{dx_{1,u}}{dt} = r_u x_{1,u} \left(1 - \frac{x_{1,u}}{K_u}\right) - \frac{a_u \hat{x}_{2} x_{1,u}}{1 + a_u h_u x_{1,u}} \qquad (21)$$

where the consumer's search rate and handling time have been subscripted with u because both are partly governed by the consumer–resource size ratio (Eqns (10), (15), and (16)). Here again, as in the case of the consumer, we can solve for positive population growth of the mutant resource to derive the condition for successful invasion, and then substitute size scaling of parameters in the quantities needed (or that must be maximized) to achieve an ESS (Pawar et al., in preparation). Specifically, to make the analyses more relevant to aquatic ecosystems, let us substitute $3D$ size scaling of search rate a (Eqn (10)). Figure 7 shows the resulting invasion fitness landscapes of consumer as well as resource size mutants. Clearly, there is an inherent conflict in the fitness landscapes and ESS' of consumers versus resources. That is, for a fixed

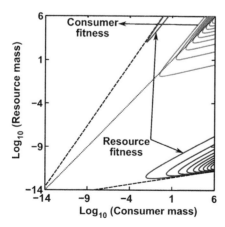

FIGURE 7 Effect of body size and size mismatch (size ratio, I) on consumer and resource fitness (of invasion of a rare mutant, given by the contours) within the feasible coexistence region of ecological coexistence (dashed lines). The ESS lies at the center of each set of contours. For a fixed consumer (or resource) size, consumer invasion fitness is maximized at minimal size ratios (size mismatch) (along the dotted diagonal I = 1 line) where its search rate is maximal. In contrast, resource fitness is maximized at extreme size mismatch, where consumers have low search rates.

consumer (or resource) size, consumer invasion fitness is maximized at intermediate size ratios where its search rate is maximal, while the resource fitness landscape has two peaks at extreme body size ratios. This means that resources can evolve to escape consumption by a consumer of fixed size by becoming much larger than the consumer if they start above the $I = 1$ line, or much smaller if they start below the $I = 1$ line. This suggests a mechanism for evolutionary size diversification in consumer—resource interactions in aquatic ecosystems.

Analogous analyses can be performed by considering consumer or resource fitness when thermal response parameters can evolve. Indeed, recent empirical evidence and preliminary theoretical studies suggest that natural selection through consumer resource interactions may play an important role in driving the evolution of activation energies (E) (Angilletta, 2009; Dell et al., 2011). In particular, Dell et al. (2011) have suggested a thermal life-dinner principle—stronger selection pressure on prey to escape capture and death results in maintenance of better performance across a range of temperatures, which means that prey escape traits are expected to show lower thermal sensitivity (lower E) than predator attack traits do. Of course, this selection has to be strong enough to overcome the energetic cost of maximizing metabolic effort at low temperatures. Note that the resulting difference between consumer and resource activation energies is by definition one of the thermal response mismatches (Δ) discussed above.

Furthermore, it is possible to use evolutionary invasion analysis to study the effect of warming on size evolution or the effect of size difference on evolution of thermal responses. Reuman et al. (2014) have performed such a study to examine the effect of warming on phytoplankton size evolution. They find that warming is likely to result in the prevalence of smaller phytoplankton at higher temperatures due to increased invasibility of smaller cells due to their higher nutrient uptake and growth. All these results clearly demonstrate the potential importance of coupled eco-evolutionary dynamics in driving the dynamics and function of aquatic ecosystems. Further work in this area would need to include adaptive dynamics to obtain a more complete and nuanced picture of such eco-evolutionary dynamics. Another, more explicitly evolutionary approach is to use a quantitative trait modeling approach (Abrams, 2000, 2001) wherein the evolutionary parameters of interest (e.g., E, T_{pk}, and body size) are allowed to evolve while ecological dynamics are taking place.

FROM CONSUMER—RESOURCE PAIRS TO COMMUNITY AND ECOSYSTEM DYNAMICS

Studying direct consumer—resource interactions in isolated pairings is a first, important step and empirical hurdle in understanding effects at the level of aquatic community food webs and ecosystems (Laska and Wootton, 1998).

The next step is to study more complex consumer–resource interactions including indirect competition between consumers and polyphagy—a consumer eating multiple types of resources. Future work using multitrophic food web models that are generalizations of the above mechanistic theoretical framework should reveal deeper insights into how eco-evolutionary dynamics constrain ecosystem functioning.

Some work in this direction has already been done, essentially by combining multiple species using the model presented in Eqn (17). For example, recent simulations using such models have posited that observed body sizes in communities allow stable coexistence of consumers and resources within local communities without any necessary optimization of consumption rates (Jonsson and Ebenman, 1998; Emmerson et al., 2005; Brose et al., 2006b; Otto et al., 2007; Tang et al., 2014). The mechanism being invoked (often implicitly) by these studies is species sorting, where destabilizing consumer–resource size combinations are "filtered out" through immigration–extinction dynamics during community assembly and maintenance (Bastolla et al., 2005; Pawar, 2009; Fahimipour and Hein, 2014; Pawar, 2015). However, all of these studies still use a $3/4$ power scaling for consumption and a $1/4$ power scaling for production rate. Future work should include better updated models of the biomechanical basis, such as the one we have presented above, to get deeper and more accurate insights. Given our results about differences in size ratios of 2D and 3D interactions, the dynamics of food web models in aquatic ecosystems that can have either dimensionality should be intriguing, and coupled 2D and 3D subcommunities should be interesting as well.

Moving beyond purely ecological models with interactions that respond in a fixed way to size and temperature, a few studies have also taken a multi-species evolutionary dynamics approach (Loeuille et al., 2002; Loeuille and Loreau, 2005, 2006; Stegen et al., 2012a,b). These studies have yielded interesting insights, but their simulation results are necessarily more complex than those for the purely ecological network models described above, and are much less amenable to analytical insights about the link between environmental change (such as climatic warming) and ecosystem dynamics, and function through metabolic and biomechanical constraints. More tractable alternatives for the study of coupled eco-evolutionary dynamics leading to changes in ecosystem functioning are trait-based modeling approaches (Norberg, 2004; Savage et al., 2007; Enquist et al., 2015) that can be used to directly link metabolic traits distributions to ecosystem functioning metrics. In this approach, the effect of functional trait variation *across* species' populations on ecosystem processes can be studied using the general model

$$\frac{dx_y}{dt} = f(T, x_{tot}, R, y)x_y + w_y \qquad (22)$$

Here y is a functional trait, x_y is biomass abundance of the (functional) group carrying the trait y, T is environmental temperature, x_{tot} is the total abundance across all trait variants (standing biomass), R is a limiting resource, and w_y determines immigration (or emigration) of trait variants into the local patch (which can be modeled as either density dependent or density independent). Alternatively, w_y can be used to model upregulated or downregulated (i.e., density-dependent or density-independent) gene expression. The advantage of this framework is that one can directly include biomechanical and metabolic constraints on traits. For example, by using Eqn (13) (for consumption rate) to specify $f(\cdot)$, we can map temperature and body size directly onto system dynamics, in which case y could be either E or body mass m. One technique that can simplify subsequent analysis is to Taylor-expand the function $f(\cdot)$ around the average trait value $\bar{y}$ to predict how system dynamics will change as a function of temperature for a given size class, including the effects of (1) total immigration (or up/downregulation) across all trait variants, (2) rate of change and static values at temporal snapshots of trait mean ($\bar{y}$), (3) rate of change and static values at temporal snapshots of trait variance, and (4) total trait ecosystem standing biomass x_{tot} as well as its rate of change. This framework can also be extended to simultaneous changes in multiple traits (e.g., $y_1 = E$ and $y_2 = m$) to ascertain the importance of functional trait redundancy and complementarity for maintaining ecosystem functioning (Savage et al., 2007). Multispecies, multitrait modeling is particularly important for aquatic ecosystems, because recent empirical studies have revealed a number of coupled systematic changes in body size structure and the temperature dependence of ecosystems functioning in freshwater as well as marine environments (Yvon-Durocher et al., 2011; O'Gorman and Emmerson, 2011; Dossena et al., 2012; Yvon-Durocher and Allen, 2012).

Thus, irrespective of the multispecies ecosystem level of model that is used, the mechanistic theory for pairwise trophic interactions we have outlined above is key to building a truly predictive theory for aquatic ecosystem dynamics and functioning. In particular, because this framework makes explicit the role of parameters such as body size, foraging strategy, shape of the thermal response, and mismatches in thermal responses of interacting species, it has the potential to allow a delineation of potentially important functional organismal and trait groups in complex ecosystems and to gradually build up the dynamics of consumer–resource pairs to whole networks of interactions.

CONCLUSIONS

We have sought to demonstrate how a general theoretical framework for the metabolic and biomechanical bases of trophic interactions has the potential to yield general predictions about the effects of individual-level metabolic and biomechanical constraints on the eco-evolutionary dynamics and functioning of aquatic ecosystems. In particular, by assuming that ectotherm body temperature is equivalent or proportional to ambient temperature, this theoretical framework can be used to make predictions about how consumer–resource interactions, and eventually how whole ecosystem dynamics, will be affected by climate change. This theory also has the potential to be applied to other types of ecological interactions that rely on metabolism and biomechanics, such as pollination, parasitism, and competitive interactions. At the same time, a number of important challenges exist in the ongoing development of the theory that need to be met in order to make it truly general and robust:

1. Although we have shown above that a random movement assumption (Eqn (6)) performs quite well in predicting search and consumption rates, the effect of more complex, nonrandom movement strategies (such as Levy flights combined with neighborhood searching) merits thorough investigation (Cantrell et al., 2006; Chen et al., 2008; Smouse et al., 2010; Viswanathan et al., 2011; Hein and McKinley, 2012).
2. A more continuous definition of spatial dimensionality (e.g., fractal-like) may be needed to better predict consumption rates in the field, especially if resources are immobile (i.e., in grazing interactions) and dispersed nonrandomly (Ritchie, 2009).
3. While the current theory uses a visual detection model, other sensory modalities including auditory, olfactory, and chemosensory need to be considered (e.g., Dusenbery and Snell, 1995 and Blake and Domenici, 2000). In particular, our speculation above that these other sensory modalities will also yield a scaling relationship of reaction distance with size needs to be tested theoretically and empirically. Also, while visual acuity is largely independent of temperature, other modes of detection are not. For example, chemosensory detection is important in aquatic environments (Atema, 1995; Brönmark and Hansson, 2007; Ferland-Raymond et al., 2010) and depends on turbulence and the $3D$ diffusion of signal molecules. At small spatial scales relevant to small organisms, diffusion is likely more important, and rates of diffusion depend upon the square root of temperature (Gainer, 1970; Krynicki et al., 1978). In addition, the biological mechanisms for detection (such as signal transduction pathways) may themselves be temperature dependent (Pantazelou and Dames, 1995).
4. The biomechanics of attack, escape, subjugation, and ingestion (Blake and Domenici, 2000; Domenici, 2001, 2002, 2010) need to be incorporated to derive truly mechanistic models of handling time and attack success. At the most fundamental level, such models should be able to explain why handling times and attack success probability tends to be unimodal with respect to body size ratio (Persson et al., 1998; Aljetlawi et al., 2004; Vucic-Pestic et al., 2010; Rall et al., 2012). In particular, this theory will need to incorporate the scaling of acceleration and angular velocity during attack—escape maneuvers (Figure 3).
5. The effect of ontogenic changes in the metabolic and biomechanical constraints on population consumption rates needs to be considered (Amarasekare and Savage, 2012; Giacomini et al., 2013; Pawar et al.,

2013). In particular, the timescale differences between biomass consumption, allocation, and production need to be reconciled. This should help explain the apparent difference in scaling of consumption rate (rate of energy intake) and metabolic rate (rate of energy use) that we report above. One way to reconcile it may be by considering differences in basal and active metabolic rates.

As these issues are addressed, concurrently incorporating them into interaction and interaction-network levels of ecosystem organization (Figure 1) will be important to produce incrementally robust and accurate predictions that can be confronted with the burgeoning data on aquatic ecosystem functioning in the face of environmental change.

ABBREVIATIONS AND MATHEMATICAL SYMBOLS

(Subscript i refers to a parameter for the ith species)
Δ Boltzmann–Arrhenius factor ratio, or thermal sensitivity mismatch (Eqn (12))
a Consumer–resource search rate (area or volume·time^{-1})
A Probability of attack success (conditional on attack)
a_0 Normalization constant for scaling of search rate with body mass
b Scaling exponent of metabolic rate (unitless)
c Biomass consumption rate (mass·time^{-1})
d Reaction distance—maximum distance at which the consumer and resource can sense and react to each other
D Euclidean dimension of consumer–resource interaction space (= 1, 2, or 3 for biological interactions)
d_0 Normalization constant for scaling of reaction distance with body mass
e Conversion efficiency of resource to consumer biomass (unitless)
E Average activation energy (eV) of rate-limiting steps in biochemical reactions (1 eV = 96.49 kJ·mol^{-1})
E_D The average energy constant at which proteins denature (deactivation energy), or more phenomenologically, a parameter that controls steepness of decline of metabolic rate beyond T_{pk}
$f(x_R)$ Prey risk function
h Handling time, or the total time taken from detection up to subjugation and ingestion of the resource by a unit consumer
h_0 Normalization constant for the size-and-temperature dependence of handling time (h)
I Size ratio, or size mismatch (Eqn (11))
k The Boltzmann constant (8.62 × 10^{-5} eV·K^{-1})
m Average individual body mass (kg)
P Whole-organism (individual) metabolic rate (units of J·s^{-1}), which may be resting, active, or field
P_0 A body mass- and temperature-independent normalization constant for metabolic rate
p_d Exponent of the scaling of reaction distance with body mass
PTR (physiological temperature range) The environmental temperature range over which organisms typically operate, here defined to lie below T_{pk} (Figure 2)

p_v Exponent of the scaling of velocity with body mass
r_0 Normalization constant for the size and temperature dependence of r_{max}
r_{max} Malthusian parameter or maximal growth rate (also known as intrinsic growth rate) (units of time^{-1}), equivalent to $g_i(\cdot)$ in certain consumer–resource model specifications (Eqn (17))
S_D Consumer–resource detection region (length^{D-1}, where D is euclidean dimension)
T Body temperature (°C or K)
T_{pk} Temperature (°C or K) for peak performance, at which metabolic rate or a functional trait value reaches its peak (P_{pk})
v Whole-organism body velocity (length·time^{-1})
v_0 Normalization constant for the size-and-temperature dependence of body velocity v
v_r Average relative velocity (length·time^{-1}) between individuals of two species
w_y Immigration (or emigration) of functional trait variants into a local patch (or alternatively, upregulation or downregulation of gene expression)
x Population biomass density (mass·area^{-1} or volume^{-1})
y Value of a functional trait (e.g., m or E)
z Intrinsic biomass loss rate due to respiration and mortality (time^{-1})

ACKNOWLEDGMENTS

We would like to thank Priyanga Amarasekare, Janice Chan, and Gabriel Yvon-Durocher for stimulating discussions and comments that helped improve this paper.

REFERENCES

Abramowitz, M., Stegun, I.A., 1964. Handbook of Mathematical Functions with Formulas, Graphs, and Mathematical Tables. U.S. Govt. Print. Off., Washington. Page xiv, 1046 p.

Abrams, P., 2000. The evolution of predator-prey interactions: theory and evidence. Ann. Rev. Ecol. Systemat. 31, 79–105.

Abrams, P.A., 2001. Modelling the adaptive dynamics of traits involved in inter- and intraspecific interactions: an assessment of three methods. Ecol. Lett. 4, 166–175.

Alexander, R.M., Goldspink, G., 1977. Mechanics and Energetics of Animal Locomotion. Chapman and Hall; Distributed by Halsted Press, London, New York. Page xii, 346 p.

Aljetlawi, A.A., Sparrevik, E., Leonardsson, K., 2004. Prey-predator size-dependent functional response: derivation and rescaling to the real world. J. Anim. Ecol. 73, 239–252.

Allen, A.P., Gillooly, J.F., Brown, J.H., 2005. Linking the global carbon cycle to individual metabolism. Funct. Ecol. 19, 202–213.

Allen, C.R., Garmestani, A.S., Havlicek, T.D., Marquet, P.A., Peterson, G.D., Restrepo, C., Stow, C.A., Weeks, B.E., 2006. Patterns in body mass distributions: sifting among alternative hypotheses. Ecol. Lett. 9, 630–643.

Amarasekare, P., Savage, V., 2012. A framework for elucidating the temperature dependence of fitness. Am. Nat. 179, 178–191.

Angilletta, M., 2009. Thermal Adaptation: A Theoretical and Empirical Synthesis. Oxford University Press, Oxford, New York. Page 289.

Angilletta, M.J., Huey, R.B., Frazier, M.R., 2010. Thermodynamic effects on organismal performance: is hotter better? Physiol. Biochem. Zool. 83, 197–206.

Atema, J., 1995. Chemical signals in the marine environment: dispersal, detection, and temporal signal analysis. Proc. Natl. Acad. Sci. USA 92, 62–66.

Bastolla, U., Lassig, M., Manrubia, S.C., Valleriani, A., 2005. Biodiversity in model ecosystems, II: species assembly and food web structure. J. Theor. Biol. 235, 531–539.

Beke-Somfai, T., Lincoln, P., Nordén, B., 2010. Mechanical control of ATP synthase function: activation energy difference between tight and loose binding sites. Biochemistry 49, 401–403.

Belgrano, A., Allen, A.P., Enquist, B.J., Gillooly, J.F., 2002. Allometric scaling of maximum population density: a common rule for marine phytoplankton and terrestrial plants. Ecol. Lett. 5, 611–613.

Bernacchi, C.J., Singsaas, E.L., Pimentel, C., Portis Jr., A.R., Long, S.P., 2001. Improved temperature response functions for models of Rubisco-limited photosynthesis. Plant Cell Environ. 24, 253–259.

Blake, R.W., Domenici, P., 2000. Biomechanics in Animal Behaviour. Experimental Biology Reviews. BIOS Scientific, Oxford. Page xv, 344 p.

Brönmark, C., Hansson, L.-A., 2007. Chemical Ecology in Aquatic Systems. Oxford University Press. Page 312.

Brose, U., Ehnes, R.B., Rall, B.C., Vucic-Pestic, O., Berlow, E.L., Scheu, S., 2008. Foraging theory predicts predator-prey energy fluxes. J. Anim. Ecol. 77, 1072–1078.

Brose, U., Jonsson, T., Berlow, E.L., Warren, P., Banasek-Richter, C., Bersier, L.F., Blanchard, J.L., Brey, T., Carpenter, S.R., Blandenier, M.F.C., Cushing, L., Dawah, H.A., Dell, T., Edwards, F., Harper-Smith, S., Jacob, U., Ledger, M.E., Martinez, N.D., Memmott, J., Mintenbeck, K., Pinnegar, J.K., Rall, B.C., Rayner, T.S., Reuman, D.C., Ruess, L., Ulrich, W., Williams, R.J., Woodward, G., Cohen, J.E., 2006a. Consumer-resource body-size relationships in natural food webs. Ecology 87, 2411–2417.

Brose, U., Williams, R.J., Martinez, N.D., 2006b. Allometric scaling enhances stability in complex food webs. Ecol. Lett. 9, 1228–1236.

Brown, J.H., Gillooly, J.F., Allen, A.P., Savage, V.M., West, G.B., 2004. Toward a metabolic theory of ecology. Ecology 85, 1771–1789.

Brown, J.H., Marquet, P.A., Taper, M.L., 1993. Evolution of body-size: consequences of an energetic definition of fitness. Am. Nat. 142, 573–584.

Cantrell, R.S., Cosner, C., Lou, Y., 2006. Movement toward better environments and the evolution of rapid diffusion. Math. Biosci. 204, 199–214.

Chase, J.M., 2000. Are there real differences among aquatic and terrestrial food webs? Trends Ecol. Evol. 15, 408–412.

Chen, B., Landry, M.R., Huang, B., Liu, H., 2012. Does warming enhance the effect of microzooplankton grazing on marine phytoplankton in the ocean? Limnol. Oceanogr. 57, 519–526.

Chen, I.-C., Hill, J.K., Ohlemüller, R., Roy, D.B., Thomas, C.D., 2011. Rapid range shifts of species associated with high levels of climate warming. Science 333, 1024–1026.

Chen, X., Hambrock, R., Lou, Y., 2008. Evolution of conditional dispersal: a reaction-diffusion-advection model. J. Math. Biol. 57, 361–386.

Chown, S.L., Gaston, K.J., 1997. The species-body size distribution: energy, fitness and optimality. Funct. Ecol. 11, 365–375.

Christian, K.A., Tracy, C.R., 1981. The effect of the thermal environment on the ability of hatchling Galapagos land iguanas to avoid predation during dispersal. Oecologia 49, 218–223.

Cohen, J.E., 2008. Body sizes in food chains of animal predators and parasites. In: Hildrew, A.G., Raffaelli, D.G., Edmonds-Brown, R. (Eds.), Body Size: The Structure and Function of Aquatic Ecosystems. Cambridge University Press, Cambridge, UK, pp. 306–325.

Cohen, J.E., Fenchel, T., 1994. Marine and continental food webs: three paradoxes? Philos. Trans. R. Soc. Lond. B Biol. Sci. 343, 57–69.

Coleman, J.R., Colman, B., 1980. Effect of oxygen and temperature on the efficiency of photosynthetic carbon assimilation in two microscopic algae. Plant Physiol. 65, 980–983.

Corkrey, R., McMeekin, T.A., Bowman, J.P., Ratkowsky, D.A., Olley, J., Ross, T., 2014. Protein thermodynamics can be predicted directly from biological growth rates. PLoS One 9, e96100.

Corkrey, R., Olley, J., Ratkowsky, D., McMeekin, T., Ross, T., 2012. Universality of thermodynamic constants governing biological growth rates. PLoS One 7, e32003.

Cyr, H., Downing, J., Peters, R., 1997a. Density-body size relationships in local aquatic communities. Oikos 79, 333–346.

Cyr, H., Peters, R.R.H., Downing, J.J.A., 1997b. Population density and community size structure: comparison of aquatic and terrestrial systems. Oikos 80, 139–149.

Damuth, J., 1981. Population density and body size in mammals. Nature 290, 170699–170700.

Dell, A.I., Pawar, S., Savage, V.M., 2011. Systematic variation in the temperature dependence of physiological and ecological traits. Proc. Natl. Acad. Sci. USA 108, 10591–10596.

Dell, A.I., Pawar, S., Savage, V.M., 2014. Temperature dependence of trophic interactions are driven by asymmetry of species responses and foraging strategy. J. Anim. Ecol. 83, 70–84.

DeLong, J.P., Okie, J.G., Moses, M.E., Sibly, R.M., Brown, J.H., 2010. Shifts in metabolic scaling, production, and efficiency across major evolutionary transitions of life. Proc. Natl. Acad. Sci. USA 107, 12941–12945.

DeLong, J., Vasseur, D., 2012. A dynamic explanation of size-density scaling in carnivores. Ecology 93, 470–476.

Deutsch, C.A., Tewksbury, J., Huey, R.B., Sheldon, K.S., Ghalambor, C.K., Haak, D.C., Martin, P.R., 2008. Impacts of climate warming on terrestrial ectotherms across latitude. Proc. Natl. Acad. Sci. 105, 6668–6672.

Domenici, P., 2001. The scaling of locomotor performance in predator-prey encounters: from fish to killer whales. Comp. Biochem. Physiol. A Mol. Integr. Physiol. 131, 169–182.

Domenici, P., 2002. The visually mediated escape response in fish: predicting prey responsiveness and the locomotor behaviour of predators and prey. Mar. Freshw. Behav. Physiol. 35, 87–110.

Domenici, P., 2010. Context-dependent variability in the components of fish escape response: integrating locomotor performance and behavior. J. Exp. Zool. A Ecol. Genet. Physiol. 313, 59–79.

Dossena, M., Yvon-Durocher, G., Grey, J., Montoya, J.M., Perkins, D.M., Trimmer, M., Woodward, G., 2012. Warming alters community size structure and ecosystem functioning. Proc. R. Soc. B Biol. Sci. 279, 3011–3019.

Dusenbery, D.B., Snell, T.W., 1995. A critical body size for use of pheromones in mate location. J. Chem. Ecol. 21, 427–438.

Emmerson, M.C., Raffaelli, D., 2004. Predator-prey body size, interaction strength and the stability of a real food web. J. Anim. Ecol. 73, 399–409.

Emmerson, M.E., Montoya, J.M., Woodward, G., 2005. Body size, interaction strength and food web dynamics. In: de Ruiter, P.C., Wolters, V., Moore, J.C. (Eds.), Dynamic Food Webs: Multispecies Assemblages, Ecosystem Development, and Environmental Change. Academic Press, San Diego, pp. 167–178.

Enquist, B.J., Norberg, J., Bonsor, S.P., Violle, C., Webb, C.T., Henderson, A., Sloat, L.L., Savage, V.M., 2015. Scaling from traits to ecosystems: developing a general Trait Driver Theory via integrating trait-based and metabolic scaling theories. Adv. Ecol. Res. 52, in press.

Fahimipour, A.K., Hein, A.M., 2014. The dynamics of assembling food webs. Ecol. Lett. 17, 606–613.

Farquhar, G.D., Caemmerer, S., Berry, J.A., 1980. A biochemical model of photosynthetic CO_2 assimilation in leaves of C3 species. Planta 149, 78–90.

Ferland-Raymond, B., March, R.E., Metcalfe, C.D., Murray, D.L., 2010. Prey detection of aquatic predators: assessing the identity of chemical cues eliciting prey behavioral plasticity. Biochem. Syst. Ecol. 38, 169–177.

Frazier, M.R., Huey, R.B., Berrigan, D., 2006. Thermodynamics constrains the evolution of insect population growth rates: "warmer is better". Am. Nat. 168, 512–520.

Gainer, J.L., 1970. Concentration and temperature dependence of liquid diffusion coefficients. Ind. Eng. Chem. Fund. 9, 381–383.

Geritz, S.A.H., Metz, J.A.J., Kisdi, É., Meszéna, G., Kisdi, E., Meszena, G., 1997. Dynamics of adaptation and evolutionary branching. Phys. Rev. Lett. 78, 2024–2027.

Giacomini, H.C., Shuter, B.J., de Kerckhove, D.T., Abrams, P.A., 2013. Does consumption rate scale superlinearly? Nature 493, E1–E2.

Gibert, J.P., Dell, A.I., DeLong, J.P., Pawar, S., 2015. Scaling-up Trait Variation from Individuals to Ecosystems. Adv. Ecol. Res. 52 in press.

Gillooly, J.F., Brown, J.H., West, G.B., Savage, V.M., Charnov, E.L., 2001. Effects of size and temperature on metabolic rate. Science 293, 2248–2251.

Gillooly, J.F., Charnov, E.L., West, G.B., Savage, V.M., Brown, J.H., 2002. Effects of size and temperature on developmental time. Nature 417, 70–73.

Hallegraeff, G.M., 2010. Ocean climate change, phytoplankton community responses, and harmful algal blooms: a formidable predictive challenge. J. Phycol. 46, 220–235.

Hein, A.M., Hou, C., Gillooly, J.F., 2012. Energetic and biomechanical constraints on animal migration distance. Ecol. Lett. 15, 104–110.

Hein, A.M., McKinley, S.A., 2012. Sensing and decision-making in random search. Proc. Natl. Acad. Sci. USA 109, 12070–12074.

Higham, T.E., 2007. The integration of locomotion and prey capture in vertebrates: morphology, behavior, and performance. Integr. Comp. Biol. 47, 82–95.

Hochachka, P.W., Somero, G.N., 1984. Biochemical Adaptation. Princeton University Press, Princeton, NJ. Page xx, 537 p.

Holling, C.S., 1959a. The components of predation as revealed by a study of small-mammal predation of the European pine sawfly. Can. Entomol. 91, 293–320.

Holling, C.S., 1959b. Some characteristics of simple types of predation and parasitism. Can. Entomol. 91, 385–398.

Holling, C.S., 1966. The functional response of invertebrate predators to prey density. Mem. Entomol. Soc. Can. 98, 1–86.

Huey, R.B., Kingsolver, J.G., 2011. Variation in universal temperature dependence of biological rates. Proc. Natl. Acad. Sci. USA 108, 10377–10378.

Huey, R., Berrigan, D., 2001. Temperature, demography, and ectotherm fitness. Am. Nat. 158, 204–210.

IPCC, I. P. O. C. C., 2007. Climate change 2007: synthesis report. In: Core Writing Team, Pachauri, R.K., Reisinger, A. (Eds.), Contribution of Working Groups I, II and III to the Fourth Assessment Report of the Intergovernmental Panel on Climate Change. IPCC, Geneva, Switzerland, Page 104.

Jeschke, J.M., Kopp, M., Tollrian, R., 2004. Consumer-food systems: why type I functional responses are exclusive to filter feeders. Biol. Rev. Camb. Philos. Soc. 79, 337–349.

Johnson, F., Eyring, H., Stover, B., 1974. The Theory of Rate Processes in Biology and Medicine. Wiley, New York. Page 703.

Johnson, F.H., Lewin, I., 1946. The growth rate of *E. coli* in relation to temperature, quinine and coenzyme. J. Cell. Comp. Physiol. 28, 47–75.

Jonsson, T., Ebenman, B., 1998. Effects of predator-prey body size ratios on the stability of food chains. J. Theor. Biol. 193, 407–417.

Kalinkat, G., Schneider, F.D., Digel, C., Guill, C., Rall, B.C., Brose, U., 2013. Body masses, functional responses and predator-prey stability. Ecol. Lett. 16, 1126–1134.

Kareiva, P., Kingsolver, J., Huey, R., 1993. Biotic Interactions and Global Change. Sinauer Associates, Inc., Sunderland, MA. Page 559.

Kiltie, R.A., 2000. Scaling of visual acuity with body size. Funct. Ecol. 14, 226–234.

Kingsolver, J.G., 2009. The well-temperatured biologist. American Society of Naturalists Presidential Address Am. Nat. 174, 755–768.

Kirschfeld, K., 1976. The resolution of lens and compound eyes. In: Zettler, F., Weiler, R. (Eds.), Neural Principles in Vision. Springer-Verlag, Berlin, Heidelberg, pp. 354–370.

Kleiber, M., 1961. The Fire of Life: An Introduction to Animal Energetics. Wiley, New York, 454 p.

Knies, J.L., Kingsolver, J.G., Burch, C.L., 2009. Hotter is better and broader: thermal sensitivity of fitness in a population of bacteriophages. Am. Nat. 173, 419–430.

Kooijman, S.A.L.M., 2000. Dynamic Energy and Mass Budgets in Biological Systems, second ed. Cambridge University Press, Cambridge, UK; New York, NY, USA Page xviii, 424 p.

Krynicki, K., Green, C.D., Sawyer, D.W., 1978. Pressure and temperature dependence of self-diffusion in water. Faraday Discuss. Chem. Soc. 66, 199.

Laska, M.S., Wootton, T.J., 1998. Theoretical concepts and empirical approaches to measuring interaction strength. Ecology 79, 461–476.

Lavergne, S., Mouquet, N., Thuiller, W., Ronce, O., 2010. Biodiversity and climate change: integrating evolutionary and ecological responses of species and communities. Annu. Rev. Ecol. Evol. Systemat. 41, 321–350.

Leaper, R., Raffaelli, D., Letters, E., 1999. Defining the abundance body-size constraint space: data from a real food web. Ecol. Lett. 2, 191–199.

Loeuille, N., Loreau, M., 2005. Evolutionary emergence of size-structured food webs. Proc. Natl. Acad. Sci. USA 102, 5761–5766.

Loeuille, N., Loreau, M., 2006. Evolution of body size in food webs: does the energetic equivalence rule hold? Ecol. Lett. 9, 171–178.

Loeuille, N., Loreau, M., Ferriere, R., 2002. Consequences of plant-herbivore coevolution on the dynamics and functioning of ecosystems. J. Theor. Biol. 217, 369–381.

López-Urrutia, A., San Martin, E., Harris, R.P., Irigoien, X., 2006. Scaling the metabolic balance of the oceans. Proc. Natl. Acad. Sci. USA 103, 8739–8744.

Manila, M., Reiger, H.A., Holems, J.A., Pauly, D., 1990. Influence of temperature changes on aquatic ecosystems: an interpretation of empirical data. Trans. Am. Fish. Soc. 119, 374–389.

Marrow, P., Dieckmann, U., Law, R., 1996. Evolutionary dynamics of predator-prey systems: an ecological perspective. J. Math. Biol. 34, 556–578.

Martin, T.L., Huey, R.B., 2008. Why "suboptimal" is optimal: Jensen's inequality and ectotherm thermal preferences. Am. Nat. 171, E102–E118.

Matthews, B., Narwani, A., Hausch, S., Nonaka, E., Peter, H., Yamamichi, M., Sullam, K.E., Bird, K.C., Thomas, M.K., Hanley, T.C., Turner, C.B., 2011. Toward an integration of evolutionary biology and ecosystem science. Ecol. Lett. 14, 690–701.

May, R.M., 1974. Stability and Complexity in Model Ecosystems. Princeton University Press, Princeton, NJ.

McArdle, B.H., Lawton, J.H., 1979. Effects of prey-size and predator-instar on the predation of Daphnia by Notonecta. Ecol. Entomol. 4, 267–275.

McGill, B.J., Mittelbach, G.G., 2006. An allometric vision and motion model to predict prey encounter rates. Evol. Ecol. Res. 8, 691–701.

Nagy, K.A., 2005. Field metabolic rate and body size. J. Exp. Biol. 208, 1621–1625.

Norberg, J., 2004. Biodiversity and ecosystem functioning: a complex adaptive systems approach. Limnol. Oceanogr. 49, 1269–1277.

O'Connor, M.I., Gilbert, B., Brown, C.J., O'Connor, M.I., Del Rio, A.E.C.M., Bronstein, E.J.L., 2011. Theoretical predictions for how temperature affects the dynamics of interacting herbivores and plants. Am. Nat. 178, 626–638.

O'Connor, M.I., Piehler, M.F., Leech, D.M., Anton, A., Bruno, J.F., 2009. Warming and resource availability shift food web structure and metabolism. PLoS Biol. 7.

O'Gorman, E.J., Emmerson, M.C., 2011. Body mass-abundance relationships are robust to cascading effects in marine food webs. Oikos 120, 520–528.

Okubo, A., 1980. Diffusion and Ecological Problems: Mathematical Models. In: Biomathematics, vol. 10. Springer-Verlag, Berlin, New York. Page xiii, 254 p.

Okubo, A., Levin, S.A., 2001. Diffusion and Ecological Problems: Modern Perspectives In: Interdisciplinary Applied Mathematics, second ed., vol. 14. Springer, New York. Page xx, 467 p.

Otto, S.B., Rall, B.C.B.C., Brose, U., 2007. Allometric degree distributions facilitate food-web stability. Nature 450, 1226–1229.

Pantazelou, E., Dames, C., 1995. Temperature dependence and the role of internal noise in signal transduction efficiency of crayfish mechanoreceptors. Int. J. Bifurc. Chaos 05, 101–108.

Parmesan, C., Yohe, G., 2003. A globally coherent fingerprint of climate change impacts across natural systems. Nature 421, 37–42.

Pawar, S., 2009. Community assembly, stability and signatures of dynamical constraints on food web structure. J. Theor. Biol. 259, 601–612.

Pawar, S., 2015. The role of body size variation in community assembly. Adv. Ecol. Res. 52 (in press).

Pawar, S., Dell, A.I.A., Savage, V.M.V., 2012. Dimensionality of consumer search space drives trophic interaction strengths. Nature 486, 485–489.

Pawar, S., Dell, A.I., Savage, V.M. Trophic interaction dimensionality drives bimodal size-ratio distributions in local communities. (in preparation).

Pawar, S., Dell, A.I., Savage, V.M., 2013. Pawar et al. reply. Nature 493, E2–E3.

Persson, L., Leonardsson, K., de Roos, A.M., Gyllenberg, M., Christensen, B., 1998. Ontogenetic scaling of foraging rates and the dynamics of a size-structured consumer-resource model. Theor. Popul. Biol. 54, 270–293.

Petchey, O.L., Brose, U., Rall, B.C., 2010. Predicting the effects of temperature on food web connectance. Philos. Trans. R. Soc. B Biol. Sci. 365, 2081–2091.

Peters, R., 1986. The Ecological Implications of Body Size. Cambridge University Press, Cambridge. Page 344 Cambridge studies in ecology.

Pörtner, H.O., Bennett, A.F., Bozinovic, F., Clarke, A., Lardies, M.A., Lucassen, M., Pelster, B., Schiemer, F., Stillman, J.H., 2012. Trade-offs in thermal adaptation: the need for a molecular to ecological integration. Physiol. Biochem. Zool. PBZ 79, 295–313.

Pörtner, H.O.H., Farrell, A.A.P., 2008. Physiology and climate change. Science 322, 690–692.

Pörtner, H.O., Knust, R., 2007. Climate change affects marine fishes through the oxygen limitation of thermal tolerance. Science (New York, N.Y.) 315, 95–97.

Rall, B.C., Brose, U., Hartvig, M., Kalinkat, G., Schwarzmüller, F., Vucic-Pestic, O., Petchey, O.L., 2012. Universal temperature and body-mass scaling of feeding rates. Philos. Trans. R. Soc. Lond. B Biol. Sci. 367, 2923–2934.

Ratkowsky, D.A., Olley, J., Ross, T., 2005. Unifying temperature effects on the growth rate of bacteria and the stability of globular proteins. J. Theor. Biol. 233, 351–362.

Reid, P.C., Johns, D.G., Edwards, M., Starr, M., Poulin, M., Snoeijs, P., 2007. A biological consequence of reducing Arctic ice cover: arrival of the Pacific diatom *Neodenticula seminae* in the North Atlantic for the first time in 800,000 years. Glob. Chang. Biol. 13, 1910–1921.

Reuman, D.C., Holt, R.D., Yvon-Durocher, G., 2014. A metabolic perspective on competition and body size reductions with warming. J. Anim. Ecol. 83, 59–69.

Reuman, D.C., Mulder, C., Banasek-Richter, C., Blandenier, M.F.C., Breure, A.M., Den Hollander, H., Kneitel, J.M., Raffaelli, D., Woodward, G., Cohen, J.E., 2009. Allometry of body size and abundance in 166 food webs. Adv. Ecol. Res. 41, 1–44.

Rip, J.M.K., McCann, K.S., 2011. Cross-ecosystem differences in stability and the principle of energy flux. Ecol. Lett. 14, 733–740.

Ritchie, M.E., 2009. Scale, heterogeneity, and the structure and diversity of ecological communities. In: Levin, S.A., Horn, H.S. (Eds.), Monographs in Population Biology. Princeton University Press, Princeton, p. 232.

Saikkonen, K., Taulavuori, K., Hyvönen, T., Gundel, P.E., Hamilton, C.E., Vänninen, I., Nissinen, A., Helander, M., 2012. Climate change-driven species' range shifts filtered by photoperiodism. Nat. Clim. Change 2, 233–234.

Savage, V.M., Gilloly, J.F., Brown, J.H., Charnov, E.L., Gillooly, J.F., West, G.B., 2004. Effects of body size and temperature on population growth. Am. Nat. 163, 429–441.

Savage, V.M., Webb, C.T., Norberg, J., 2007. A general multi-trait-based framework for studying the effects of biodiversity on ecosystem functioning. J. Theor. Biol. 247, 213–229.

Schmidt-Nielsen, K., 1972. Locomotion: energy cost of swimming, flying, and running. Science 177, 222–228.

Schmidt-Nielsen, K., 1984. Scaling, Why Is Animal Size So Important? Cambridge University Press, Cambridge, New York. Page xi, 241 p.

Schoolfield, R.M., Sharpe, P.J., Magnuson, C.E., 1981. Non-linear regression of biological temperature-dependent rate models based on absolute reaction-rate theory. J. Theor. Biol. 88, 719–731.

Schramski, J.R., Dell, A.I., Grady, J.M., Sibly, R.M., Brown, J.H., 2015. Metabolic theory predicts whole-ecosystem properties. PNAS 112, 2617–2622.

Schulte, P.M., Healy, T.M., Fangue, N.A., 2011. Thermal performance curves, phenotypic plasticity, and the time scales of temperature exposure. Integr. Comp. Biol. 51, 691–702.

Sharpe, P.J., DeMichele, D.W., 1977. Reaction kinetics of poikilotherm development. J. Theor. Biol. 64, 649–670.

Shurin, J.B., Gruner, D.S., Hillebrand, H., 2006. All wet or dried up? Real differences between aquatic and terrestrial food webs. Proc. Biol. Sci./Royal Soc. 273, 1–9.

Skellam, J.G., 1958. The mathematical foundations underlying the use of line transects in animal ecology. Biometrics 14, 385–400.

Smouse, P.E., Focardi, S., Moorcroft, P.R., Kie, J.G., Forester, J.D., Morales, J.M., 2010. Stochastic modelling of animal movement. Philos. Trans. R. Soc. B Biol. Sci. 365, 2201–2211.

Stegen, J.C., Ferriere, R., Enquist, B.J., 2012a. Evolving ecological networks and the emergence of biodiversity patterns across temperature gradients. Proc. R. Soc. B Biol. Sci. 279, 1051–1060.

Stegen, J., Enquist, B., Ferrière, R., 2012b. Eco-evolutionary community dynamics: covariation between diversity and invasibility across temperature gradients. Am. Nat. 180, E110–E126.

Tang, S., Pawar, S., Allesina, S., 2014. Correlation between interaction strengths drives stability in large ecological networks. Ecol. Lett. 17, 1094–1100.

Thuiller, W., Münkemüller, T., Lavergne, S., Mouillot, D., Mouquet, N., Schiffers, K., Gravel, D., 2013. A road map for integrating eco-evolutionary processes into biodiversity models. Ecol. Lett. 16 (Suppl. 1), 94–105.

Tilman, D., 1977. Resource competition between plankton algae: an experimental and theoretical approach. Ecology 58, 338–348.

Vasseur, D.A., DeLong, J.P., Gilbert, B., Greig, H.S., Harley, C.D.G., McCann, K.S., Savage, V., Tunney, T.D., O'Connor, M.I., 2014. Increased temperature variation poses a greater risk to species than climate warming. Proc. R. Soc. B Biol. Sci. 281, 20132612.

Vasseur, D.A., McCann, K.S., 2005. A mechanistic approach for modeling temperature-dependent consumer-resource dynamics. Am. Nat. 166, 184–198.

Viswanathan, G.M., da Luz, M.G.E., Raposo, E.P., Stanley, H.E., 2011. The Physics of Foraging: An Introduction to Random Searches and Biological Encounters. Cambridge University Press. Page 178.

Vucic-Pestic, O., Rall, B.C., Kalinkat, G., Brose, U., 2010. Allometric functional response model: body masses constrain interaction strengths. J. Anim. Ecol. 79, 249–256.

Walker, B., Ariza, L.S., Kaines, S., Badger, M.R., Cousins, A.B., 2013. Temperature response of in vivo Rubisco kinetics and mesophyll conductance in *Arabidopsis thaliana*: comparisons to *Nicotiana tabacum*. Plant Cell Environ. 36, 2108–2119.

Weitz, J.S., Levin, S.A., 2006. Size and scaling of predator-prey dynamics. Ecol. Lett. 9, 548–557.

West, G.B., Brown, J.H., 2005. The origin of allometric scaling laws in biology from genomes to ecosystems: towards a quantitative unifying theory of biological structure and organization. J. Exp. Biol. 208, 1575–1592.

West, G.B., Brown, J.H., Enquist, B.J., 1997. A general model for the origin of allometric scaling laws in biology. Science (New York, N.Y.) 276, 122–126.

Wilken, S., Huisman, J., Naus-Wiezer, S., Van Donk, E., 2013. Mixotrophic organisms become more heterotrophic with rising temperature. Ecol. Lett. 16, 225–233.

Yodzis, P., Innes, S., 1992. Body size and consumer-resource dynamics. Am. Nat. 139, 1151–1175.

Yvon-Durocher, G., Allen, A.P., 2012. Linking community size structure and ecosystem functioning using metabolic theory. Philos. Trans. R. Soc. Lond. B Biol. Sci. 367, 2998–3007.

Yvon-Durocher, G., Caffrey, J.M., Cescatti, A., Dossena, M., del Giorgio, P., Gasol, J.M., Montoya, J.M., Pumpanen, J., Staehr, P.A., Trimmer, M., Woodward, G., Allen, A.P., 2012. Reconciling the temperature dependence of respiration across timescales and ecosystem types. Nature 487, 472–476.

Yvon-Durocher, G., Montoya, J.M., Trimmer, M., Woodward, G., 2011. Warming alters the size spectrum and shifts the distribution of biomass in freshwater ecosystems. Glob. Chang. Biol. 17, 1681–1694.

Chapter 2

Ecological Effects of Intraspecific Consumer Biodiversity for Aquatic Communities and Ecosystems

Eric P. Palkovacs[1], David C. Fryxell[1], Nash E. Turley[2] and David M. Post[3]

[1]*Department of Ecology and Evolutionary Biology, University of California, Santa Cruz, CA, USA;*
[2]*Department of Biology, University of Toronto at Mississauga, Mississauga, ON, Canada;*
[3]*Department of Ecology and Evolutionary Biology, Yale University, New Haven, CT, USA*

INTRODUCTION

Consumer identity and biodiversity have important impacts on aquatic communities and ecosystems (Steiner, 2001; Cardinale et al., 2002; Jonsson and Malmqvist, 2003; Woodward, 2009; Narwani and Mazumder, 2010). Researchers have traditionally focused on the importance of species-level biodiversity, but biodiversity within species can also have important consequences for ecology. Studies performed in aquatic systems have played a key role in revealing the ecological importance of intraspecific consumer biodiversity. Experiments have shown that divergent populations sharing a recent common ancestor can differ markedly in their effects on communities and ecosystems (Harmon et al., 2009; Palkovacs et al., 2009; Palkovacs and Post, 2009; Bassar et al., 2010). Field studies in nature show that recent diversification can be important for the ecology of whole ecosystems (Post et al., 2008). The ecological importance of intraspecific consumer biodiversity has been examined for diverse aquatic systems including lakes, ponds, and streams, as well as for a wide array of ecological processes including community structure, trophic interactions, primary production, nutrient recycling, and decomposition.

A number of recent reviews have highlighted the importance of recent and ongoing evolution for ecological dynamics (Hairston et al., 2005; Fussmann et al., 2007; Post and Palkovacs, 2009; Matthews et al., 2011; Schoener, 2011).

However, the importance of intraspecific consumer biodiversity relative to traditional ecological drivers such as species removal or replacement is not well known (Ellner et al., 2011). Here we describe aquatic study systems that have revealed important insights into the ecological effects of intraspecific biodiversity of consumers. We synthesized data from studies that evaluated both the effects of genetic or phenotypic diversity within consumer species (intraspecific effects) and the effects of species removal or replacement (species effects) (Table 1), and performed a meta-analysis to evaluate the relative importance of these types of effects for community and ecosystem responses. Results show that species effects are stronger than intraspecific effects for community responses, but that intraspecific effects are equal in magnitude to species effects for ecosystem responses. These findings show that intraspecific consumer biodiversity can have important consequences for the ecology of aquatic systems.

CASE STUDIES

Aquatic study systems have revealed important effects of intraspecific consumer biodiversity for communities and ecosystems. We are particularly interested in the magnitude of intraspecific biodiversity effects relative to the effects of species removal or replacement, which has been a traditional focus of ecologists. We first describe study systems that have evaluated both intraspecific effects and species effects (Figure 1). We then report the results of a meta-analysis in which we evaluated the relative importance of these effects across studies and ecological responses.

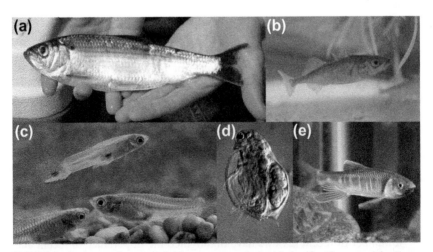

FIGURE 1 Study organisms included in our meta-analysis: (a) alewife, *Alosa pseudoharengus*, (b) threespine stickleback, *Gasterosteus aculeatus*, (c) Trinidadian guppy, *Poecilia reticulata*, (d) *Daphnia*, (e) pale chub, *Zacco platypus*.

TABLE 1 Studies Included in Meta-analysis Comparing the Effects of Intraspecific Consumer Biodiversity to the Effects of Species Removal or Replacement for Aquatic Communities and Ecosystems

Study System	Trait Variation	Experimental Venue	Responses Measured
Alewife			
Post et al. (2008)	Migratory life history (anadromous-landlocked)	Whole lake contrasts	Zooplankton size Zooplankton biomass Phytoplankton biomass
Palkovacs and Post (2009)	Migratory life history (anadromous-landlocked)	Lake mesocosms (18,850-L)	Zooplankton size Zooplankton biomass Zooplankton diversity Phytoplankton biomass
Guppy			
Palkovacs et al. (2009)	Predator-driven life history (low predation–high predation)	Stream mesocosms (200-L)	Benthic invertebrate biomass Periphyton biomass Decomposition Nutrient excretion
Bassar et al. (2010)	Predator-driven life history (low predation–high predation)	Stream mesocosms (200-L)	Benthic invertebrate biomass Periphyton biomass Decomposition Gross primary production Community respiration Nutrient excretion Benthic organic material

Continued

TABLE 1 Studies Included in Meta-analysis Comparing the Effects of Intraspecific Consumer Biodiversity to the Effects of Species Removal or Replacement for Aquatic Communities and Ecosystems—cont'd

Study System	Trait Variation	Experimental Venue	Responses Measured
Daphnia			
Hargrave et al. (2011)	Standing genetic variation (feeding differences among clones)	Laboratory microcosms (2.5-L)	Algae clearance rate
Walsh et al. (2012)	Predator-driven life history (anadromous–landlocked–no alewife)	Laboratory mesocosms (56-L)	Net primary production
Chislock et al. (2013)	Response to toxic prey (sensitive–tolerant)	Lake mesocosms (3100-L)	Phytoplankton biomass Gross primary production
Stickleback			
Ingram et al. (2012)	Response to intraguild predator (sculpin present–sculpin absent)	Pond mesocosms (1136-L)	Zooplankton biomass Benthic invertebrate biomass
Des Roches et al. (2013)	Foraging habitat (generalist–benthic–limnetic)	Pond mesocosms (1136-L)	Zooplankton biomass Phytoplankton biomass Periphyton biomass Dissolved oxygen
Pale chub			
Katano (2011)	Within-population foraging behavior(bottom–surface–mixed)	Pond mesocosms (839-L)	Benthic invertebrate density Periphyton biomass

Migration and Foraging Trait Divergence in Alewife

The alewife (*Alosa pseudoharengus*) is an anadromous (migratory sea-run) zooplanktivorous fish native to the Atlantic coast of North America. In Connecticut (United States), dam construction beginning in the seventeenth century (Twining and Post, 2013; Twining et al., 2013) appears to have caused some populations to evolve a landlocked (freshwater-resident) life history (Palkovacs et al., 2008). The important ecological effects of landlocked alewife populations in Connecticut lakes have been known for a long time due to the classic work of Brooks and Dodson (1965), which showed that lakes with landlocked populations are dominated by small-bodied zooplankton species, whereas lakes without alewife populations are dominated by large-bodied zooplankton species. These effects led to the designation of the alewife as a keystone species in temperate lakes (Power et al., 1996). Post et al. (2008) revisited many of the lakes sampled by Brooks and Dodson (1965), plus lakes with anadromous alewife populations that were not sampled previously. This study revealed that anadromous alewives have effects on lake ecosystems that are strong but distinct from those of landlocked alewives. In lakes with anadromous alewife populations, large-bodied zooplankton are present in early spring, before the annual alewife spawning migration. However, during summer large-bodied species are absent and average zooplankton body size is reduced below that seen in landlocked alewife lakes. Consequently, anadromous alewives cause a stronger midsummer trophic cascade than landlocked alewives (Post et al., 2008).

Landlocked alewife populations show parallel divergence in foraging traits from their anadromous ancestors (Palkovacs et al., 2008; Palkovacs and Post, 2008). Anadromous alewife populations have large gapes, widely spaced gill rakers, and are highly selective for large-bodied prey items, whereas landlocked populations have small gapes, narrowly spaced gill rakers, and show reduced prey selectivity. Palkovacs and Post (2008) hypothesized that ecoevolutionary feedbacks were responsible for foraging trait divergence between anadromous and landlocked alewife populations, which then cause differences in midsummer zooplankton and phytoplankton communities. Palkovacs and Post (2009) conducted a mesocosm experiment to test the strength of contemporary divergence in foraging traits relative to the effects of alewife removal. This experiment revealed significant effects of alewife divergence on zooplankton size, biomass, species richness, and species diversity. Subsequent analyses of phytoplankton data from this experiment have revealed additional cascading effects on phytoplankton biomass (Palkovacs and Dalton, 2012) and community structure (Weis and Post, 2013). Continued investigations in this study system have shown trait divergence between anadromous and landlocked alewife populations in body shape and dietary niche (Schielke et al., 2011; Jones et al., 2013), and additional effects on zooplankton and phytoplankton metacommunity structure (Howeth et al., 2013). Contemporary trait divergence among alewife populations also drives contemporary evolution in populations of *Daphnia ambigua* (Walsh and Post, 2011, 2012), which in turn creates an additional layer of intraspecific effects in *D. ambigua* that further shapes phytoplankton production (Walsh et al., 2012).

Life History Divergence in the Trinidadian Guppy

The Trinidadian guppy (*Poecilia reticulata*) is a classic study system in evolutionary biology. Work by Reznick and Endler (1982) documented life history divergence among guppy populations in response to fish predators. Guppy populations from high-predation streams containing fish predators (*Crenicichla alta*, *Hoplias malabaracus*) display earlier age and smaller size at maturation compared with populations from low-predation streams containing *Rivulus hartii*. Subsequent work established the evolutionary basis of this life history divergence (Reznick and Bryga, 1996; Reznick et al., 1996), and experimental introductions showed that such evolution can occur on contemporary timescales (Reznick et al., 1990, 1997). Evidence for contemporary evolution in response to predators prompted researchers to ask whether life history divergence in guppies has impacts on community and ecosystem processes.

A series of mesocosm experiments have explored the ecological effects of guppy life history evolution. Palkovacs et al. (2009) established treatments containing different combinations of guppy and *Rivulus* populations to test the effects of guppy invasion, guppy evolution and guppy–*Rivulus* coevolution on community and ecosystem processes. Results show that divergence between high-predation and low-predation guppies in diet and excretion rates cause differences in nutrient dynamics and periphyton biomass and accrual rates. Guppy–*Rivulus* coevolution appears to impact benthic invertebrate biomass, but the trait differences underlying this effect are not known. Subsequent work has revealed additional differences between high-predation and low-predation guppy populations in feeding rates and diet selectivity (Palkovacs et al., 2011; Zandonà et al., 2011). Bassar et al. (2010) conducted an experiment in which they examined the interacting effects of guppy evolution and guppy density. The findings of this study support the results of Palkovacs et al. (2009) for nutrients and periphyton and show additional effects on decomposition rates, gross primary productivity, and one component of the benthic invertebrate community (chironomids). These responses were modified by differences in guppy density. Subsequent reanalysis of the data from Bassar et al. (2010) has revealed important complexities that may help explain why intraspecific biodiversity effects have been underappreciated until recently. Different ecological pathways may be opposing and thus offsetting (Bassar et al., 2012), and the direction of evolutionary changes may often mask (rather than exacerbate) community and ecosystem changes (Ellner et al., 2011). These factors may contribute to what Yoshida et al. (2007) have called the "cryptic" effects of rapid evolution on trophic interactions.

Divergence Due to Predators and Toxic Prey in *Daphnia*

The large-bodied cladoceran zooplankter *Daphnia* is the dominant grazer on phytoplankton in many aquatic ecosystems. *Daphnia* is a preferred prey item for many zooplanktivorous fishes, and predation on *Daphnia* can cause strong trophic cascades in lakes and ponds (Hurlbert et al., 1972; Carpenter et al., 1987). *Daphnia* is also vulnerable to toxic cyanobacteria, which often increase in abundance due to cultural eutrophication. *Daphnia* exhibit rapid evolutionary responses to both predators and toxic prey (Hairston et al., 1999, 2001; Cousyn et al., 2001). The potential for contemporary evolution and its important role in aquatic food webs suggest that *Daphnia's* intraspecific biodiversity effects may be strong in many ecosystems (Miner et al., 2012).

Several mesocosm studies have investigated the strength of intraspecific effects in *Daphnia*. Studies by Walsh and Post (2011, 2012) examined evolutionary responses of *D. ambigua* to predation by alewives, showing that *D. ambigua* from lakes with anadromous alewives grow faster, mature earlier, and produce more offspring than do *D. ambigua* from lakes with landlocked alewives or lakes without alewives. A subsequent study used laboratory mesocosms to test the effect of this evolutionary response on consumer–resource dynamics, showing that *D. ambigua* from lakes with anadromous alewife significantly reduce phytoplankton abundance and net primary productivity relative to *D. ambigua* from lakes with landlocked alewives or lakes without alewives (Walsh et al., 2012). Whether such "cascades" of eco-evolutionary effects are common is unknown.

Another experiment testing the effects of *Daphnia* genetic variation on phytoplankton consumption rates manipulated the number of *Daphnia* species and clones, chosen from cultures with similar recent histories of laboratory selection, within micrososms (Hargrave et al., 2011). This study found some variation among clones in consumption rates, yet no net effect of clonal richness on overall algal consumption. In another study, Chislock et al. (2013) examined the ecological effects of adaptation in response to toxic cyanobacteria using lake mesocosms. Results show that toxin-resistant genotypes of *Daphnia pulicaria* reduced phytoplankton biomass and gross primary productivity compared with toxin-sensitive genotypes. This result suggests that the evolution of toxin resistance is what allows *Daphnia* in eutrophic habitats to control phytoplankton abundance despite the prevalence of toxic cyanobacteria in these environments.

Foraging Habitat Divergence in Threespine Stickleback

The threespine stickleback (*Gasterosteus aculeatus*) is a model system for understanding adaptation and ecological diversification (Bell and Foster, 1994). Marine stickleback repeatedly invaded freshwater lakes after the last glaciation 10,000–12,000 years ago. After entering freshwater, populations evolve as a foraging habitat generalist or adapt to specialize on either limnetic or benthic habitats (Schluter and McPhail, 1992). Limnetic stickleback display fusiform bodies, narrow gapes, and densely packed gill rakers, and also prefer zooplankton, whereas benthic stickleback display deep bodies, wide gapes, and few short gill rakers, and also have a preference for benthic invertebrates (Schluter, 1993, 1994).

Several mesocosm studies have examined the effects of foraging-habitat specialization in stickleback on community and ecosystem processes. Harmon et al. (2009) established treatments that included generalist, benthic, limnetic, and benthic-plus-limnetic stickleback (this study is not included in our meta-analysis because it did not include a species removal or replacement treatment). Results show that different stickleback forms shaped communities and ecosystems in different ways, but ecological responses were not predictable based on feeding preferences. For example, habitat specialists would be predicted to have the greatest impact on their preferred prey, but the generalist form reduced some benthic (e.g., ostracods) and some limnetic (e.g., *Daphnia*) prey items more than either specialized form. Differences in dissolved organic carbon and light transmission were observed among treatments, but the causal mechanisms driving these differences are unknown. A study conducted by Des Roches et al. (2013) added a species-removal (no stickleback) treatment and a stickleback-density treatment to the Harmon et al. (2009) design. This study found relatively weak effects of phenotypic diversity relative to the effects of density. Des Roches et al. (2013) found differences in components of the zooplankton community, but no significant differences among treatments (including the no-fish control) in the biomass of large zooplankton. This result is surprising given a long history of work showing major impacts of planktivorous fishes on large zooplankton (Lazzaro, 1987). Ingram et al. (2012) examined the ecological effects of stickleback adaptation in response to the presence of an intraguild predator, the prickly sculpin (*Cottus asper*). Sculpin are a benthic predator and appear to drive selection on stickleback to adopt a more limnetic lifestyle. Results of a mesocosm experiment show that stickleback that have evolved in the presence of sculpin consume more zooplankton and reduce zooplankton biomass relative to stickleback that have evolved without sculpin. This result shows that stickleback habitat specialization can have predictable impacts on community and ecosystem processes, but the strength of such effects differ among experiments and evolutionary scenarios.

Within-Population Variation in Feeding Behavior in Pale Chub

The above examples describe study systems where genetic and phenotypic differences are the results of adaptive divergence among populations. However, trait differences within populations can also have important ecological effects (Bolnick et al., 2011). To examine the ecological effects of feeding specialization within populations, Katano (2011) conducted an experiment involving different behavioral feeding types of the pale chub (*Zacco platypus*). This study characterized different stereotypical feeding behaviors that differed among individuals but appeared consistent within individuals over time. These behavioral types included bottom feeders, surface feeders, and intermediate feeders that displayed both behaviors. The differential ecological effects of these feeding types were examined in experimental mesocosms. Surface feeders and intermediate feeders caused trophic cascades, increasing periphyton biomass by reducing the abundance of benthic invertebrates by consuming them when they entered the drift. In contrast, bottom feeders did not cause a strong trophic cascade; although they consumed benthic invertebrates, they also directly consumed periphyton, and this direct consumption of algae appears to have compensated for the reduction of invertebrate grazers. The results of this study show that within-population variation can have important ecological effects. Heritable within-population trait variation is the raw material upon which selection acts, making within-population variation important for current and future eco-evolutionary dynamics (Carlson et al., 2011).

META-ANALYSIS

Recent and ongoing evolution creates phenotypic variation within species that may be important for communities and ecosystems. We examined the effects of biodiversity within consumer species (intraspecific effects) relative to the effects of species removal or replacement (species effects) and performed a meta-analysis to evaluate the relative importance of these effects for community and ecosystem responses (Table 1). We calculated intraspecific effects as mean of pairwise differences among within-species treatments. For most studies, the intraspecific effect was the pairwise contrasts that best represented the authors interpretation of the study system (e.g., anadromous versus landlocked alewives (Post et al., 2008; Palkovacs and Post, 2009), high predation guppies at low-density versus low-predation guppies at high density (Bassar et al., 2010), benthic versus limnetic stickleback (Des Roches et al., 2013)). For Hargrave et al. (2011), the intraspecific effect was the mean contrast among the four clones of *Daphnia pulex* and *D. pulicaria*. For Katano (2011), the intraspecific effect was the mean contrast among bottom, surface, and intermediate feeding forms of pale chub. We estimated species effects as the mean pairwise differences between treatments that removed or replaced the target species of interest and the mean of all other treatments that included the target species. For the Hargrave et al. (2011) study, we estimated the species effect as the mean contrast among the four species of *Daphnia* included in the study.

We calculated Hedge's *g*, a bias corrected version of Cohen's *d*, as our standardized effect size (Hedges, 1981). Hedge's *g* was calculated as (intraspecific effect–species effect)/pooled standard deviation. A negative Hedge's *g* indicates that species effects were larger than intraspecific effects. We estimated Hedge's *g* for each of the response variables measured in each study, and we classified response variables as a community or ecosystem, and direct or indirect, and as influenced by consumption, nutrient excretion, or both consumption and nutrient excretion. We dropped the nutrient excretion classification from our final analysis because there were only two response variables that could be unambiguously classified as nutrient excretion only. We performed all calculations with the Metafore package version 1.9.2 in R (Viechtbauer, 2010). We estimated Hedge's *g* and the sampling variance using the *escalc* function, and we performed our meta-analysis using the *rma* function, with effect size estimates weighted by the inverse of the sampling variance.

Overall, species effects were significantly greater than intraspecific effects (mean Hedge's $g = -0.44$, SE = 0.12, $z = -3.56$, $n = 54$, $p < 0.001$; Figure 2). Specifically, removing a species or replacing it with a different species had an effect that was on average 0.44 standard deviation units greater than manipulating variation within species. Although species effects were larger than intraspecific effects, there was considerable variation among response variables (Figure 2).

Following the rule of thumb proposed by Cohen (1988), we defined effects as large intraspecific effects if $g > 0.3$, large species effects if $g < -0.3$, and roughly equal if $-0.3 < g < 0.3$. For 20% of the response variables, intraspecific effects were large; for 20% of the response variables, intraspecific and

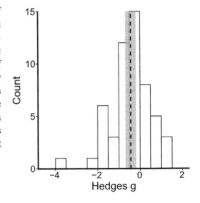

FIGURE 2 The overall ecological effects of replacing or removing consumer species (species effects) were larger than replacing forms, populations, or genotypes within species (intraspecific effects). The histogram shows the distribution of effects sizes across all response variables in our database. Negative values indicate that species effects were greater than intraspecific effects. The vertical dashed line is a predicted mean effect from a inverse sampling variance-weighted meta-analysis and the gray shaded area show 95% CI around that estimate.

species effects were about equal; and for about 60% of the response variables, species effects were large. Large intraspecific effects (relative to species effects) include those for alewive foraging trait divergence on small-bodied zooplankton biomass and phytoplankton biomass (Post et al., 2008; Palkovacs and Post, 2009), guppy life history divergence on periphyton biomass, nitrogen, and decomposition rates (Palkovacs et al., 2009; Bassar et al., 2010), and *Daphnia* cyanotoxin tolerance on phytoplankton biomass and gross primary productivity (Chislock et al., 2013). Large species effects (relative to intraspecific effects) include those for alewife presence on large-bodied zooplankton biomass and zooplankton length distributions (Post et al., 2008; Palkovacs and Post, 2009), stickleback presence on zooplankton biomass (Ingram et al., 2012) and phytoplankton biomass (Des Roches et al., 2013), *Daphnia* presence or species identity on net primary productivity (Hargrave et al., 2011; Walsh et al., 2012), and guppy presence on invertebrate biomass, phosphorus, and periphyton production (Palkovacs et al., 2009; Bassar et al., 2010).

We found that species effects were significantly larger for community response variables than for ecosystem response variables (estimated difference = 0.70, $z = 3.03$, $N = 52$, p = 0.024; Figure 3). For community response variables, the overall species effect was 0.75 standard deviation units greater than the overall intraspecific effect. In contrast, for ecosystem response variables, intraspecific effects and species effects were not significantly different (Figure 3). Ecosystem and community response variables were represented in all studies, suggesting that this difference is a result of the response variables themselves rather than variation among studies or study systems. We found that for studies with aquatic consumers, community variables were nearly all driven by the effects of consumption, whereas ecosystem variables

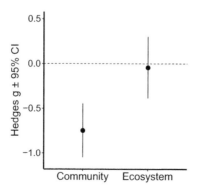

FIGURE 3 Replacing or removing consumer species (species effects) had a larger impact on community response variables than replacing forms, populations, or genotypes within species (intraspecific effects). Species effects and intraspecific effects had equal impacts on ecosystem responses. Graph shows predicted means from a formal meta-analysis and 95% CI.

were driven by both consumption and nutrient excretion. Thus, our analysis shows that the effects of species removal or replacement were strongest on community variables driven directly by top-down effects, but that intraspecific effects and species effects were equal in magnitude for ecosystem variables driven by both top-down and bottom-up effects.

CONCLUSIONS

Community and ecosystem ecology have a long history of focusing on the species-level biodiversity, implicitly assuming that intraspecific biodiversity is unimportant for ecological dynamics. This perspective has been challenged by recent studies in aquatic systems that have directly examined the importance of intraspecific consumer biodiversity relative to species removal or replacement. Our meta-analysis reveals that, while species effects are stronger on average, intraspecific effects can be important, especially for ecosystem responses driven by the combined effects of consumption and nutrient excretion. Our meta-analysis reveals such effects across a diversity of organisms (fishes and invertebrates) and aquatic study systems (lakes, ponds, and streams), suggesting that intraspecific consumer biodiversity is broadly important for aquatic communities and ecosystems.

ACKNOWLEDGMENTS

We thank the Quebec Center for Biodiversity Science working group on eco-evolutionary dynamics for many thoughtful discussions that helped us develop this work.

REFERENCES

Bassar, R.D., Ferriere, R., Lopez-Sepulcre, A., Marshall, M.C., Travis, J., Pringle, C.M., Reznick, D.N., 2012. Direct and indirect ecosystem effects of evolutionary adaptation in the trinidadian guppy (*Poecilia reticulata*). Am. Nat. 180, 167–185.

Bassar, R.D., Marshall, M.C., López-Sepulcre, A., Zandonà, E., Auer, S., Travis, J., Pringle, C.M., Flecker, A.S., Thomas, S.A., Fraser, D.F., Reznick, D.N., 2010. Local adaptation in Trinidadian guppies alters ecosystem processes. Proc. Natl. Acad. Sci. USA 107, 3616–3621.

Bell, M.A., Foster, S.A. (Eds.), 1994. The Evolutionary Biology of the Threespine Stickleback. Oxford University Press, Oxford.

Bolnick, D.I., Amarasekare, P., Araujo, M.S., Burger, R., Levine, J.M., Novak, M., Rudolf, V.H.W., Schreiber, S.J., Urban, M.C., Vasseur, D.A., 2011. Why intraspecific trait variation matters in community ecology. Trends Ecol. Evol. 26, 183–192.

Brooks, J.L., Dodson, S.I., 1965. Predation, body size, and composition of plankton. Science 150, 28–35.

Cardinale, B.J., Palmer, M.A., Collins, S.L., 2002. Species diversity enhances ecosystem functioning through interspecific facilitation. Nature 415, 426–429.

Carlson, S.M., Quinn, T.P., Hendry, A.P., 2011. Eco-evolutionary dynamics in Pacific salmon. Heredity 106, 438–447.

Carpenter, S.R., Kitchell, J.F., Hodgson, J.R., Cochran, P.A., Elser, J.J., Elser, M.M., Lodge, D.M., Kretchmer, D., He, X., Vonende, C.N., 1987. Regulation of lake primary productivity by food web structure. Ecology 68, 1863–1876.

Chislock, M.F., Sarnelle, O., Olsen, B.K., Doster, E., Wilson, A.E., 2013. Large effects of consumer offense on ecosystem structure and function. Ecology 94, 2375–2380.

Cohen, J., 1988. Stastical Power Analysis for the Behavioral Sciences, second ed. Lawrence Erlbaum Associates, Hillsdale, NJ.

Cousyn, C., De Meester, L., Colbourne, J.K., Brendonck, L., Verschuren, D., Volckaert, F., 2001. Rapid, local adaptation of zooplankton behavior to changes in predation pressure in the absence of neutral genetic changes. Proc. Natl. Acad. Sci. USA 98, 6256–6260.

Des Roches, S., Shurin, J.B., Schluter, D., Harmon, L.J., 2013. Ecological and evolutionary effects of stickleback on community structure. PLoS ONE 8.

Ellner, S.P., Geber, M.A., Hairston, N.G., 2011. Does rapid evolution matter? measuring the rate of contemporary evolution and its impacts on ecological dynamics. Ecol. Lett. 14, 603–614.

Fussmann, G.F., Loreau, M., Abrams, P.A., 2007. Eco-evolutionary dynamics of communities and ecosystems. Funct. Ecol. 21, 465–477.

Hairston, N.G.J., Ellner, S.P., Geber, M.A., Yoshida, T., Fox, J.A., 2005. Rapid evolution and the convergence of ecological and evolutionary time. Ecol. Lett. 8, 1114–1127.

Hairston, N.G.J., Holtmeier, C.L., Lampert, W., Weider, L.J., Post, D.M., Fischer, J.M., Caceres, C.E., Fox, J.A., Gaedke, U., 2001. Natural selection for grazer resistance to toxic cyanobacteria: evolution of phenotypic plasticity? Evolution 55, 2203–2214.

Hairston, N.G.J., Lampert, W., Caceres, C.E., Holtmeier, C.L., Weider, L.J., Gaedke, U., Fischer, J.M., Fox, J.A., Post, D.M., 1999. Lake ecosystems—rapid evolution revealed by dormant eggs. Nature 401, 446.

Hargrave, C.W., Hambright, K.D., Weider, L.J., 2011. Variation in resource consumption across a gradient of increasing intra- and interspecific richness. Ecology 92, 1226–1235.

Harmon, L.J., Matthews, B., Des Roches, S., Chase, J.M., Shurin, J.B., Schluter, D., 2009. Evolutionary diversification in stickleback affects ecosystem functioning. Nature 458, 1167–1170.

Hedges, L.V., 1981. Distribution theory for Glass's estimator of effect size and related estimators. J. Educ. Stat. 6, 107–128.

Howeth, J.G., Weis, J.J., Brodersen, J., Hatton, E.C., Post, D.M., 2013. Intraspecific phenotypic variation in a fish predator affects multitrophic lake metacommunity structure. Ecol. Evol. 3, 5031–5044.

Hurlbert, S.H., Zedler, J., Fairbank, D., 1972. Ecosystem alteration by mosquitofish (*Gambusia affinis*) predation. Science 175, 639–641.

Ingram, T., Svanbäck, R., Kraft, N.J.B., Kratina, P., Southcott, L., Schluter, D., 2012. Intraguild predation drives evolutionary niche shift in threespine stickleback. Evolution 66, 1819–1832.

Jones, A.W., Palkovacs, E.P., Post, D.M., 2013. Recent parallel divergence in body shape and diet source of alewife life history forms. Evol. Ecol. 27, 1175–1187.

Jonsson, M., Malmqvist, B., 2003. Importance of species identity and number for process rates within different stream invertebrate functional feeding groups. J. Anim. Ecol. 72, 453–459.

Katano, O., 2011. Effects of individual differences in foraging of pale chub on algal biomass through trophic cascades. Environ. Biol. Fishes 92, 101–112.

Lazzaro, X., 1987. A review of planktivorous fishes: their evolution, feeding behaviours, selectivities, and impacts. Hydrobiologia 146, 97–167.

Matthews, B., Narwani, A., Hausch, S., Nonaka, E., Peter, H., Yamamichi, M., Sullam, K.E., Bird, K.C., Thomas, M.K., Hanley, T.C., Turner, C.B., 2011. Toward an integration of evolutionary biology and ecosystem science. Ecol. Lett. 14, 690–701.

Miner, B.E., De Meester, L., Pfrender, M.E., Lampert, W., Hairston, N.G., 2012. Linking genes to communities and ecosystems: Daphnia as an ecogenomic model. Proc. R. Soc. B-Biol. Sci. 279, 1873—1882.

Narwani, A., Mazumder, A., 2010. Community composition and consumer identity determine the effect of resource species diversity on rates of consumption. Ecology 91, 3441—3447.

Palkovacs, E.P., Dalton, C.M., 2012. Ecosystem consequences of behavioral plasticity and contemporary evolution. In: Wong, B.B.M., Candolin, U. (Eds.), Behavioral Responses to a Changing World: Mechanisms and Consequences. Oxford University Press, Oxford, U.K.

Palkovacs, E.P., Dion, K.B., Post, D.M., Caccone, A., 2008. Independent evolutionary origins of landlocked alewife populations and rapid parallel evolution of phenotypic traits. Mol. Ecol. 17, 582—597.

Palkovacs, E.P., Marshall, M.C., Lamphere, B.A., Lynch, B.R., Weese, D.J., Fraser, D.F., Reznick, D.N., Pringle, C.M., Kinnison, M.T., 2009. Experimental evaluation of evolution and coevolution as agents of ecosystem change in Trinidadian streams. Philos. Trans. R. Soc. B-Biol. Sci. 364, 1617—1628.

Palkovacs, E.P., Post, D.M., 2008. Eco-evolutionary interactions between predators and prey: can predator-induced changes to prey communities feed back to shape predator foraging traits? Evol. Ecol. Res. 10, 699—720.

Palkovacs, E.P., Post, D.M., 2009. Experimental evidence that phenotypic divergence in predators drives community divergence in prey. Ecology 90, 300—305.

Palkovacs, E.P., Wasserman, B.A., Kinnison, M.T., 2011. Eco-evolutionary trophic dynamics: loss of top predators drives trophic evolution and ecology of prey. PLoS ONE 6, e18879.

Post, D.M., Palkovacs, E.P., 2009. Eco-evolutionary feedbacks in community and ecosystem ecology: interactions between the ecological theatre and the evolutionary play. Philos. Trans. R. Soc. B-Biol. Sci. 364, 1629—1640.

Post, D.M., Palkovacs, E.P., Schielke, E.G., Dodson, S.I., 2008. Intraspecific variation in a predator affects community structure and cascading trophic interactions. Ecology 89, 2019—2032.

Power, M.E., Tilman, D.E., Estes, J.A., Menge, B.A., Bond, W.J., Mills, L.S., Daily, G.S., Castilla, J.C., Lubchencho, J., Paine, R.T., 1996. Challenges in the quest for keystones. BioScience 46, 609—620.

Reznick, D., Endler, J.A., 1982. The impact of predation on life history evolution in Trinidadian guppies (*Poecilia reticulata*). Evolution 36, 160—177.

Reznick, D.A., Bryga, H., Endler, J.A., 1990. Experimentally induced life-history evolution in a natural population. Nature 346, 357—359.

Reznick, D.N., Bryga, H.A., 1996. Life-history evolution in guppies (*Poecilia reticulata*: Poeciliidae). 5. Genetic basis of parallelism in life histories. Am. Nat. 147, 339—359.

Reznick, D.N., Butler, M.J., Rodd, F.H., Ross, P., 1996. Life-history evolution in guppies (*Poecilia reticulata*). 6. Differential mortality as a mechanism for natural selection. Evolution 50, 1651—1660.

Reznick, D.N., Shaw, F.H., Rodd, F.H., Shaw, R.G., 1997. Evaluation of the rate of evolution in natural populations of guppies (*Poecilia reticulata*). Science 275, 1934—1937.

Schielke, E.G., Palkovacs, E.P., Post, D.M., 2011. Eco-evolutionary feedbacks drive dietary niche differences in alewives. Biol. Theory 6, 211—219.

Schluter, D., 1993. Adaptive radiation in sticklebacks - size, shape, and habitat use efficiency. Ecology 74, 699—709.

Schluter, D., 1994. Experimental-evidence that competition promotes divergence in adaptive radiation. Science 266, 798—801.

Schluter, D., McPhail, J.D., 1992. Ecological character displacement and speciation in sticklebacks. Am. Nat. 140, 85–108.

Schoener, T.W., 2011. The newest synthesis: understanding the interplay of evolutionary and ecological dynamics. Science 331, 426–429.

Steiner, C.F., 2001. The effects of prey heterogeneity and consumer identity on the limitation of trophic-level biomass. Ecology 82, 2495–2506.

Twining, C.W., Post, D.M., 2013. Cladoceran remains reveal presence of a keystone size-selective planktivore. J. Paleolimnol. 49, 253–266.

Twining, C.W., West, D.C., Post, D.M., 2013. Historical changes in nutrient inputs from humans and anadromous fish in New England's coastal watersheds. Limnol. Oceanogr. 58, 1286–1300.

Viechtbauer, W., 2010. Conducting meta-analyses in R with the metafor package. J. Stat. Software 36, 1–48.

Walsh, M.R., DeLong, J.P., Hanley, T.C., Post, D.M., 2012. A cascade of evolutionary change alters consumer-resource dynamics and ecosystem function. Proc. R. Soc. B-Biol. Sci. 279, 3184–3192.

Walsh, M.R., Post, D.M., 2011. Interpopulation variation in a fish predator drives evolutionary divergence in prey in lakes. Proc. R. Soc. B-Biol. Sci. 278, 2628–2637.

Walsh, M.R., Post, D.M., 2012. The impact of intraspecific variation in a fish predator on the evolution of phenotypic plasticity and investment in sex in *Daphnia ambigua*. J. Evol. Biol. 25, 80–89.

Weis, J.J., Post, D.M., 2013. Intraspecific variation in a predator drives cascading variation in primary producer community composition. Oikos 122, 1343–1349.

Woodward, G., 2009. Biodiversity, ecosystem functioning and food webs in fresh waters: assembling the jigsaw puzzle. Freshwater Biol. 54, 2171–2187.

Yoshida, T., Ellner, S.P., Jones, L.E., Bohannan, B.J.M., Lenski, R.E., Hairston, N.G., 2007. Cryptic population dynamics: rapid evolution masks trophic interactions. PLoS Biol. 5, 1868–1879.

Zandonà, E., Auer, S.K., Kilham, S.S., Howard, J.L., Lopez-Sepulcre, A., O'Connor, M.P., Bassar, R.D., Osorio, A., Pringle, C.M., Reznick, D.N., 2011. Diet quality and prey selectivity correlate with life histories and predation regime in Trinidadian guppies. Funct. Ecol. 25, 964–973.

Chapter 3

How Does Evolutionary History Alter the Relationship between Biodiversity and Ecosystem Function?

David A. Vasseur and Susanna M. Messinger
Department of Ecology and Evolutionary Biology, Yale University, New Haven, CT, USA

INTRODUCTION

The relationship between biodiversity and ecosystem function (B-EF) has been of great interest recently (Cardinale et al., 2002, 2006; Hooper et al., 2005; Loreau, 1998a, b; Loreau et al., 2001; Reich et al., 2012; Reiss et al., 2009; Tilman, 1996, 2000; Wardle et al., 1997) due in part to the increasing rate of species loss in many ecosystems (Pimm and Raven, 2000) and the need to sustain the human-derived goods and services from these systems (Reiss et al., 2009; Thompson and Starzomski, 2007). A wealth of experiments conducted between 1994 and 2014 has led to general acceptance of the hypothesis that ecosystem function increases to a saturating maximum as species richness increases in communities (Cardinale et al., 2006; Loreau et al., 2001; Tilman et al., 2001); however, the multiple mechanisms that potentially drive this relationship have allowed researchers to question the true importance of diversity for ecosystem function (Jiang et al., 2008; Loreau, 1998a; Reiss et al., 2009).

A positive relationship between B-EF can arise when species are able to use complementary niches in space or time or when species interactions enhance the capture of resources when species co-occur. Together, these processes are known as "complementarity" (Cardinale et al., 2002) and this mechanism relating diversity to ecosystem function underlies much of our intuition about the role of diversity in B-EF studies (Hooper and Dukes, 2004). However, when assembling communities with greater species richness, it is increasingly likely that dominant species (those which contribute more to ecosystem function) are included, therefore artificially generating a relationship between diversity and ecosystem function. This is known as the "selection" (or sampling) effect (Huston, 1997; Wardle, 1999). Various experimental designs and comparative measures have been suggested in order to partition the relative contributions of these two mechanisms to the B-EF relationship (Loreau and Hector, 2001; Loreau, 1998a, 2000).

Central to partitioning the contributions of complementarity and sampling effects are comparisons of the yield of a particular ecosystem function (e.g., biomass, carbon storage, resource depletion, etc.) between a diverse polyculture of species and monocultures of the same species, grown under the same environmental conditions (Hector et al., 2002). Nontransgressive overyielding occurs when the yield of a polyculture is greater than an average or weighted average of the yields of individual species in monoculture. Nontransgressive overyielding is common in nature but can arise due to either complementarity or selection effects (Cardinale et al., 2006; Hector et al., 2002). In order for complementarity to underlie a positive effect of diversity on ecosystem function, polycultures should exhibit transgressive overyielding, which occurs when the polyculture yield exceeds that of all of its constituent species grown in monoculture (Cardinale et al., 2004; Hooper and Dukes, 2004; Tilman et al., 1997). In a summary of 111 field, greenhouse, and laboratory experiments on the B-EF relationship, Cardinale et al. (2006) found significant nontransgressive overyielding when considering either the standing stocks (biomass) or resource depletion of communities; however, transgressive overyielding was nonsignificant for both measures and when data were further partitioned by aquatic or terrestrial habitat and by trophic group. This result suggested only a limited role for complementarity as a driver of positive B-EF relationships and highlighted the lack of concordance between empirical data and theory.

One of the shortcomings befalling many experiments is that their limited duration may not allow communities to reach the equilibrium conditions imposed by most theoretical models. Hooper and Dukes (2004) demonstrated a shift from selection-dominated to complementarity dominated overyielding over a period of 6 years in a long-term grassland experiment; however, these effects were not strong enough to cause transgressive overyielding. Recently, Cardinale et al. (2007) summarized a set of 44 independent experiments and demonstrated a significant time effect on the B-EF relationship for plants. Longer duration experiments exhibited increased polyculture yield (biomass) relative to those with shorter duration and furthermore, the probability of transgressive overyielding tended to increase with experiment duration, with the trend fit to metadata suggesting that it takes approximately 5 years for the most diverse community to exhibit transgressive overyielding (Cardinale et al., 2007). The authors suggested that the delayed impact of diversity results from temporal changes to resource use and interspecific interactions that yield greater species' complementarity, and that many experiments may simply be too short for plants to develop, for example, differentiated rooting depths to allow access to complementary soil resources. Reich et al. (2012) analyzed two long-term (>13 years) studies on plants and similarly found that the effects of diversity on productivity increased through time. Here, species-rich plots were found to accumulate more soil nitrogen over time, suggesting an important effect of diversity on complementarity in N-cycling and availability (Reich et al., 2012).

For aquatic systems, the timescales and traits involved in generating complementarity likely differ from those estimated for terrestrial plants. For example, a study conducted with cladocerans did not find a consistent effect of diversity on resource depletion over the course of a 5 week study, yet demonstrated that a positive B-EF relationship was possible when polycultures were assembled with species that were known a priori to have complementary feeding habits (Norberg, 2000). Although the experiment incorporated approximately five generations of the study organisms, it is unlikely that any sexual reproduction occurred, thus limiting the ability of the species to develop new phenotypic traits that could increase complementary. Phytoplankton have been shown to exhibit complementarity based on photosynthetic pigments (Striebel et al., 2009). Undoubtedly, the development of such specialized molecular machinery has occurred over a long coevolutionary history, limiting its potential to drive overyielding in short-term experiments. Understanding both the timescale of adaptation and the traits involved is necessary for identifying when and where complementarity may emerge in experimental assemblages.

In conjunction with the issue of experimental timescale is the ancestral history of the organisms used in the experiment. At the same time that diverse polycultures are adapting to local environmental conditions and the suite of species sharing and impacting this environment, monocultures of species may also be adapting to the environment generated by the absence of interspecific competitors. For example, a species that is a strong competitor for a particular resource type or size class may, in the absence of interspecific competitors, deplete resources to a greater extent and generate further selection on resource uptake traits. Thus the ability or inability of a diverse community to overyield the average or most productive monoculture may in fact depend on the extent to which species are adapted to both monoculture and polyculture conditions. Given that some experiments utilize species with a long history of sympatric conditions and others assemble species with limited or no history of sympatry, short-term assessments of overyielding may be expressing the variability introduced by the community's coevolutionary history.

In this study, we seek to identify the importance of the ancestral state of the community for the B-EF relationship. Using two dynamic models of resource competition, which represent autotrophic and heterotrophic competitors and for which evolutionary stable states have been previously identified under allopatric and sympatric (herein two-competitor) conditions, we demonstrate how experiments initiated from different ancestral states differ in their ability to generate net and transgressive overyielding. We find that nontransgressive overyielding and transgressive overyielding have a much greater likelihood and extent when experiments use species that are adapted to sympatry ("sympatric ancestry"; i.e., they have coevolved) rather than allopatry ("allopatric ancestry"; i.e., they have independently evolved). Although an empirical test generally supports this result, statistical comparisons are not significant. Furthermore our results suggest that short-term overyielding results best reflect long-term expectations when species with a sympatric ancestry are used to initiate experiments.

METHODS

To study the importance of what we herein refer to as the "ancestral state" of species in B-EF relationships, we begin by introducing a pair of models where two competitors (referred to as consumers because they occupy the second trophic level of our models) compete explicitly for two dynamically available resources. Both models are based on the premise that consumers have a fixed stoichiometric demand for nonliving nutrients (e.g., nitrogen and phosphorus), but, over evolutionary time, can affect their ability to take up these nutrients from their surroundings. Although stoichiometry can be flexible in some circumstances (Diehl et al., 2005), this flexibility has limits (Klausmeier et al., 2004), making this a popular approach for describing competitive relationships and deriving optimality theory (Fox and Vasseur, 2008; Grover, 2002; Klausmeier et al., 2007; Tilman, 1976; Vasseur and Fox, 2011).

In the first model we describe, nutrients are packaged in fixed ratios in the tissues of two different living resources (e.g., autotrophs), making the resources "partially substitutable" in the diets of our focal competitors. This implies that consumers do not require both resources in their diet in order to grow, but can achieve a greater rate of growth by consuming a mixed diet (Abrams, 1987a). For example, insect herbivores have been shown to select a mixed diet of partially substitutable plants that maximizes their fitness (Behmer and Joern, 2008). In the second model we describe, abiotic nutrients are taken up directly by the consumers, making the resources "essential" in the diets of the focal competitors. This implies that consumers cannot grow unless both resources are present in their diet. Most models of autotroph population growth are based on the premise that multiple resources are essential and this premise has permeated our understanding of phytoplankton competition and coexistence (e.g., Tilman, 1976).

We analyze this set of models by (1) solving the allopatric and sympatric optimum uptake rates as a function of internal consumer stoichiometries, (2) determining equilibrium abundances for our focal competitors at these different optima, and (3) contrasting three scenarios in which we measure nontransgressive and transgressive overyielding: (a) monocultures and polycultures at their allopatric evolutionary stable strategies (ESS) (constructed communities); (b) monocultures and polycultures at their sympatric ESS (deconstructed communities); monocultures at their allopatric ESS and polycultures at their sympatric ESS. Respectively these represent our "allopatric ancestry," "sympatric ancestry," and "expected long-term" cases.

Resource Competition Models

Our models of resource competition use a common differential equation framework to describe the ecological dynamics of two resource and two consumer populations:

$$\frac{dR_i}{dt} = D(S_i - R_i) - R_i \sum_j f_i(u_j) C_j$$

$$\frac{dC_j}{dt} = C_j(g_j(u_j) - d_j)$$

for $i,j = 1,2$. (1)

Here resources (R) are supplied in a chemostat fashion given by the dilution rate D, and the supply concentrations S_i. Resource consumption is given by the functional responses f_i and these are in turn dependent on the uptake rates of consumers u_j. In the consumer (C) equations, ingested resources are converted to biomass in the growth functions $g_j(u_j)$, and consumers die or are washed out of the system at a rate d_j. Importantly, we assume throughout that a consumer's uptake rates of the two resources are constrained to sum to one. This assumption implies a linear trade-off between a consumer's ability to ingest the two resource types; it does not, however, imply that the two types will have equal value for consumer growth as we describe further below.

Model 1: Partially Substitutable Resources

Building on the work of Hsu et al. (1977) and Abrams (1987a), Vasseur and Fox (2011) analytically solved the ESS uptake rates for a pair of consumers that compete for essential nutrients bound in fixed ratios in the edible tissue of two resources types (Figure 1(a)). Here we reiterate the basic framework for this model.

Two resources (R_i) each contain two chemical nutrients a and b (e.g., these might be nitrogen and phosphorus), which are essential for consumer growth. These nutrients are maintained at fixed "compositional" ratios ($\alpha_i = b/a$) in the two resource populations. For simplicity, we assume that resources are entirely composed of these two nutrients, such that

$$k_i = \frac{1}{\alpha_i + 1}$$

(2)

represents the proportion of each unit of resource i that is composed of nutrient a. Since resources with identical compositional ratios would be

58 SECTION | I Theoretical Background

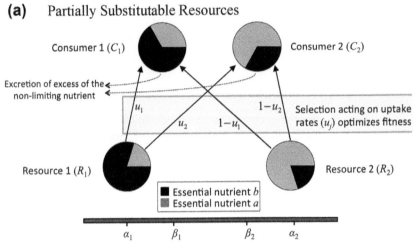

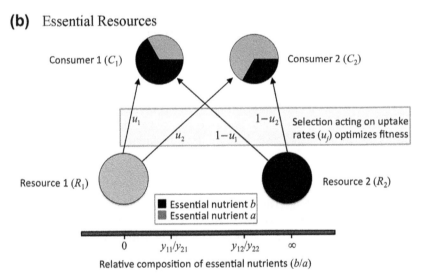

FIGURE 1 Conceptual basis for the two competition models used herein. (a) There are two resources each composed of two essential nutrients (a, b) in the ratios α_1 and α_2. Consumers can exist by consuming only one resource, but benefit by consuming a mixed diet; the two resources are therefore partially substitutable. (b) There are two resources each composed of only a single essential nutrient. Consumers can only exist if they consume a mixed diet; the two resources are therefore essential. In both models, we assume that stoichiometry is fixed while uptake rates are plastic on evolutionary timescales but constrained to sum to one.

perfectly substitutable in a consumer's diet, we limit analysis to the case where $\alpha_1 < \alpha_2$.

Consumers (C_j) each require nutrients a and b at fixed "demand ratios" ($\beta_j = b/a$) that differ between consumers ($\beta_1 \neq \beta_2$) and these ratios are fixed throughout evolutionary time. For convention we will assume that $\beta_1 < \beta_2$, but as these values diverge, consumers become increasingly differentiated in their stoichiometric demands. In this model, consumers achieve optimal fitness (ESS) when their intake of nutrients matches their internal stoichiometry (Abrams, 1987b; Vasseur and Fox, 2011). The case of "partially substitutable resources" requires that the consumer demand ratios be bracketed by the resource compositional ratios, which implies here that $\alpha_1 < \beta_j < \alpha_2$. The functional responses of consumers,

$$f_1(u_j) = u_j \\ f_2(u_j) = 1 - u_j, \tag{3}$$

are linear, and reflect the trade-off in ability to acquire the two resource types. The g-function of consumer j, which reflects its ability to assimilate ingested biomass, is given by

$$g_j(u_j) = (1 - \delta_j)(Ia_j + Ib_j), \tag{4}$$

where Ia_j and Ib_j represent the amount of ingested nutrients a and b, and $1 - \delta_j$ determines the fraction of ingestion that is actually assimilated, taking into account the differences between consumer and diet stoichiometry. Vasseur and Fox (2011) showed that if excess nutrients are excreted at no cost to the consumer, Ia_j, Ib_j, and $1 - \delta_j$ are given by

$$Ia_j = k_1 u_j R_1 + k_2 (1 - u_j) R_2 \\ Ib_j = (1 - k_1) u_j R_1 + (1 - k_2)(1 - u_j) R_2 \\ 1 - \delta_j = 1 - \frac{\text{Max}\left[Ib_j - Ia_j \beta_j, Ia_j - Ib_j/\beta_j\right]}{Ia_j + Ib_j} \tag{5}$$

Two consumers can always coexist under this type of resource competition provided that their demand ratios β_j are neither both below or above the ratio at which nutrients are supplied when both resources are considered together at equilibrium (Vasseur and Fox, 2011). Herein we consider only the case where resource compositional ratios are reciprocal and supply rates S_1 and S_2 are equal; this ensures that coexistence is possible whenever $\beta_1 < 1 < \beta_2$.

For this model the uptake rates corresponding to the ESS, those values that optimize the per-capita growth rate of their respective consumer populations, are analytically tractable; however, the equilibrium densities of the two consumers cannot always be analytically determined (Vasseur and Fox, 2011). We therefore resort to a numerical approach to solve consumer densities at the ESS (see below). When two consumers coexist in sympatry, their ESS uptake rates are given by (Vasseur and Fox, 2011):

$$u_j^S = \frac{1 - k_2(1 + \beta_j)}{(k_1 - k_2)(1 + \beta_j)} \qquad (6)$$

When either of the consumers exists in allopatry, their ESS uptake rates are a discontinuous function of their demand ratio β_j and the supply ratios of essential nutrients provided by the two resources. This discontinuity occurs because the optimal uptake rate may select a diet where either or both of the nutrients are limiting for growth. Solutions for the allopatric ESS uptake rates are shown in Figure 2 and are given in Appendix A of Vasseur and Fox (2011).

Case 2: Essential Resources

Building on the work of Leon and Tumpson (1975) and Abrams (1987b), Fox and Vasseur (2008) solved the ESS uptake rates of two consumers that compete for two resources which are both essential for consumer growth and therefore nonsubstitutable in their diet. Under this scenario the consumer growth function is given by

$$g_j(u_j) = \min\left[y_{1j}u_jR_1, y_{2j}(1 - u_j)R_2\right] \qquad (7)$$

and the functional responses are given by

$$f_i(u_j) = \frac{g_j(u_j)}{y_{ij}R_i}. \qquad (8)$$

Here the term y_{ij} represents the yield of consumer j that is generated from ingestion of a unit of resource i. Thus, the growth rate of consumer j is set in Eqn (7) by the lesser of the growth rates given by ingestion of R_1 and R_2. The functional response reflects the impact of the actual consumer growth rate on resource i. As above, we assume that consumers have a fixed internal stoichiometry that is evolutionarily inflexible, such that the yield terms y_{ij} are fixed and represent static differences between the two competing consumers.

Leon and Tumpson (1975) derived the conditions for the stable coexistence of two consumers competing for essential resources and these are reiterated for the model, structured as above in Fox and Vasseur (2008). In brief, under the assumption of equal rates of resource supply ($S_1 = S_2$) and equal consumer death rates ($d_1 = d_2$), it is sufficient to assume that the stable coexistence of two consumers is possible at some combination of uptake rates provided that $y_{11} \leq y_{21}$ and $y_{22} \leq y_{12}$. Herein we set $y_{21} = y_{12} = 1$ and examine the effects of varying the remaining yield coefficients. As y_{11} and y_{22} decrease the consumers become increasingly differentiated in their stoichiometric demands.

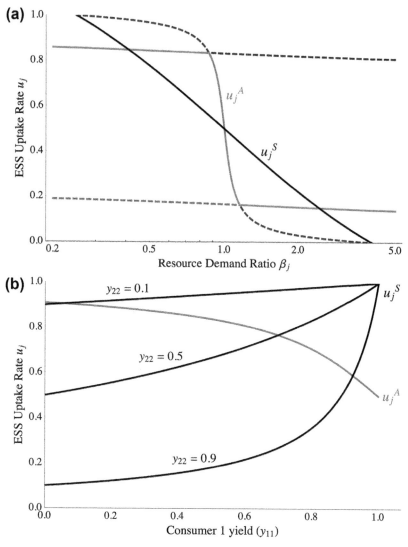

FIGURE 2 The evolutionary stable strategy for the models of competition for (a) partially substitutable resources and (b) essential resources when consumer 1 exists in allopatry (u_j^A) or in sympatry (u_j^S) as a function of specialization on resource 1. Small values on the abscissa represent a high degree of specialization on resource 1. In panel b three curves are shown for the sympatric ESS, each representing a different degree of specialization for the second consumer. In panel b the sympatric ESS does not depend on the second consumer's degree of specialization.

The ESS uptake rates for consumers of essential resources in sympatry are given by

$$u_j^S = \frac{y_{1j}(y_{22}-1)}{y_{11}y_{22}-1} \quad (9)$$

and in allopatry they are given by

$$u_j^A = \frac{x_j + 2 - \sqrt{x_j^2 + 4}}{2x_j}, \quad (10)$$

where $x_1 = (y_{11}S_1 - S_2)/d_1$ and $x_2 = (y_{22}S_2 - S_1)/d_2$. Fox and Vasseur (2008) showed that only one of the roots of Eqn (10) falls within the domain of feasible parameter space, ensuring a single ESS for any parameter combination. In this type of competition the sympatric ESS depends on the yields of both consumers (unlike the case for partially substitutable resources); thus Figure 2 shows the allopatric and sympatric optima for one consumer paired with three different competitors ($y_{22} = 0.01, 0.1,$ and 0.5). As above, we resort to a numerical approach for solving the equilibrium densities of consumers at the various ESSs.

Model Analysis

To examine the role of adaptation on overyielding under the three scenarios outlined in the Introduction (allopatric ancestry, sympatric ancestry, expected long-term outcome), we used numerical simulations to determine the steady state abundances of consumers:

1. in *allopatry* at the *allopatric* ESS,
2. in *allopatry* at the *sympatric* ESS,
3. in *sympatry* at the *allopatric* ESS, and
4. in *sympatry* at the *sympatric* ESS.

Under each scenario, we ran numerical simulations in Mathematica v9.0 for 100,000 time steps using a Runge–Kutta integration method and we recorded the equilibrium values (C_j^*). For the model of essential resources we examined a two-dimensional range of consumer yield coefficients ($0 < y_{11} < 1$ and $0 < y_{22} < 1$) using a grid with discrete steps valued at 0.01. Because the allopatric equilibrium is neutrally stable for this model (Fox and Vasseur, 2008), we took the average densities from 100 replicates where the initial conditions were taken randomly on the interval (0,1). For the model of partially substitutable resources, we examined a two-dimensional range of consumer demand ratios ($\alpha_1 < \beta_1 < \alpha_S$ and $\alpha_S < \beta_2 < \alpha_2$), in discrete steps of 0.01. Parameter values are given in Table 1.

We determined the extent of nontransgressive overyielding (OY) and transgressive overyielding (OY_{max}) (Loreau, 1998a; Wardle et al., 1997) by

TABLE 1 Model Parameter Definitions and Values

Common Parameters		
D	Dilution rate	0.1
S_i	Resource i supply density	1.0
d_j	Consumer j death rate	0.1
u_j	Consumer j uptake rate of R_1	Plastic
$1 - u_j$	Consumer j uptake rate of R_2	Plastic
Partially Substitutable Resources		
k_i	Proportion of R_1 composed of nutrient a	0.8, 0.2
α_i	Ratio of nutrients b/a in R_i	0.25, 4.0
β_j	Ratio of nutrients b/a in C_j	Between 0.25 and 4.0
$1 - \delta_j$	Fraction of biomass ingested by C_j that is assimilated	
Essential Resources		
y_{ij}	Yield of C_j from one unit of R_i	y_{11} between 0 and 1 $y_{12} = 1.0$ $y_{21} = 1.0$ y_{22} between 0 and 1

comparing the polyculture yields to the average or maximum monoculture yields of consumers according to

$$\mathrm{OY} = \mathrm{Log}_e \left[\frac{\left(C^*_{1(S)} + C^*_{2(S)} \right)}{0.5 \left(C^*_{1(A)} + C^*_{2(A)} \right)} \right] \quad (11)$$

$$\mathrm{OY}_{\max} = \mathrm{Log}_e \left[\frac{\left(C^*_{1(S)} + C^*_{2(S)} \right)}{\mathrm{Max} \left(C^*_{1(A)}, C^*_{2(A)} \right)} \right] \quad (12)$$

Using this formalism, values less than zero indicate underyielding of the diverse community relative to monocultures while values greater than zero indicate overyielding.

Reanalysis of Empirical Data

We reviewed a previously published summary of 58 studies from the literature that experimentally manipulated biodiversity at a particular trophic level and measured the direct effect on standing biomass or resource depletion,

TABLE 2 Summary of Experiments and Analysis

Trophic level	Source of organisms	Number of experiments	P value nontransgressive overyielding	P value transgressive overyielding
Predators (C)	Allopatric	12	0.0507[a]	0.118
	Sympatric	9		
Detritivores (D)	Allopatric	8	0.142	0.340
	Sympatric	23		
Heterotrphs (H)	Allopatric	2	0.487	0.817
	Sympatric	4		
Primary producers (P)	Allopatric	3	0.0402[b]	0.111
	Sympatric	7		

[a]Significant at $\alpha = 0.1$.
[b]Significant at $\alpha = 0.05$.

quantified by the log response ratio (LRR = $\ln(\text{yield}_{\text{polyoculture}}/\text{yield}_{\text{monoculture}})$) (Cardinale et al., 2006). For each study, we determined whether the organisms used in the experiment had previously developed in allopatry or sympatry. We were able to clearly identify the source of organisms as either sympatric or allopatric for 28 studies yielding data for 68 experiments (Table 2). We compared the LRR between experiments assembled from allopatric or sympatric organisms using a Mann–Whitney U test. We additionally compared the LRR between experiments grouped by the focal trophic level (predators, detritivores, heterotrophs, or primary producers). For each test, we compared the LRR measured as the proportional difference in yield between the most species-rich polyculture and the average species monoculture (a positive value indicates overyielding) or the species with the highest yield in monoculture (a positive value indicates transgressive overyielding). We used the data published by Cardinale et al. (2006) except in three cases, where we had to infer the highest yield value from figures in the original study.

RESULTS

Figures 3–6 show the extent of nontransgressive and transgressive overyielding in the models of competition for partially substitutable resources and for essential resources. The axes on these figures, while in different units for the two models, are formatted such that increasing values represent a greater need for one of the essential nutrients relative to the other (increased resource specialization). Thus, at the origin the two competitors have identical

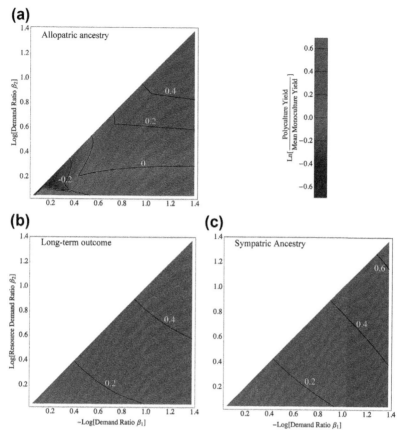

FIGURE 3 Nontransgressive overyielding (polyculture yield/average monoculture yield) under competition for complementary resources. Increasing values on the axes represent increased (fixed) resource specialization. (a) Both underyielding and overyielding are possible under allopatric ancestry, depending on the degree of resource specialization. Under (b) sympatric ancestry and (c) over the long term, when both monocultures and polycultures are at their ESS, only overyielding occurs, with larger values at higher degrees of resources specialization.

stoichiometric requirements (functionally redundant) while at the upper right of the panels the two competitors have very different stoichiometric requirements. The extent to which their uptake rates reflect these differences depends on the extent to which they are adapted to their current environment. In the upper left panel of these figures (panel a) the two competitors are adapted to allopatry and therefore exhibit suboptimal fitness in polycultures. In the lower right panel of these figures (panel c) the two competitors are adapted to sympatry and therefore exhibit suboptimal fitness in monocultures. In the lower left panel (panel b) fitness is optimized in both poly- and monocultures reflecting the long-term expectation for overyielding in these two models.

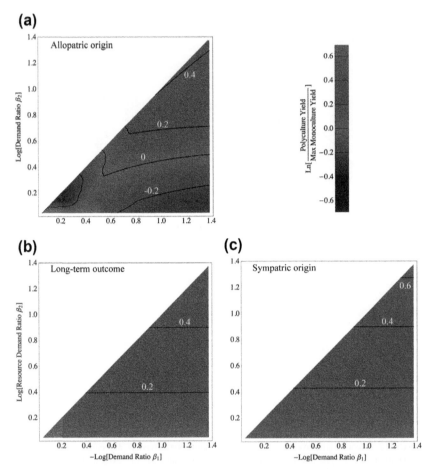

FIGURE 4 Transgressive overyielding (polyculture yield/highest monoculture yield) under competition for complementary resources. Increasing values on the axes represent increased (fixed) resource specialization. (a) Both underyielding and overyielding are possible under allopatric ancestry, depending on the degree of resource specialization in both species. Under (b) sympatric ancestry and (c) over the long-term, when both monocultures and polycultures are at their ESS, only overyielding occurs, with larger values occurring only when both competitors show strong (divergent) resource specialization.

For the model of partially substitutable resources, we find that non-transgressive and transgressive overyielding predominate our results (Figures 3 and 4). Under sympatric ancestry (where monocultures exhibit suboptimal fitness) both nontransgressive and transgressive overyielding show a positive effect of diversity on yield (Figures 3(c) and 4(c)). Nontransgressive overyielding increases as either competitor becomes more specialized but transgressive overyielding only changes in response to the less specialized of the two competitors (Figure 4(c)). Under allopatric ancestry both under- and

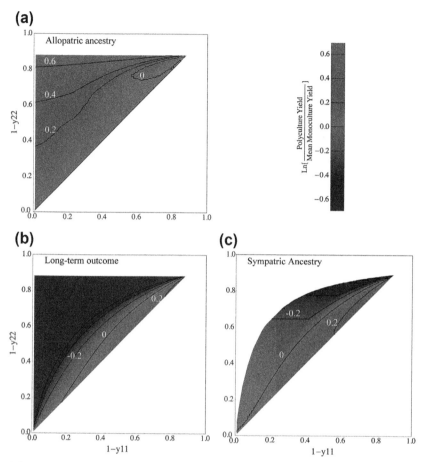

FIGURE 5 Nontransgressive overyielding (polyculture yield/average monoculture yield) under competition for essential resources. Increasing values on the axes represent increased (fixed) resource specialization. (a) Under allopatric ancestry overyielding typically occurs because sympatric coexistence is not possible without adaptation. (c) Under sympatric ancestry and (b) over the long term, when both monocultures and polycultures are at their ESS, overyielding only occurs where both competitors are specialized on different resources to a relatively equal extent.

overyielding are apparent (Figures 3(a) and 4(a)). Overyielding requires a moderate amount of specialization in both competitors and tends to be limited by the more generalist competitor. The long-term overyielding patterns tend to resemble those of sympatric ancestry, indicating that the extent to which monoculture yields suffer when competitors are adapted to sympatric conditions is small relative to the extent to which polyculture yields suffer when competitors are adapted to allopatry.

In contrast to the above model, under competition for essential resources nontransgressive and transgressive underyielding predominate (Figures 5 and 6).

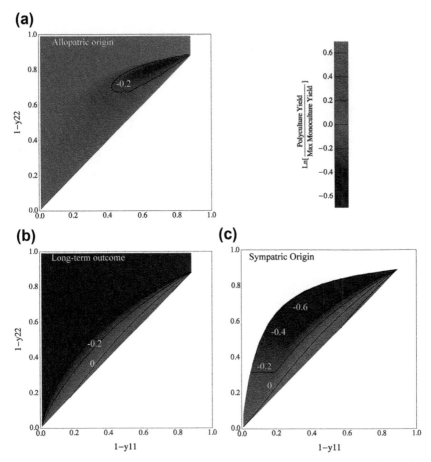

FIGURE 6 Transgressive overyielding (polyculture yield/average monoculture yield) under competition for essential resources. Increasing values on the axes represent increased (fixed) resource specialization. (a) Under allopatric ancestry transgressive overyielding never occurs. (c) Under sympatric ancestry and at the long-term outcome (b) overyielding only occurs where competitors are specialized on different resources, but to a generally equal extent.

Although nontransgressive overyielding is apparent under allopatric ancestry in this model (Figure 5(a)), this result is driven by the inability of competitors to coexist in polyculture when adapted to allopatric conditions (Fox and Vasseur, 2008). This is apparent in the lack of transgressive overyielding in Figure 6(a). Under sympatric ancestry, the competitors show a small amount of non-transgressive and transgressive overyielding when their resource requirements are strongly and near-equally differentiated. As the extent to which competitors are differentiated becomes too large, sympatric coexistence is strained leading to underyielding and eventually competitive exclusion. As in the model of partially

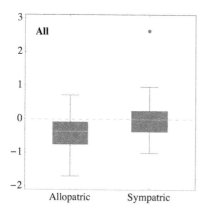

FIGURE 7 Comparison of transgressive overyielding between experiments assembled from allopatric or sympatric organisms. The gray point indicates a far outlier. $N = 25$ (allopatry) and 43 (sympatry).

substitutable resources, we find that long-term patterns are similar to those of sympatric ancestry.

We found no overall significant difference in nontransgressive (result not shown) and transgressive overyielding (Figure 7) between experiments assembled from allopatric or sympatric organisms ($p = 0.789$ and $p = 0.056$, respectively). However, the trend suggests that transgressive overyielding is more likely to occur in experiments assembled from sympatric organisms (Figure 7). When partitioned into groups by trophic status, we find only one significant difference in nontransgressive overyielding among primary producers (Figure 8(a)).

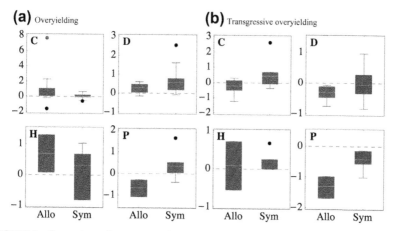

FIGURE 8 Comparison of nontransgressive and transgressive overyielding between experiments assembled from allopatric or sympatric organisms by trophic level. C, consumer (predator); D, detritivore; H, heterotroph; P, primary producer. The left four graphs show nontransgressive overyielding calculated using the average monoculture, and the right four transgressive overyielding calculated using the most productive monoculture. A positive log response ratio indicates overyielding (transgressive overyielding in the case of LRR2). Black points indicate outliers, and gray points indicate far outliers.

Although no groups show significant differences in transgressive overyielding we find the same trend in all cases, with experiments classified as sympatric ancestry tending to generate greater transgressive overyielding than those with allopatric ancestry (Figure 8(b)).

DISCUSSION

Our results demonstrate that both nontransgressive and transgressive overyielding are more likely to occur when the species used to conduct experiments have coevolved in polyculture (sympatric ancestry) rather than in monoculture (allopatric ancestry). Because our measures of overyielding are represented as a log ratio of polyculture to monoculture yield, it is not surprising that sympatric ancestry leads to increased overyielding because species that are well adapted to polyculture conditions will have optimal fitness in polyculture and suboptimal fitness in monocultures. Although there is not an exact correspondence between fitness and biomass yield in general (Leimu et al., 2006) and in these models (unpublished result), fitness and biomass are strongly correlated (Leimu et al., 2006). Thus suboptimal fitness in monocultures increases the extent of overyielding in communities with sympatric ancestry. Under allopatric ancestry, the suboptimal fitness of species in polyculture weakens the ability of the community to overyield. Although empirical tests of the difference between ancestral states of the experimental community are nonsignificant, their trend matches the predictions made by the two models described here.

As the duration of experiments increases, it has been shown that complementarity tends to strengthen (Hooper and Dukes, 2004; Reich et al., 2012; Tilman et al., 2001), leading to an increase in nontransgressive and transgressive overyielding (Cardinale et al., 2007). While increases in complementary can arise as species sorting occurs in competitive communities, it may also be driven by selection on traits to allow greater niche differentiation. In our models, the long-term outcome depicted in panel b of Figure 7 shows the extent of overyielding when species have reached optimal fitness in all settings. Here the extent to which communities overyield is intermediate between the scenarios of sympatric and allopatric ancestry because there are no longer any treatments where populations are at suboptimal fitness. Somewhat surprisingly, the long-term outcome does not represent a midpoint between the two scenarios, but more closely resembles the results of sympatric ancestry. This suggests that the effects of evolutionary suboptimality are not equal across monocultures and polycultures; species yield much less biomass in polyculture when not optimally adapted, whereas biomass yield is less affected by adaptation in monoculture.

The closer match between long-term outcomes and the sympatric ancestry scenario described herein suggests that temporal increases in overyielding should be apparent only when organisms have an allopatric ancestry in B-EF experiments. Although we assign the majority of experiments to the first category (Table 2) because they draw species from a local pool, small-scale heterogeneity under environmental conditions and species interactions may ensure that experimental conditions provide novel environments for both mono- and polycultures. Explicit tests of the importance of coevolutionary history on monoculture and polyculture biomass yield, or other ecosystem function, may provide important insight into the tendency for complementarity to drive B-EF relationships. Given recent acknowledgment that evolutionary changes are sufficiently rapid as to interact with ecological processes (Jones et al., 2009; Palkovacs and Post, 2009; Yoshida et al., 2003), the importance of adaptation should be explicitly considered.

The models described here represent a simplified view of competition in natural systems that include a greater diversity of species and a greater dimension of resources that potentially limit growth. Moreover, there is a potential for multiple evolutionary stable states to emerge as the complexity of the community increases, challenging our ability to predict the importance of evolutionary history for B-EF relationships. As molecular methods continue to provide greater power to resolve evolutionary changes, we will become better equipped to deal with this complexity. Ultimately, improved understanding of the ecoevolutionary feedback in competitive communities will allow us to better conserve and restore ecosystems and to better construct functional and resilient ecosystems that are optimized with respect to particular functions.

ABBREVIATION

B—EF Biodiveristy—Ecosystem Function

ACKNOWLEDGMENTS

D.A.V. acknowledges support of Yale University. S.M.M. was supported by a Yale Institute for Biospheric Studies Donnelley Postdoctoral Fellowship. Jesse Korman assisted with the collection of empirical data.

REFERENCES

Abrams, P.A., 1987a. The functional responses of adaptive consumers of two resources. Theor. Popul. Biol. 32, 262—288. http://dx.doi.org/10.1016/0040-5809(87)90050-5.

Abrams, P.A., 1987b. Alternative models of character displacement and niche shift. I. Adaptive shifts in resource use when there is competition for nutritionally nonsubstitutable resources. Evolution 651—661.

Behmer, S.T., Joern, A., 2008. Coexisting generalist herbivores occupy unique nutritional feeding niches. Proc. Natl. Acad. Sci. USA 105, 1977—1982. http://dx.doi.org/10.1073/pnas.0711870105.

Cardinale, B.J., Ives, A.R., Inchausti, P., 2004. Effects of species diversity on the primary productivity of ecosystems: extending our spatial and temporal scales of inference. Oikos 104, 437—450.

Cardinale, B.J., Palmer, M.A., Collins, S.L., 2002. Species diversity enhances ecosystem functioning through interspecific facilitation. Nature 415, 426–429.

Cardinale, B.J., Srivastava, D.S., Duffy, J.E., Wright, J.P., Downing, A.L., Sankaran, M., Jouseau, C., 2006. Effects of biodiversity on the functioning of trophic groups and ecosystems. Nature 443, 989–992.

Cardinale, B.J., Wright, J.P., Cadotte, M.W., Carroll, I.T., Hector, A., Srivastava, D.S., Loreau, M., Weis, J.J., 2007. Impacts of plant diversity on biomass production increase through time because of species complementarity. Proc. Natl. Acad. Sci. USA 104, 18123–18128. http://dx.doi.org/10.1073/pnas.0709069104.

Diehl, S., Berger, S., Wöhrl, R., 2005. Flexible nutrient stoichiometry mediates environmental influences on phytoplankton and its resources. Ecology 86, 2931–2945.

Fox, J.W., Vasseur, D.A., 2008. Character convergence under competition for nutritionally essential resources. Am. Nat. 172, 667–680.

Grover, J.P., 2002. Stoichiometry, herbivory and competition for nutrients: simple models based on planktonic ecosystems. J. Theor. Biol. 214, 599–618.

Hector, A., Bazeley-White, E., Loreau, M., Otway, S., Schmid, B., 2002. Overyielding in grassland communities: testing the sampling effect hypothesis with replicated biodiversity experiments. Ecol. Lett. 5, 502–511.

Hooper, D.U., Chapin Iii, F.S., Ewel, J.J., Hector, A., Inchausti, P., Lavorel, S., Lawton, J.H., Lodge, D.M., Loreau, M., Naeem, S., et al., 2005. Effects of biodiversity on ecosystem functioning: a consensus of current knowledge. Ecol. Monogr. 75, 3–35.

Hooper, D.U., Dukes, J.S., 2004. Overyielding among plant functional groups in a long-term experiment. Ecol. Lett. 7, 95–105. http://dx.doi.org/10.1046/j.1461-0248.2003.00555.x.

Hsu, S.B., Hubbell, S., Waltman, P., 1977. A mathematical theory for single-nutrient competition in continuous cultures of micro-organisms. SIAM J. Appl. Math. 32, 366–383.

Huston, M.A., 1997. Hidden treatments in ecological experiments: re-evaluating the ecosystem function of biodiversity. Oecologia 110, 449–460.

Jiang, L., Pu, Z., Nemergut, D.R., 2008. On the importance of the negative selection effect for the relationship between biodiversity and ecosystem functioning. Oikos 117, 488–493.

Jones, L.E., Becks, L., Ellner, S.P., Hairston, N.G., Yoshida, T., Fussmann, G.F., 2009. Rapid contemporary evolution and clonal food web dynamics. Philos. Trans. R. Soc. B Biol. Sci. 364, 1579–1591.

Klausmeier, C.A., Litchman, E., Levin, S.A., 2004. Phytoplankton growth and stoichiometry under multiple nutrient limitation. Limnol. Oceanogr. 1463–1470.

Klausmeier, C.A., Litchman, E., Levin, S.A., 2007. A model of flexible uptake of two essential resources. J. Theor. Biol. 246, 278–289. http://dx.doi.org/10.1016/j.jtbi.2006.12.032.

Leimu, R., Mutikainen, P., Koricheva, J., Fischer, M., 2006. How general are positive relationships between plant population size, fitness and genetic variation? J. Ecol. 94, 942–952. http://dx.doi.org/10.1111/j.1365-2745.2006.01150.x.

Leon, J., Tumpson, D., 1975. Competition between two species for two complementary or substitutable resources. J. Theor. Biol. 50, 185–201. http://dx.doi.org/10.1016/0022-5193(75)90032-6.

Loreau, M., 1998a. Separating sampling and other effects in biodiversity experiments. Oikos 82, 600–602.

Loreau, M., 1998b. Biodiversity and ecosystem functioning: a mechanistic model. Proc. Natl. Acad. Sci. USA 95, 5632–5636.

Loreau, M., 2000. Biodiversity and ecosystem functioning: recent theoretical advances. Oikos 91, 3–17. http://dx.doi.org/10.1034/j.1600-0706.2000.910101.x.

Loreau, M., Hector, A., 2001. Partitioning selection and complementarity in biodiversity experiments. Nature 412, 72–76.

Loreau, M., Naeem, S., Inchausti, P., Bengtsson, J., Grime, J.P., Hector, A., Hooper, D.U., Huston, M.A., Raffaelli, D., Schmid, B., 2001. Biodiversity and ecosystem functioning: current knowledge and future challenges. Science 294, 804–808.

Norberg, J., 2000. Resource-niche complementarity and autotrophic compensation determines ecosystem-level responses to increased cladoceran species richness. Oecologia 122, 264–272.

Palkovacs, E.P., Post, D.M., 2009. Experimental evidence that phenotypic divergence in predators drives community divergence in prey. Ecology 90, 300–305. http://dx.doi.org/10.1890/08-1673.1.

Pimm, S.L., Raven, P., 2000. Biodiversity: extinction by numbers. Nature 403, 843–845.

Reich, P.B., Tilman, D., Isbell, F., Mueller, K., Hobbie, S.E., Flynn, D.F.B., Eisenhauer, N., 2012. Impacts of biodiversity loss escalate through time as redundancy fades. Science 336, 589–592. http://dx.doi.org/10.1126/science.1217909.

Reiss, J., Bridle, J.R., Montoya, J.M., Woodward, G., 2009. Emerging horizons in biodiversity and ecosystem functioning research. Trends Ecol. Evol. 24, 505–514. http://dx.doi.org/10.1016/j.tree.2009.03.018.

Striebel, M., Behl, S., Diehl, S., Stibor, H., 2009. Spectral niche complementarity and carbon dynamics in pelagic ecosystems. Am. Nat. 174, 141–147.

Thompson, R., Starzomski, B.M., 2007. What does biodiversity actually do? A review for managers and policy makers. Biodiversity Conserv. 16, 1359–1378.

Tilman, D., 1976. Ecological competition between algae: experimental confirmation of resource-based competition theory. Science 192, 463–465.

Tilman, D., 1996. Biodiversity: population versus ecosystem stability. Ecology 77, 350–363.

Tilman, D., 2000. Causes, consequences and ethics of biodiversity. Nature 405, 208–211. http://dx.doi.org/10.1038/35012217.

Tilman, D., Lehman, C.L., Thomson, K.T., 1997. Plant diversity and ecosystem productivity: theoretical considerations. Proc. Natl. Acad. Sci. USA 94, 1857–1861.

Tilman, D., Reich, P.B., Knops, J., Wedin, D., Mielke, T., Lehman, C., 2001. Diversity and productivity in a long-term grassland experiment. Science 294, 843–845.

Vasseur, D.A., Fox, J.W., 2011. Adaptive dynamics of competition for nutritionally complementary resources: character convergence, displacement, and parallelism. Am. Nat. 178, 501–514.

Wardle, D.A., 1999. Is "sampling effect" a problem for experiments investigating biodiversity-ecosystem function relationships? Oikos 403–407.

Wardle, D.A., Bonner, K.I., Nicholson, K.S., 1997. Biodiversity and plant litter: experimental evidence which does not support the view that enhanced species richness improves ecosystem function. Oikos 247–258.

Yoshida, T., Jones, L.E., Ellner, S.P., Fussmann, G.F., Hairston, N.G., 2003. Rapid evolution drives ecological dynamics in a predator–prey system. Nature 424, 303–306.

Chapter 4

Effects of Metacommunity Networks on Local Community Structures: From Theoretical Predictions to Empirical Evaluations

Ana Inés Borthagaray, Verónica Pinelli, Mauro Berazategui, Lucía Rodríguez-Tricot and Matías Arim
Departamento de Ecología y Evolución, Facultad de Ciencias and Centro Universitario Regional Este (CURE), Universidad de la República, Montevideo, Uruguay

INTRODUCTION

Significant losses of species, homogenization of biotas, changes in climatic conditions, and reduction or fragmentation of ecosystems will likely occur through the current century (D'Antonio et al., 2001; May et al., 1995; Scheffer et al., 2009). In addition to the direct effect on biodiversity, several ecosystem processes that support human societies will be impacted (Loreau, 2010; Nicholson et al., 2009). The discipline of ecology will have a central role in this changing world by accounting for mechanisms involved in these ongoing processes (Loreau, 2010; Loreau et al., 2001). Ecological theory has focused primarily on local determinants of species coexistence as a main approach to explaining biodiversity patterns (Diamond, 1975; Gotelli et al., 1997; MacArthur, 1970; MacArthur and Levins, 1964, 1967). At the same time, Lotka—Volterra models and small-scale experiments with few species have dominated the literature (Hutchinson, 1959; MacArthur and Levins, 1964; Pianka, 1974; Ritchie, 2010). In spite of the significant advances achieved by these approaches (Chase and Leibold, 2003; Chesson, 2000), they alone do not account for biodiversity patterns and functions across diverse ecological and evolutionary scales (Maurer, 1999; Morin, 2010; Ricklefs and Schluter, 1993). In consequence, new theoretical and methodological approaches to the study of biodiversity have emerged in recent decades.

A groundbreaking set of novel theories with new methodologies and a focus on alternative mechanisms has been formulated in recent years (Ritchie, 2010; p. 199). Among them, we emphasize the unified neutral theory of biodiversity and biogeography (Hubbell, 2001; Rosindell et al., 2011), the metabolic theory of ecology (Brown et al., 2004), the spatial scaling law (Ritchie, 2010), and two theories based on maximum entropy (MaxEnt) formalisms (Harte, 2011; Shipley, 2010a). To some extent, macroecology, with its focus on basic principles structuring biodiversity, prepared the groundwork for these new theories (Brown and Maurer, 1989). Perspectives such as those embedded in the neutral theory avoid considerations of niche-based processes and helped in the formulation of a novel synthesis of niche theory (Chase and Leibold, 2003). With such a plethora of new ideas, a search for unified theories combining some or all of them and unifying new theories with more classic concepts has become the holy grail of ecology (Chase, 2005; McGill, 2010). The discipline of metacommunity ecology may provide a framework for the unification of the different theories.

Metacommunity concepts have a long history in the science of ecology (Chase and Bengtsson, 2010); they represent natural extensions of the metapopulation approach to multispecies systems (Hanski and Gilpin, 1991; Loreau, 2010). The metacommunity concept connects primary biogeographical and community theories, such as those relating to species co-occurrences among communities (Holt, 1997; Leibold and Mikkelson, 2002; Ulrich and Gotelli, 2007), the link between local and regional diversity (Holt, 1993), and the neutral theory of ecology, which considered metacommunity—community immigration as a basic determinant of community structure (Hubbell, 2001). Some of the most important contributions to biological thinking have emerged in the introduction of new concepts (Mayr, 1997). There is no doubt that the metacommunity concept, together with the growing formalization of related theories, is producing important changes in the discipline of community ecology. Local communities are no longer considered isolated units in which local processes determine their structure and function (Baiser et al., 2013; Brown and Swan, 2010; Carrara et al., 2012; Loreau et al., 2003; Moritz et al., 2013). The theoretical and empirical evidence available indicates that the effects of large-scale processes on community structure and diversity may be equally as important as or more important than long-studied local determinants (Borthagaray et al., 2012; Economo and Keitt, 2008; Leibold et al., 2004; Mouquet and Loreau, 2003). Furthermore, the classic local—regional dichotomy might well be an oversimplification (Cadotte, 2006).

Dissociation between theoretical predictions and their empirical evaluation is a chronic problem in ecology (Abrams, 2001; Arim et al., 2007; Hanski, 1999). While foundational publications on ecological theories have been cited thousands of times (e.g., Leibold et al., 2004), reports of their empirical testing are orders of magnitude fewer (Logue et al., 2011). Metacommunity ecology is rapidly advancing through theoretical considerations, but the empirical counterpart is in its infancy (Baiser et al., 2013; Dorazio et al., 2010; Driscoll and Lindenmayer, 2009; Logue et al., 2011). In order to contribute to the construction of a strong theoretical framework, the empirical approach must be able to discriminate among potential mechanisms even when several of these drivers may seem to underlay observed patterns (Platt, 1964). Analyses of the purely neutral model, partitioning of variance, and site-by-species incidence matrices have been foci of interest in empirical evaluations of metacommunity theory (Meynard et al., 2013). Spatial autocorrelation in community compositions has also been used to analyze dispersal patterns (Shurin et al., 2009; Soininen et al., 2007). Baiser et al. (2013) and Dorazio et al. (2010) have proposed additional new methods for empirical testing. Here, we argue that MaxEnt theory has a major role in the analysis of metacommunity mechanisms (Shipley, 2010b; Shipley et al., 2006) and that graph theory provides a robust procedure for the quantification of metacommunity network structure and local community isolation. The use of graph theory for metacommunity studies has been emphatically recommended in recent years (Altermatt, 2013; Economo and Keitt, 2010; Gonzalez et al., 2011; Peterson et al., 2013). However, the methods by which networks may be empirically determined are not always obvious; here, we review potential alternatives.

In this chapter we contribute to reducing the distance between metacommunity theory and empirical evaluation by reviewing predictions from metacommunity theory and proposing tools for their testing using field observational data. In the subsequent sections, we briefly introduce the main conceptual frameworks related to metacommunity theory. Following this introduction to mechanisms and predictions, we focus on methodological procedures for their evaluation.

FOUR PARADIGMS

Outstanding reviews and presentations of metacommunity theory have been published previously (Chase and Bengtsson, 2010; Holyoak et al., 2005a; Leibold et al., 2004). Here, we highlight basic mechanisms and predictions that should be a focus of empirical analyses. We start by presenting four nonexclusive metacommunity paradigms: patch dynamics, mass effect, species sorting, and neutral dynamics (see Figure 1). The central message here is that a wide range of predictions emerges from metacommunity mechanisms, and closely similar predictions frequently arise from different theoretical mechanisms. In addition, analyzing the role of organismal traits in the assembly of communities and in environment–trait associations is crucial to the disentanglement of competing theoretical mechanisms predicting similar community patterns.

Species Sorting

Differences in functional attributes of species determine differential success among different environments

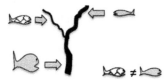

Environment ⇔ spp traits
Environment 1 ≠ Environment 2

Patch Dynamics

Species coexistence determined by a compromise between colonization and extinction rates among species

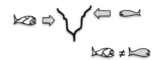

Tradeoff: colonization ⇔ extinction
Environment 1 = Environment 2

Mass Effect

Individual immigration precludes species extinction and promotes recolonization after extinction

Neutral Model

Individual traits have no role in species success. Environmental conditions determine community abundance and systematic trends in diversity

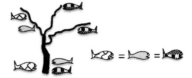

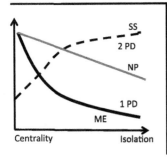
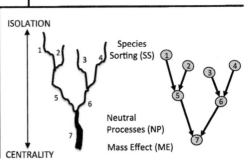

FIGURE 1 Four nonexclusive metacommunity mechanisms. This example considers a fish metacommunity in a river. Fish patterns represent observed traits, and a different environment is assumed among river headers, and between headers and river mouths. Each box contains one of the four "paradigms" of metacommunity ecology: species sorting, patch dynamics, mass effect, and neutral processes. These paradigms are expected to systematically rise or decrease with community isolation. The trend in patch dynamics with isolation is contingent on the relative effects of isolation on dominant and subordinate competitors. The lower right side panel presents a graph representation of a river and the regions where the different mechanisms are expected to be more relevant.

Patch Dynamics and Mass Effect

We focus here on two of the four main perspectives in metacommunity ecology: patch dynamics and mass effect. These mechanisms may be conceptualized by different analytical and conceptual approaches (Holyoak et al., 2005a). Among them, metapopulation theory has the advantage of connecting population ecology with biogeography and landscape ecology through simple and general models (Gotelli, 1991). We concentrate on the derivation of patch dynamics and mass effect from metapopulation models. This approach identifies basic mechanisms beyond each of the paradigms and provides further predictions on the potential operation of specific mechanisms along a gradient of community centrality—isolation (Figure 1).

Metapopulation models represent three contrasting scenarios of scale and ecological level. First, a model may represent the occurrence of populations across a landscape (Levins, 1969). Second, it may represent space occupancy by individuals from a single population (Tilman, 1994), and the model may have analogies to the logistic equation (Hanski, 1999). Third, a model may connect to the community level because the fraction of occupied patches is the probability that a species is present in a single patch. Assuming that species occurrences are mutually independent, as in the MacArthur and Wilson (1967) model, the expected richness in a local community is the product of species occurrence probability and the number of species in the regional pool (Gotelli, 1991; Hanski, 1999, 2010).

Patch Dynamics

In a basic metapopulation model, local patches are colonized through immigration (i) and evacuated through extinction (e). The dynamics of occupied patches (P) may be modeled as follows (Levins, 1969):

$$dP/dt = iP(1 - P) - eP \quad (1)$$

an expression implying that the fraction of occupied patches at equilibrium is $P^* = 1 - e/i$. A main consequence of this relationship is that no matter how good a species may be as a competitor, if it does not persist forever in a patch, a fraction of free environment will always remain unused.

It is instructive to consider the extreme example of one superior and one inferior competitor (Tilman, 1994). The superior competitor species always excludes the inferior one on arrival in a patch. The dynamics of the superior competitor species P_1 are not affected by the presence in the landscape of the competitively inferior species: $dP_1/dt = i_1 P_1(1-P_1) - e_1 P_1$. However, the inferior species experiences a reduction in the fraction of available habitat and an increase in local extinction once the superior competitor species arrives in the patch. The dynamics of the inferior competitor species P_2 will be determined by:

$$dP_2/dt = i_2 P_2(1 - P_1 - P_2) - e_2 P_2 - i_1 P_1 P_2 \quad (2)$$

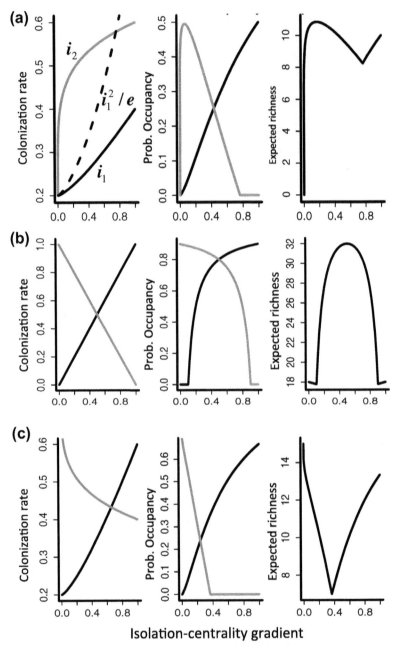

FIGURE 2 Predictions of metapopulation models. Gray and black lines represent the trends in colonization rates and probabilities of patch occupancy of inferior and superior competitor species across an isolation gradient. (a): *Patch dynamic*—coexistence depends on the inferior competitors more rapidly capturing resources than the superior competitors. The dashed line represents the

which is a relationship implying that in order to maintain a positive population growth rate when the superior competitor species occupies its equilibrium proportion of the environment, the following relationship among demographic parameters should be satisfied:

$$i_2 > \frac{i_1(i_1 + e_2 - e_1)}{e_1} \qquad (3)$$

When extinction rates are equal, the inequality simplifies to $i_2 > i_1^2/e$. This inequality implies that coexistence depends on the inferior competitor species having a faster rate of empty patch colonization, an outcome that may be generalized to many species (Tilman, 1994). This constraint is relaxed when the mortality rate of the inferior competitor species is lower than that of the superior competitor species, and the reverse is also true. These relationships determine patch dynamics in which species coexistence depends on a trade-off between species' rates of empty patch colonization versus species' abilities to colonize and defend a patch from future immigrants (Holyoak et al., 2005b).

In general, the combination of demographic parameters of inferior and superior competitor species should ensure that the former may persist in the patches left empty by the latter. The proportion of free patches for colonization when the superior competitor species is at equilibrium is: $1 - \hat{P} = 1 - 1 + e_1/i_1 = e_1/i_1$. Community isolation is considered to be a direct determinant of the rate of individual arrival and, consequently, of the colonization rate i (Altermatt, 2013; Economo and Keitt, 2010; MacArthur and Wilson, 1967). Since an isolation gradient determines an immigration gradient, more isolated communities should have reduced colonization rates. Patch dynamics involve inferior competitor species with enhanced dispersal abilities. The reduction in colonization rate among increasingly isolated patches should be greatest among better competitors, thus determining a reduction in the proportion of patches occupied by the dominant species (Figure 2(a)). The relative increase in the inferior species colonization rate in comparison with the superior competitor species rate relaxes the constraint of Eqn (1). In addition, free space should also foster opportunities for isolated patch specialist species. As a consequence, the more isolated a community is, the more opportunities there are for the

---

constraint $i_2 > i_1^2/e$, assuming equal extinction rates. Two pools of 10 dominant and 10 subordinate species are considered. A large turnover between subordinate and dominant species is expected. (b): *Species mixing*—a pool of isolation-specialists and a pool of central patch species are considered. Each species pool follows Levins dynamics with equal extinction rates but inverse trends in colonization with increasing community isolation. The coexistence of species from the two pools at intermediate isolation levels determines a humped pattern of diversity. (c): Similar to (b), but isolation-specialists are competitive subordinates. The more rapid decrease in the probability (Prob.) of isolation-specialist occurrence may determine minimum richness at intermediate isolation levels.

operation of patch dynamics. However, this trend may reverse if superior competitor species are able to reach more distant patches; for example, if those with large body sizes are able to travel longer distances between patches (Arim et al., 2010; Borthagaray et al., 2012; McCann, 2012). The propensity for patch dynamics in isolated communities (in contrast to the mass effect described below) is a main prediction of metacommunity theory (Brown and Swan, 2010). Metapopulation models emphasize the dependence of the main predictions of functional attributes present in the species pool (e.g., specialists in isolated environments) and/or the differential effect of isolation on the immigration rates of superior and inferior competitors.

Mass Effect

Metapopulation models also permit evaluation of expected trends in the role of an additional metacommunity mechanism, the "mass effect." The individual flow from "source" populations in suitable environments to "sink" populations in less propitious environments may enhance species persistence under unfavorable conditions (Hanski, 1999; Shmida and Wilson, 1985). This source—sink dynamic may determine a rescue effect, whereby immigration prevents species extinction (Brown and Kodric-Brown, 1977). Even when extinction takes place, the inflow of individuals promotes a rapid colonization and persistence of the species in the environment. The enhanced species viability and community richness promoted by source—sink dynamics and rescue effects determine the second metacommunity mechanism termed "mass effect" (Holyoak et al., 2005a; Shmida and Wilson, 1985).

One advantage of metapopulation models is their flexibility in incorporating additional ecological processes (Gotelli, 1991). The Levins model (Eqn (1)) may be modified to include mass effect caused by the processes of source—sink dynamics (propagule rain) and the rescue effect (Gotelli, 1991). Propagule rain was previously conceptualized to account for the arrival of propagules in local communities (sink or nonsink communities) from a mainland or a very large population source (MacArthur and Wilson, 1967). This rain has the effect of making the production of propagules by the metapopulation of little consequence for colonization rates (Gotelli, 1991). In addition, the rescue effect involves a reduction in extinction probability because of the immigration of individuals from other patches, which reduces extinction rates as the landscape becomes occupied. The Levins model modified to incorporate these two components of mass effect is expressed as (Gotelli, 1991):

$$dP/dt = i(1-P) - ep(1-P) \tag{5}$$

A metapopulation dynamic incorporating the rescue effect and the propagule rain will have an equilibrium landscape occupancy of $P^* = i/e$. For the same extinction and colonization rates, this equilibrium is much larger than that expected without the mass effect. Furthermore, species can persist even when the extinction rate is greater than the colonization rate; when immigration is greater than extinction, species tend to occupy most of the space available (Gotelli, 1991). As a consequence, the mass effect represents a strong force promoting species viability in local communities, raising local richness, and reducing beta diversity because the species are able to persist in more communities (Holyoak et al., 2005b). Immigration to local communities is determined by the influx of individuals from the metacommunity. Thus, the larger the centrality (i.e., lower the isolation) of a local community becomes, the greater the importance of immigration and the mass effect (Figure 3). When considering a single species pool, the probability of patch occupancy may be translated into an expected richness by multiplying this probability by the number of species in the pool (Hanski, 1999, 2010). The expected pattern is a poor representation of species across isolated communities—that is, a nonlinear increase in species richness with all species of the pool present among more central communities (Figure 3). This result is congruent with predictions of nonneutral models considering species-specific demographic parameters (Loreau, 2010; Loreau and Mouquet, 1999). The metacommunity may be considered as comprising two pools of species, one with attributes that promote better performance in central communities and the other with attributes that promote better performance in isolated communities (e.g., Chase and Shulman, 2009; Welsh and Hodgson, 2011). Both groups of species have inverse relationships in colonization and/or extinction rates across an isolation gradient. In the framework of a Levins metapopulation model for each pool and, for simplicity, considering only trends in colonization rates, the combination of species from the two pools at intermediate levels of isolation contribute to a humped pattern of diversity (Figure 2(b) and Arim et al., 2002). When we consider species interactions for the dominant/subordinate species

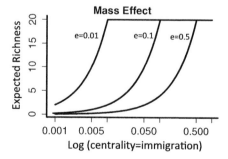

FIGURE 3 Expected trend in richness across an isolation gradient under the influence of *mass effect*. The metapopulation model incorporating the rescue effect and propagule rain captures the mass effect and has a fraction of occupied patches at an equilibrium of $P = i/e$. The rise in immigration with increasing centrality ensures that the expected richness quickly rises to the total number of species in the pool (20 species).

discussed previously, we see that each species in the metapopulation model represents a species pool, and that total richness is determined by the sum of expected richness from each pool. When isolation has a similar effect on dominant and subordinate species, subordinate species would need better colonization capabilities in order to coexist (see Figure 2(a)). At the metacommunity level, this may involve a transition across the isolation gradient, with subordinate species in isolated patches and dominant species in more central patches (Figure 2(a)). This transition would be more pronounced (even producing a U-shaped trend in species richness) when subordinate species are better colonizers in isolated patches and the dominant species better colonizers in central patches (Figure 2(c)).

Two main messages emerge from these metacommunity models. First, a wide range of richness—isolation patterns may be expected even from the simplest considerations of mass effect and patch dynamics mechanisms. These patterns include monotonic, humped, and U-shaped trends. Second, trait-mediated species responses to isolation and biotic interactions determine expected patterns. The operation of this second mechanism is reinforced by "species sorting," the next metacommunity process for consideration below.

Species Sorting

At one extreme, neutral theories (considered in the next section) predict expected patterns when ecological processes operate among individuals or species independently of their traits (Hubbell, 2001; Rosindell et al., 2011). The *species sorting* view of metacommunity processes presents the other extreme (Holyoak et al., 2005b). The basic premise of species sorting is that the combination of attributes carried by a species determines its colonization and growth success in local communities (Leibold et al., 2004; Mouquet and Loreau, 2003). Trade-offs in species performances under different conditions are at the heart of species sorting and most explanations of species coexistence (Kneitel and Chase, 2004). Immigration to local communities should ensure recruitment of those species with attributes well fitted to the environment of the regional pool, but immigration rates should not be so high that they produce a significant mass effect that erases trait—environment associations (Leibold et al., 2004).

Loreau and Mouquet (1999) introduced a metacommunity model with the potential to incorporate species sorting and mass effect, considering both as local and regional dynamics, and a dispersal parameter determining the fraction of locally produced individuals that migrate to a regional pool. In summary, this is a model that considers key components of metacommunity theory. Local communities are represented by a Levins's metapopulation model (Tilman, 1994) in which parameters change among and within the same species across communities. Each species has local reproductive and mortality rates. As different local communities have different conditions, different species in each community will have the largest reproductive potential.

Consequently, in the absence of dispersal, each community is inhabited by a single species with the best environment—trait linkage, and regional coexistence depends on interpatch heterogeneity (Mouquet and Loreau, 2002). An intuitive and formal result of this model is that the number of species that can coexist at the metacommunity level is equal to the number of local communities (Mouquet and Loreau, 2002). In this context, species sorting determines a large variation of trait performance among communities, with a hypothetical extreme of single local communities selecting for unique combinations of species traits. A further outcome that emerges from this concept is the possibility of local coexistence when the reproductive rates of different species are different in local communities but similar across the whole metacommunity (Mouquet and Loreau, 2002).

When there is a gradient of dispersal rates, such as the continuum that may exist across the space between isolated and central communities, species sorting is predicted to function as a strong structuring mechanism at low dispersal rates, progressively weakening when higher dispersal rates enhance local abundance of the best competitors at the metacommunity level—for example, by mass effect. Mouquet and Loreau (2003) validated this prediction by showing that local diversity (α) tracks a humped trend with dispersal, while beta diversity (β) and gamma diversity (γ) decay with dispersal rate (Mouquet and Loreau, 2003). The analysis of local food webs connected by dispersal suggests that food web branching also tracks a humped association with dispersal rate, corroborating the humped association between local richness and dispersal (Pillai et al., 2012). However, the inclusion of stochasticity into dispersal rate eliminates this association, moving the richness mode to higher dispersal rates (Matias et al., 2013). The humped trend may be reinforced when keystone predators enhance local coexistence or weaken when predation or competition promotes species extinction (Kneitel and Miller, 2003).

Neutral Mechanisms

Neutral theories are based on the assumption that organismal traits have no selective role. Thus, whatever the variation in phenotypic attributes observed among species or individuals, all have equal chances of survival and reproduction (Hubbell, 2001). Neutral theories may incorporate processes of competition (Hubbell, 2001), disturbance (Kadmon and Benjamini, 2006), landscape structure (Borthagaray et al., 2014b; Economo and Keitt, 2008), or any sound ecological/evolutionary mechanism. It is not the simplicity or complexity of a theory that determines its neutral status, but rather the fact that organism and/or species traits have no role in the operation of these processes.

The island biogeography model of MacArthur and Wilson (MW) (1967) assumes a species pool on a mainland that serves as a source for island colonization. It further assumes that colonization and extinction rates are determined by island isolation and area, but for each island these rates are equal among all species (MacArthur and Levins, 1967; MacArthur and Wilson, 1967). As a consequence, the MW model represents a neutral theory at the species level (Hubbell, 2001). The expected species richness on an island is a result of the colonization–extinction balance, which determines a dynamic equilibrium whereby the number of species on the island remains fixed, but the composition changes at a constant rate (Gotelli, 2008; MacArthur and Levins, 1967). The model considers both local (area) and landscape (isolation) determinants of local community richness and species turnover. Two main predictions on isolation emerge from this model: (1) species richness decreases with isolation, and (2) species turnover also decreases with isolation. As a consequence of these two predictions, a third may be proposed: (3) as a result of the sampling effect, beta diversity between communities that are similarly isolated increases at higher isolation levels. When there is a progressive colonization process with immigrant species coming not only from the mainland but also from other islands, a nested pattern can emerge at the metacommunity scale. Across a gradient of community isolation, poorer communities tend to be subsamples of richer ones (Patterson and Atmar, 1986).

Throughout the 10 years following the publication of *The Unified Neutral Theory of Biodiversity and Biogeography* (Hubbell, 2001), responses to the concept changed from virulent attacks to acceptance of neutral processes as main components of ecological theory (Holyoak et al., 2005b; Rosindell et al., 2011). However, debate about the role of neutral processes and the approach to neutrality in Hubbell's formulation remains ongoing (Clark, 2012; Clarke and Johnston, 1999; Shipley, 2010b). The focus of Hubbell's theory on the analysis of ecological processes (with an underlying premise of neutrality at the individual level) confirmed previous predictions and significantly expanded the range of ecological patterns and processes for consideration (Rosindell et al., 2011). The accurate prediction of the species abundance distribution (SAD) by neutral models was previously considered a validation of the theory. However, it became evident that the same SAD pattern may be expected under the operation of neutral, quasi-neutral, or nonneutral mechanisms (Mouquet and Loreau, 2003; Rosindell et al., 2011). Consequently, the reproduction of SAD patterns may actually be considered a necessary but insufficient condition for the validation of alternative ecological mechanisms. Simultaneous focus on SAD and other patterns, such as species–area relationships and temporal or spatial turnover, is considered a more appropriate approach to the analysis of neutral theories (Rosindell et al., 2011). Major predictions of neutral models, including MacArthur and Wilson's, are summarized in Figure 4.

Neutral and niche theories may be considered two ends of a conceptual continuum rather than contrasted alternatives (Gravel et al., 2006; Shipley, 2010a). When the immigration rate is sufficiently high that it surpasses the signal of local community selection for organismal traits, a neutrality domain emerges (Ai et al., 2013). Under these circumstances, it may be predicted that more central communities will be influenced by a mass effect than those that

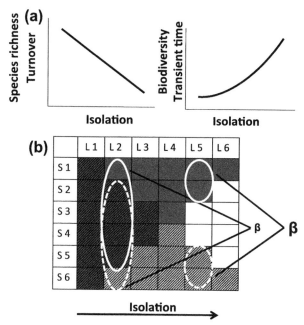

FIGURE 4 Main predictions of neutral models on the effect of isolation on community structure. (a) Isolation is expected to reduce species richness and slow temporal turnover (left panel). Beta diversity among isolated communities is expected to increase with isolation and transient time to community attractors (right panel). (b) At the metacommunity level, a nested pattern is expected to emerge when immigrants to isolated communities come from closer and less-isolated communities. More-isolated communities contain a subset of species compositions of those nearer to the species pool. Two possible nested patterns (white with cross-hatching as well as gray) are depicted in a species (S) × localities (L) matrix. Black and crosshatched cells indicate the presence of species S in locality L, whereas white cells indicate absences. Two of any of the communities (encircled by white dashed or continuous lines) are more likely to share species (low beta diversity) when they are in more-central locations.

are isolated, which reduces the role of traits on community assembly and enhances neutral dynamics. Thus, for neutral dynamics to operate in some local communities, niche-mediated species assemblage should be operating at the metacommunity level.

Significant improvements have been made in the estimation of neutral model parameters from field data (Munoz and Couteron, 2011; Rosindell et al., 2011). Modeling and empirical parameterization of neutral dynamics should become relatively straightforward with novel approaches (Munoz and Couteron, 2011; Rosindell et al., 2011). Furthermore, as the niche–neutral continuum becomes more widely recognized, it will be necessary to develop a methodology that is able to identify systems at any point on this gradient. The analysis of community assembly through MaxEnt formalisms may well provide the appropriate procedure for making these determinations across the continuum (Shipley et al., 2006).

Theory Data

Empirical support for the construction of ecological theory depends on clear assumptions and predictions, and testing with procedures that either support them or do not. The four paradigms of metacommunity ecology (also termed "views" or "mechanisms") have explicit assumptions and predictions (Figure 1). However, empirical approaches using field data (Cottenie, 2005), and even experimental analyses (Logue et al., 2011), are limited in their ability to provide adequate information to make choices among mechanisms. The isolation of a single paradigm operating independently of the others may be impossible in many cases. While this complexity is part of the natural world, it is nevertheless rewarding to seek approaches that would disentangle the paradigms. In this context, we promote the view that further application of graph theory for the estimation of proxies of local and regional dispersal (Altermatt, 2013; Economo and Keitt, 2010), and the incorporation of MaxEnt analysis of functional diversity (Shipley et al., 2006) will likely markedly enlarge the metacommunity ecologist's toolbox and contribute to the construction of robust theory.

Several methods are currently used to explore metacommunity mechanisms through the analysis of field data. Spatial autocorrelation in community structure has been used as a proxy of dispersal, whereby the scale of correlation is assumed to be proportional to dispersal (Shurin et al., 2009; Soininen et al., 2007). Partitioning of variance dissects the total variation in a community matrix into unique environmental and spatial components, and is capable of separating each of these components independently of one or all of the others using a purely statistical procedure (Cottenie, 2005). It is usually assumed that environmental components represent local filters, that spatial components represent dispersal limitation, and that their interaction term is difficult to interpret (Meynard et al., 2013). Another approach involves the analysis of a species site-by-species incidence matrix (Meynard et al., 2013) that detects structures that may support random assembly, competition, or environmental filtering. The combined use of this matrix analysis and the partitioning of variance into components that are environmental or spatial may robustly analyze empirical data for the determination of metacommunity mechanisms (see Meynard et al., 2013). Finally, neutral models and the relative deviation between observed pattern and neutral expectations are now amenable to relatively straightforward computation (Munoz and Couteron, 2011; Rosindell et al., 2011). As a consequence, the evaluation of neutral mechanisms—for example, through analysis of a spectrum of communities with different dispersal rates—may function as a powerful empirical procedure. A potential limitation of this approach is that there are several potential neutral models for consideration. Consequently, a deviation from model expectations may represent a limitation of the model considered rather than of the magnitude of the neutral mechanisms. On the other hand, the fitting procedure of some null models may involve very flexible equations that represent a wide range of data spanning neutral and nonneutral mechanisms (McGill, 2003). MaxEnt analysis of metacommunity patterns provides an alternative measure of neutrality; it is an analysis conditioned to the set of traits observed among species, but independent of any particular neutral mechanism.

The degree of isolation of a local community is accepted as a good proxy for the potential dispersal experienced by a local community (Altermatt, 2013; Brown and Swan, 2010; Economo and Keitt, 2010). Isolation could be represented by the distance of a local community to other patches—for example, those at headwaters (Miyazono and Taylor, 2013). However, the explicit consideration of metacommunity networks makes it possible to use several alternative and complementary measures of isolation that provide quantitative and continuous estimations of centrality. In addition, the same degree of community isolation may represent a strong or a weak barrier to dispersal depending on the dispersal abilities of species (Borthagaray et al., 2012; Economo and Keitt, 2010; Keitt et al., 1997; Urban and Keitt, 2001). As a consequence, the organismal perspective has to be considered in the analysis of a metacommunity pattern. The problem of incorporating individual perspective into the definition of a metacommunity is not new, but it has seldom been evident in ecological thinking following the pioneer studies in community ecology (Holyoak et al., 2005). In the next section, we review the main network metrics that may be used to estimate the relative centrality—isolation of local communities as proxies of dispersal.

Metacommunity Networks

Exploring the role of the metacommunity network on communities and metacommunity structure will require comparisons of both (1) local communities with different patterns of insertion within the network and (2) different metacommunities with contrasting structures at the network level. Graph theory provides an exceptional procedure for progress on these two issues (Urban and Keitt, 2001). Graph theory has long been used as a framework for analysis of ecological networks considered as food webs (Cohen et al., 1990; Pimm, 1982), of mutualistic networks (Bascompte et al., 2003), and increasingly in landscape ecology (Keitt et al., 1997; Urban and Keitt, 2001). The central role of graph theory in advancing a spatially explicit approach to metacommunity ecology has been increasingly emphasized in recent works (Altermatt, 2013; Borthagaray et al., 2012, 2014b; Carrara et al., 2012; Economo and Keitt, 2010; Peterson et al., 2013). Graph theory provides a wide set of tools for representing metacommunities and quantifying their structures at the level of the whole network and at the level of individual communities. A metacommunity can be described by a graph defined as a set of nodes connected by links. Typically, nodes correspond to communities, and connections refer to some kind of structural or functional relationship among them—for example, the flow of individuals (Urban and Keitt, 2001). Here, we review network metrics used to estimate the isolation (centrality) of local communities and to provide main descriptions of whole network properties. However, to determine these properties, there must be a metacommunity network, which in most cases is not easy to estimate. Thus, we will finish this chapter by describing several methodologies for empirically estimating metacommunity networks.

Community-Level Properties

In recent years, centrality metrics have been used as main measures of community isolation (Ai et al., 2013; Altermatt, 2013; Altermatt et al., 2013; Borthagaray et al., 2012; Carrara et al., 2012; Desjardins-Proulx and Gravel, 2012; Economo and Keitt, 2010; Urban et al., 2009). The centrality index potentially reflects individual flows through local communities and then onward through the whole metacommunity. The various centrality metrics focus on different concepts and definitions of centrality in a network (Newman, 2010). Importantly, alternative centrality measures complement one another in the representation of different components of community isolation (see below). This becomes evident upon realization that different processes reflected in different metrics refer to different taxa. Following the same line of reasoning, the isolation or centrality of patches in different metacommunities may be better represented by different centrality measures. Therefore, it would be advantageous to calculate a large set of metrics to reflect community isolation in different ways; evaluating relative performances, the metrics account for the patterns of interest (Figure 5).

In a metacommunity context, the four major centrality metrics are degree, eigenvector, closeness, and betweenness (see Table 1) (Economo and Keitt, 2010; Estrada and Bodin, 2008). Larger values of centrality indicate lower levels of community isolation (Economo and Keitt, 2010). *Degree centrality* is the number of direct connections between a community and its neighbors—for example, direct links between patches (Freeman, 1979; Wasserman and Faust, 1994). The term "connectivity of a node" has also been used to refer to degree centrality. *Eigenvector centrality* ranks communities not only by the number of direct connections, but also by the number of connections that their neighbors have (Bonacich, 1972; Wasserman and Faust, 1994). In this sense, it may be seen as an extended degree centrality, since it is proportional to the sum of the degree centralities of the community neighbors. Definitions of closeness centrality and betweenness centrality are based on the length of the shortest path between communities (Freeman, 1979; Wasserman and Faust, 1994). *Closeness centrality* is the reciprocal of the average length of the shortest path between the reference community and all others; it provides a representation of how close or how far a community is from the remaining communities in the metacommunity. A main limitation of closeness is that it cannot be calculated for all communities in a disconnected metacommunity since two unconnected communities do not have a finite distance between them (Opsahl et al., 2010).

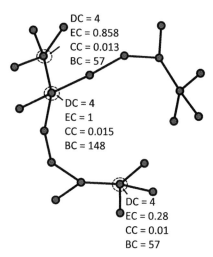

FIGURE 5 Estimation of alternative indices of centrality (the reverse of isolation) in a hypothetical metacommunity. A circle represents local communities and lines the connections among them. The dashed circles indicate the locations of three communities for which alternative centrality indices were estimated: DC: degree centrality; EC: eigenvector centrality; CC: closeness centrality; BC: betweenness centrality. Each index emphasizes a different component of community isolation; the communities considered here have the same local connections (DC), but have different isolations measured in terms of the connections between their neighborhoods and the neighborhoods of their neighborhoods (EC), relative distance to all other communities (CC), and roles as stepping stones for individual flow across the whole community (BC).

Alternatively, closeness may also be expressed by its inverse value to represent isolation, a metric termed *farness* (Altermatt, 2013; Newman, 2010). *Betweenness centrality* describes how often a community acts as a mediator on the shortest path between two other communities, identifying key connectors—that is, stepping stones of individual flow to the rest of the metacommunity. The different centrality metrics emphasize different aspects relating to the spatial scope considered (Figure 5). Degree centrality highlights the potential flow of individuals through the local neighborhood. Eigenvector centrality expands the potential flow of individuals to a larger neighborhood. Closeness centrality is a measure of how long it takes an organism to move sequentially from one patch to all other patches in the metacommunity (Altermatt, 2013). Betweenness centrality emphasizes the bridging role of the patches in maintaining metacommunity connectivity beyond the local scale.

Other centrality metrics have been suggested to emphasize metacommunity structure at intermediate scales. Among them, the *k-neighborhood* metric is defined as the number of communities that are within a k-shortest path from the focal community (Economo and Keitt, 2010). *Subgraph centrality* measures the number of times a community occurs in all subgraphs or in closed loops within the metacommunity, with weighting according to their

TABLE 1 Definition of Centrality Indices and Their Calculation with R Program

Index	Definition	R Package (Function)
Degree centrality	Number of neighboring communities	sna (degree); Igraph (degree)
Eigenvector centrality	Scores communities not only by their immediate connections but also by the degree of connection	sna (evcent); Igraph (evcent)
Closeness centrality	Calculates how close a focal community is to all other communities	sna (closeness); Igraph (closeness)
Betweenness centrality	Quantifies how frequently a community is on the shortest path between every possible pair of the other communities	sna (betweenness); Igraph (betweenness)
k-neighborhood	Number of communities within distance k of the focal community	Igraph (neighborhood.size)
Subgraph centrality	Number of closed loops in which a focal community is involved	Igraph (subgraph.centrality)

size (Estrada and Rodríguez-Velázquez, 2005). All these metrics may be calculated with R software (R Development Core Team, 2013) (see Table 1). Parameter estimation is based on an adjacency matrix in which each element is 1 if two communities are connected and otherwise 0 (Urban and Keitt, 2001).

A final caveat for mention is the fact that centrality indices are relative measures of local community importance. The value of each index as an indicator of the degree of community isolation depends on the spatial arrangement of the other communities within the metacommunity. In this sense, to make results comparable among metacommunities, it would be appropriate to standardize centrality metrics to the size of the metacommunity or to normalize them so that they sum to 1.

Metacommunity-Level Properties

A well-supported advance in the empirical association between metacommunity structure and function is based on the comparative analysis of different metacommunities (e.g., Bertuzzo et al., 2011; Carrara et al., 2012). Although slower than the development of theoretical studies (Driscoll and Lindenmayer, 2009), empirical explorations are now underway (Burns, 2007; Driscoll and Lindenmayer, 2009). A future challenge for advances in this field will be the recording and comparison of contrasting metacommunity structures and associated biological information. This procedure is of relevance because properties of network structure strongly affect metacommunity dynamics (Economo and Keitt, 2008; Gilarranz and Bascompte, 2012). Thus, we now present alternative metrics for the description of structural properties at the level of the whole network.

Three of the most widely used and characteristic structural properties of any kind of network are *linkage density*, *connectance*, and *diameter*. Connectance is the number of connections or links realized (L) divided by the maximum possible number of links (Newman, 2003; Proulx et al., 2005; Williams et al., 2002). In a metacommunity with lc local communities, connectance is calculated as $(2 \times L/lc \times (lc-1))$. A metacommunity with a high connectance probably has more redundancy in connections among local communities, which may foster robustness when link and patch removal occur (Melián and Bascompte, 2002). At the same time, higher connectance may promote disease propagation and global extinction because of the synchronization of local dynamics (Liebhold et al., 2004). Linkage density is defined as the average number of realized links per local community and is calculated as (L/lc), where (L) is the total number of links in the network and (lc) is the number of local communities. The diameter of a metacommunity represents the number of steps necessary for movement through the whole metacommunity. Estimations of diameter are based on the distribution of the shortest paths between all pairs of local communities. Thus, diameter has been defined as the length of the number of links in the longest path among shortest paths between all pairs of communities (Newman, 2003; Urban and Keitt, 2001; Wasserman and Faust, 1994). It is therefore necessary to determine all of the shortest paths between all pairs of communities in the network and then find the longest among them. Diameter has also been defined as the average of the shortest paths between all pairs of communities in the network (Proulx et al., 2005), and also as the characteristic path length (Williams et al., 2002). A good description of metacommunity structure may be obtained by considering linkage density, connectance, and diameter.

Other basic and widely used metrics to describe network structure are based on frequency distributions rather than on single parameters. For example, linkage density estimates the mean degree, which in several networks provides a good description of the connection between local patches. However, when the distribution of links per patch is markedly asymmetric—that is, most patches with few links and a few with several connections—a focus on whole distributions may be a more appropriate approach. *Degree distribution* (also called *connectivity distribution*) provides such a focus; it describes the frequency distribution of the number of links to each local community (May, 2006; Proulx et al., 2005). A first approach to estimating degree distribution is the construction of a histogram of community degree plotted on a logarithmic scale (see Newman, 2003). However, a better representation is obtained by using the inverse cumulative distribution, which indicates the fraction of communities that have a degree of k or higher (Newman, 2005). The shape of this distribution represents a description of the level of heterogeneity of the whole metacommunity and is associated with the robustness of the network in maintaining the flow of individuals when increasing numbers of communities or paths are deleted (Burns, 2007; Gilarranz and Bascompte, 2012; Melian and Bascompte, 2002). The relevance of such a structure is that it represents recolonization capability from highly connected patches (Gilarranz and Bascompte, 2012). However, this would be advantageous only for individuals with long-range dispersal ability or individuals with high rates of extinction. In other cases, a homogeneous structure may be better for maintaining population persistence (Gilarranz and Bascompte, 2012).

A topological metric that complements degree distribution in the description of metacommunity structure is *degree correlation* (Krapivsky and Redner, 2001). In a metacommunity, degree correlation (also called *connectivity correlation*) is the relationship between the number of neighbors in a community and the average connections of neighbors (Maslov and Snepen, 2002; Melián and Bascompte, 2002). The slope of this correlation indicates the level of spatial aggregation of the overall network. Links between highly connected communities generate a cohesive structure, but when such links are suppressed, a compartmentalized metacommunity structure would be expected (Melián and Bascompte, 2002).

The properties described above emphasize large-scale structures that affect the dispersal and interchange of organisms through the whole metacommunity. *Modularity* is a property of intermediate scales between local community and metacommunity levels; it is defined as the degree to which some groups of communities have a higher probability of mutual flows of organisms as opposed to flows to other communities (May, 1972; Newman and Girvan, 2004). In this sense, modularity detects groups of communities (called modules or compartments) with relatively high numbers of mutual connections favoring the movement of neighboring individuals, even when each individual is able to reach any community in the landscape (Borthagaray et al., 2014b). Although modularity was proposed as an early key metric of networks (May, 1972), the intermediate level of structure captured by modularity has rarely been considered (Borthagaray et al., 2014b). It should be noted that other metrics have been developed to identify modules and compartments (see Bodin and Norberg, 2007; Urban and Keitt, 2001), but only after the recent development of robust modularity-detecting algorithms (Guimerà and Amaral, 2005; Newman and Girvan, 2004) have modules and compartments been widely detected in ecological networks (e.g., Olesen et al., 2007; Stouffer and Bascompte, 2011).

Weighted Metacommunity Networks

To this point, all metacommunity- and community-level properties have been defined for an unweighted and undirected network and calculated from an adjacency matrix (a 0–1 matrix indicating the presence or absence of individual flows between all pairs of communities). Alternatively, distance or dispersal matrices are used to define weighted connections between local communities. A *distance matrix* is one in which each element corresponds to a distance between local communities. This distance is measured as the minimum edge-to-edge or centroid-to-centroid Euclidean distance (Urban and Keitt, 2001). In addition, the distance between patches may be a biological distance (e.g., the Jaccard index). Another alternative is the *dispersal matrix* in which each element is defined as the probability that an individual in a community moves to another community. Dispersal probability is often approximated by a negative exponential function of the distance between two communities (Bunn et al., 2000; Hanski, 1999; Urban and Keitt, 2001), but other distributions may also be used (Clark et al., 1999). Therefore, community-centrality metrics and metacommunity metrics for weighted networks are based on the sum of the link weights rather than on the number of links (e.g., Estrada and Bodin, 2008); that is, a distance is defined as the sum of the link weights instead of the number of links between two communities. For example, the length of the shortest path between communities that defines closeness centrality, betweenness centrality, or the diameter of a metacommunity is calculated by summing the link weights. Finally, community-centrality metrics may also be defined by combining both the number of communities to which a focal community is connected and the sum of link weights (Opsahl et al., 2010).

Importantly, an adjacency matrix may be generated from the distance matrix or the dispersal matrix defining a threshold distance or probability to determine a link between two communities (Bunn et al., 2000). Two communities in a network are connected by a link when the separation between them is below the threshold distance (Keitt et al., 1997; Urban and Keitt, 2001).

Moreover, metacommunity network structure may be asymmetric, implying a directed network (e.g., a river). Thus, the distance defining the connection from community i to community j is different from the reverse path from community i to community j. Accordingly, two types of degree centrality are differentiated. In-degree, defined as the number of links that end in community i, and out-degree defined as the number of links originating in community i. Similarly, two types of closeness and betweenness centrality may be estimated for a directed network. These metrics may be calculated with the functions specified in Table 1 and executed with R software (first indicating that the matrix is directed and weighted) (R Development Core Team, 2013).

Patch isolation and its effect on immigration rate have been extensively considered by practitioners of metapopulation ecology (Hanski, 1999). An interesting approach is the estimation of the potential immigration rate (formerly termed connectivity, S_i) to a local community from all other communities, taking into consideration their distances, areas, occupancy probabilities, and a kernel function of decay in dispersal with distance to the source (Hanski, 2010). With this approach, the colonization rate of a focal patch is defined as the sum of the contributions from all possible source patches. These contributions are weighted by three factors: (1) the areas of the source A_j and focal patches A_i, (2) the distance from the focal patch to the source patch d_{ij}, and (3) the probability of the source patch being occupied p_j. In addition, the effect of areas on individual flows may be elevated to an exponent to account for nonlinear effects ($A_i^{\varepsilon_{im}}$ and $A_j^{\varepsilon_{em}}$), and the distance effect may be an exponential (with an α decay parameter) or any appropriate function (Hanski, 2010):

$$S_i = A_i^{\varepsilon_{im}} \sum A_j^{\varepsilon_{em}} p_j \exp(-\alpha d_{ij}) \qquad (6)$$

In estimating local community immigration rate, potential immigration from each of the other communities is considered. Combined vectors of immigration rates for all patches can be used to estimate a migration matrix and the associated migration-weighted graph. Finally, it should be emphasized that the parameter of this metric of centrality and the immigration rate may be directly estimated from the occurrence pattern, making this procedure a powerful tool for empirical approaches (Hanski, 2010).

Methodologies for Estimating Metacommunity Networks

A main challenge in the analysis of metacommunities is identification of network configuration (Jacobson and Peres-Neto, 2010)—that is, the spatial arrangement of communities and their connections, through which individuals move. In this sense, a natural network configuration is one defined for organisms subjected to directional dispersal driven by wind and water flows (Figure 6, Altermatt, 2013; Vanschoenwinkel et al., 2008a,b). Similarly, using molecular or genetic approaches it is possible to establish the connectivity pattern between communities and determine the metacommunity structure (Becker et al., 2007; Fortuna et al., 2009).

However, in most cases the metacommunity configuration is unknown or the level of sampling effort required for its determination cannot be achieved. Two readily applicable approximations for estimating spatial networks are the minimum spanning tree (MST) and the percolation network (Figures 7 and 8). Both methodologies define fully connected networks but differ in the criteria for connecting two communities. The MST connects every local community with the shortest path length—that is, with the minimum number of links to ensure that all patches are connected to a single graph. The MST represents the backbone of the network maximizing the flow of individuals with different dispersal abilities among communities (Urban and Keitt, 2001; Urban et al., 2008). Notice that in this approximation only the number of links (not their lengths) defines the network structure. The percolation network is defined as a graph in which patches are

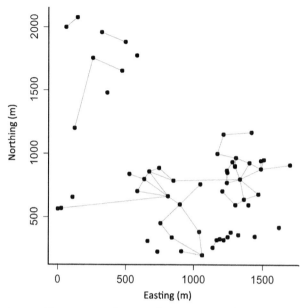

FIGURE 6 A potential metacommunity network for a system of temporary ponds located in Uruguay (34°25′47″S, 53°98′10″W). The ponds (nodes) are sporadically linked by water corridors (lines) as the system drains after heavy rainfall events. The physical connection between communities is an intuitive approach to estimating the metacommunity network. However, the functional metacommunity network may be different.

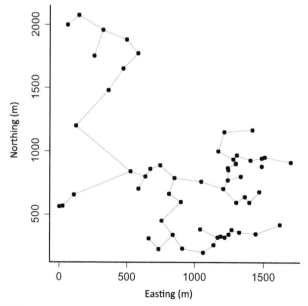

FIGURE 7 Minimum spanning tree (MST) for the temporary pond system in Figure 6. An MST is the shortest network that includes all the nodes in the graph (Urban and Keitt, 2001).

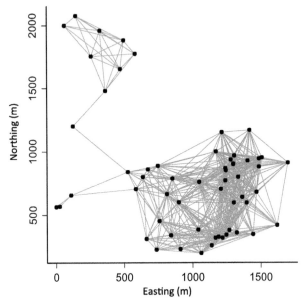

FIGURE 8 Percolation network for the temporary pond system in Figure 6. This network is obtained by linking the pairs of nodes that reach the minimum distance linking the whole patch system and form a single cluster. In this case, the percolation threshold is 539 m.

connected when the distance between them is less than a threshold distance, which is the minimum distance (i.e., the percolation point) at which all patches are connected in a single network (Rozenfeld et al., 2008; Urban and Keitt, 2001). The structure of this network with a threshold behavior is reminiscent of other landscape connectivity metrics such as connectedness or diameter (Keitt et al., 1997; With and Crist, 1995; Keitt et al., 1997). Similarly, the ability of organisms to disperse among communities typically fits a pattern of abrupt transitions from movements on local scales to movements that allow potential access to all parts of the landscape beyond the critical threshold distance (Borthagaray et al., 2012).

Although the MST and percolation network provide robust estimations of metacommunity networks, they may differ significantly from the landscape perceived by the organisms under consideration, or poorly represent the processes shaping local community structure. Local communities harbor species with a wide range of biological attributes that likely determine the spatial scales at which they experience the landscape (Borthagaray et al., 2012, 2014b; Keitt et al., 1997). Moreover, the same geographic distance between a pair of communities may be perceived as either a weak or a strong barrier to dispersal by different organisms that are part of the same metacommunity. Under these circumstances, exploration of a wide set of networks defined by a gradient of threshold distances provides an alternative approach to determining metacommunity networks (e.g., Borthagaray et al., 2014b; Urban and Keitt, 2001). This gradient is the continuum of geographic distances from the minimum to the maximum edge-to-edge Euclidean distance between two communities, which reflects the wide range of potential dispersal distances in a metacommunity pool (Figure 9). As the threshold distance increases, communities are connected into larger components that are themselves connected at even greater threshold distances (Urban and Keitt, 2001). In addition, a biological distance such as the Jaccard index, or a combination of both distance indices, may be more meaningful for connecting two communities than geographic distance alone (Figure 10). Biological indices may demonstrate that geographically close communities are isolated from one another, while distant communities are connected (Fortuna et al., 2009). Another methodology uses modeling of dispersal based on individual biological attributes to determine the putative connections that define metacommunity configuration (e.g., Moritz et al., 2013). An alternative procedure, which has been well developed in metapopulation ecology, is calculation of the incidence function; this function has the potential for application in a metacommunity context (Eqn (6); Melian and Bascompte, 2002).

Finally, we propose a new methodology for constructing a metacommunity network based on maximization of coherence with a community attribute (e.g., diversity or productivity). This approach consists of four steps: (1) constructing a set of metacommunity networks along a gradient of linkage distances to connect two communities; (2) calculating community-centrality metrics for each network; (3) determining the association between the centrality index and the community attribute of each metacommunity network; and (4) identifying the linkage distance and the related metacommunity network using a maximum community attribute−metacommunity network association. The underlying principle is that a community attribute−community centrality association will have an extreme value at a linkage distance congruent with the distance affecting organism flow (Borthagaray et al., 2014a).

Maximum Entropy

MaxEnt formalisms have performed well in niche modeling, and consequently this statistic has become a major component of the ecologist's toolbox (Phillips et al., 2006). Bill Shipley and coauthors recently proposed a novel use of MaxEnt for exploring organism trait−environment connections (Shipley, 2010a,b; Shipley et al., 2006). Unraveling of this connection permits detection of major processes in community assembly that may be used to predict community structure. Shipley has provided a short, succinct review of the MaxEnt theory (Shipley, 2010a) and has also published more extensive considerations of the topic (Shipley, 2010b). As a truly groundbreaking concept, this theory is subject to extensive debate (McGill, 2006; Petchey, 2010), but its good performance (Shipley et al., 2011) and demystification through the use of

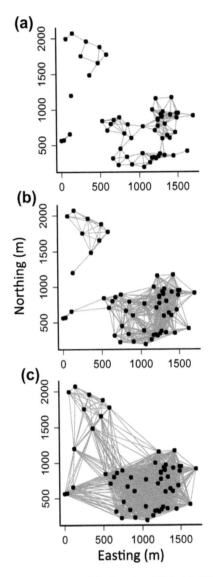

FIGURE 9 Networks obtained by three link distances: 250 m (a), 500 m (b), and 1000 m (c). All patches closer than the link distance are connected. By considering a wide range of link distances, a range of network perceptions determined by the range of dispersal abilities within the species pool may be inferred.

clear presentations (Shipley, 2010b) is consolidating its place in mainstream ecological theory.

MaxEnt translates the concept of community assembly by trait-based habitat filtering into a mathematical formalization that estimates trait roles

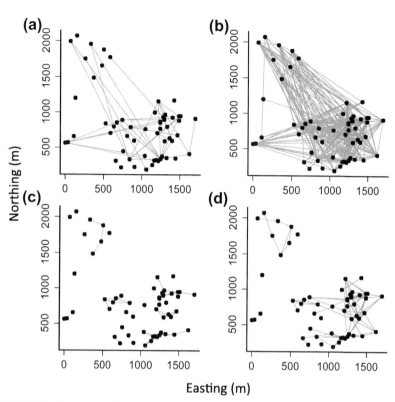

FIGURE 10 Networks obtained linking the set of ponds in Figures 6–9 based on ecological distances between vegetation of local communities. The distances were estimated as Jaccard indices for pairs of ponds. The networks were obtained by linking pairs of ponds in the whole system with index values up to 0.3 (a) and 0.4 (b), and by combinations of indices up to 0.4 and spatial distances up to 250 (c) and 500 m (d).

in observed local abundances. The underpinning logic is that each trait carried by a species (e.g., body size, dispersal mode, diet) has the potential to determine outcomes of interactions with other species and with the physical environment (Keddy, 1992; Weiher and Keddy, 1999). A first examination of MaxEnt indicates that the concept brings radical change to the methodological approach for analyzing community patterns. The objective of MaxEnt is to predict the relative abundance of a species in a local community from its combination of traits. If all individuals from a local community are pooled, an average value can be estimated for each trait (e.g., mean mass, mean proportion of aerial dispersers). In MaxEnt, these measures of community-aggregated traits are termed "empirical macroscopic constraints." An additional constraint is that relative abundances should sum to one. For any proposed combination of species abundances to conform to the local community, the values of these empirical constraints would have to be preserved.

However, consideration of the set of traits in each species identifies several combinations of species abundances that satisfy macroscopic constraints. Which of these combinations is more congruent with our knowledge of the system is as yet unknown; this knowledge gap is retained in the MaxEnt formalism, which at the same time allows selection of a particular combination of expected abundances. The selected combination is the one that maximizes entropy. Ecologists are familiar with the entropy concept in their common use of the Shannon diversity measure: $H = -\sum p_i \log(p_i)$. MaxEnt focuses on maximization of a related measure termed relative entropy (Shipley, 2010b):

$$\max\left(H = -\sum p_i \log\left(\frac{p_i}{q_i}\right)\right) \qquad (7)$$

where p_i is the predicted abundance of species i in a local community and q_i is a prior expected abundance. This prior distribution of abundances is an additional element in MaxEnt analysis. We may have no information on the distribution of abundance among species and assume accordingly that each species has the same prior relative abundance; for example, there is a uniform prior distribution. In addition, in the absence of processes shaping local community abundances, the community should be a random sample of the metacommunity (Borthagaray et al., 2012); the relative abundance in the metacommunity is then used as the prior distribution. If the predicted distribution of relative abundances is the same as the prior distribution of relative abundances, the relative entropy is zero: there is no increase in our knowledge of the system. On the other hand, the larger the discrepancy between p_i and q_i the more our knowledge increases. MaxEnt maximizes this increase in knowledge but is restricted to macroscopic constraints. The pressing issue now is application of a method for assigning a relative abundance p_i to each species in agreement with macroscopic constraints and maximizing relative entropy. "Lagrange multipliers" perform this function (Shipley, 2010a), for which there is a general solution, the "Gibbs distribution":

$$p_i = \frac{q_i \cdot \sum_{j=1}^{K} X_{ij} \cdot \lambda_j}{\sum_i q_i \cdot e^{-\sum_{j=1}^{K} X_{ij} \cdot \lambda_j}} \qquad (8)$$

This ungainly formula is a powerful tool for use in functional and metacommunity ecology. The parameter K is the number of traits observed in a species (e.g., mass, diet, defenses, fecundity). If no trait were measured, the expected relative abundance would be the prior relative abundance q_i. Parameter X_{ij} is an observation of trait j in species i, and λ_i is one of the Lagrange multipliers estimated by an iterative procedure (Shipley, 2010b). If observed traits have no role in species relative abundances, Lagrange multipliers will be zero and expected abundance will equal prior abundance. This is valuable ecological information, because as Lagrange multipliers become closer to zero, the closer to neutrality will be the community assembly in relation to the traits under consideration. Parameter λ_j represents an estimation of the selective advantage of a species with trait j in a local community. In summary, MaxEnt focuses on functional ecology and provides a method for estimating environment–trait connections that determine local abundances of species and community structure. Notably, each metacommunity paradigm may be transformed to an explicit prediction about the values of Lagrange multipliers and expected trends across the community centrality–isolation gradient.

In the *species sorting* (SS) paradigm, different communities have different environments, and different species have different combinations of traits. To support an SS concept of community assembly, the Lagrange multipliers should change among communities, with a negative correlation between the Lagrange and community attributes. This negative association determines the differential success of species with different traits among different environments. The *mass effect* involves the movement of individuals from source to sink communities. This implies that some communities have larger Lagrange multipliers (sources) than others (sinks). *Patch dynamics* is supported by the species trade-off between colonization and competition attributes. In order to identify this mechanism, those traits with a negative correlation among species attributes (trade-off) should also have large Lagrange multipliers. This is because there is a patch dynamic requirement that those traits that negatively covary among species have effects on organism performance. Finally, *neutral dynamics* predict no association between the successes of organisms and their traits. This prediction directly transforms to a Lagrange multiplier having a value of zero. Neutrality here is always conditional on the set of attributes under consideration (Shipley, 2010b). Trends in the operational strengths of these mechanisms across a gradient of community centrality–isolation will probably predict the expected trend in Lagrange multipliers. Increases in neutrality and mass effect toward central communities on the gradient (Ai et al., 2013; Brown and Swan, 2010) should be reflected in a negative association between Lagrange multipliers and centrality. However, the increase in patch dynamics toward isolated or central communities should be reflected in a positive trend in the Lagrange multipliers involved in species trade-offs. The increase in species sorting among isolated communities (Brown and Swan, 2010) should be reflected in a larger negative association between Lagrange multipliers in isolated communities than in central communities.

Our considerations to this point emphasize the wide range of diversity–isolation relationships that may be expected and the congruence of different mechanisms supporting any single pattern. MaxEnt may be further used to disentangle alternative mechanisms beyond a single pattern—for example, if a humped trend in richness across a community isolation gradient were to originate from the mixing of two species pools, one adapted to central locations on the gradient and other adapted to isolated locations (Arim et al.,

2002), MaxEnt should identify two groups of Lagrange multipliers: one increasing, and the other decreasing with community isolation. If the humped pattern originates through species sorting in isolated communities and mass effect among central communities (Mouquet and Loreau, 2003), all Lagrange multipliers should approach zero among central communities. In general, we emphasize that the incorporation of MaxEnt analysis into community structure considerations will represent a significant advance in the testing of metacommunity theory from observational data. This is likely the most powerful approach available for advancing empirical evaluations of metacommunity theory (Carrara et al., 2012; Dorazio et al., 2010; Driscoll and Lindenmayer, 2009; Logue et al., 2011).

Finally, we would like to stress that the estimation of the Lagrange multipliers is straightforward. The function "maxent" of the R software package "FD" (Laliberté and Shipley, 2011) estimates Lagrange multipliers starting from the set of macroscopic values in local communities, the set of traits values, and a prior distribution of abundances. Furthermore, the MaxEnt analysis of community assembly will become even more straightforward since the recent recognition of its equivalence to Poisson regressions (Renner and Warton, 2013; Warton et al., in press). This will expand the range of trait—environment associations for consideration, incorporate statistical methods for model validation and contrasts among models, and in general incorporate all the machinery of generalized linear models.

ACKNOWLEDGMENTS

This work was supported by a grant from Fondo Clemente Estable 2011-2-7117 and 2007-054 to M. A. A.I.B. thanks a postdoctoral fellowship CONICYT - FONDECYT N°3130360 (Chile) and CONICYT: Anillo en Complejidad Social SOC-1101. The authors thank PROBIDES and Establecimiento Barra Grande for field assistance and Anthony Chapman for English editing.

REFERENCES

Abrams, P.A., 2001. Describing and quantifying interspecific interactions: a commentary on recent approaches. Oikos 94, 209—218.

Ai, D., Gravel, D., Chu, C., Wang, G., 2013. Spatial structures of the environment and of dispersal impact species distribution in competitive metacommunities. PLoS One 8, e68927.

Altermatt, F., 2013. Diversity in riverine metacommunities: a network perspective. Aquat. Ecol. 47, 365—377.

Altermatt, F., Seymour, M., Martinez, N., 2013. River network properties shape a-diversity and community similarity patterns of aquatic insect communities across major drainage basins. J. Biogeogr. 40, 2249—2260.

Arim, M., Abades, S., Laufer, G., Loureiro, M., Marquet, P.A., 2010. Food web structure and body size: trophic position and resource acquisition. Oikos 119, 147—153.

Arim, M., Barbosa, O., Molino, J.-F., Sabatier, D., 2002. Humped pattern of diversity: fact or artifact? Science 297, 1763a.

Arim, M., Marquet, P.A., Jaksic, F.M., 2007. On the relationship between productivity and food chain length at different ecological levels. Am. Nat. 169, 62–72.

Baiser, B., Buckley, H.L., Gotelli, N.J., Ellison, A.M., 2013. Predicting food-web structure with metacommunity models. Oikos 122, 492–506.

Bascompte, J., Jordano, P., Melián, C.J., Olesen, J.M., 2003. The nested assembly of plant–animal mutualistic networks. Proc. Natl. Acad. Sci. USA 100, 9383–9387.

Becker, B.J., Levin, L.A., Fodrie, F.J., McMillan, P.A., 2007. Complex larval connectivity patterns among marine invertebrate populations. Proc. Natl. Acad. Sci. USA 104, 3267–3272.

Bertuzzo, E., Suweis, S., Mari, L., Maritan, A., Rodriguez-Iturbe, I., Rinaldo, A., 2011. Spatial effects on species persistence and implications for biodiversity. Proc. Natl. Acad. Sci. USA 108, 4346–4351.

Bodin, Ö., Norberg, J., 2007. A network approach for analyzing spatially structured populations in fragmented landscape. Landscape Ecol. 22, 31–44.

Bonacich, P., 1972. Factoring and weighting approach to clique identification. J. Math. Sociol. 2, 113–120.

Borthagaray, A.I., Arim, M., Marquet, P.A., 2012. Connecting landscape structure and patterns in body size distributions. Oikos 121, 697–710.

Borthagaray, A.I., Berazategui, M., Arim, M., 2014a. Disentangling the effects of local and regional processes on biodiversity patterns through taxon-contingent metacommunity network analysis. Oikos. http://dx.doi.org/10.1111/oik.01317.

Borthagaray, A.I., Barreneche, J.M., Abades, S.R., Arim, M., 2014b. Modularity along organism dispersal gradients challenges a prevailing view of abrupt transitions in animal landscape perception. Ecography 37, 001–008.

Brown, B.L., Swan, C.M., 2010. Dendritic network structure constrains metacommunity properties in riverine ecosystems. J. Anim. Ecol. 79, 571–580.

Brown, J.H., Gillooly, J.F., Allen, A.P., Savage, V.M., West, G.B., 2004. Toward a metabolic theory of ecology. Ecology 85, 1771–1789.

Brown, J.H., Kodric-Brown, A., 1977. Turnover rates in insular biogeography: effect of immigration on extinction. Ecology 58, 445–449.

Brown, J.H., Maurer, B.A., 1989. Macroecology: the division of food and space among species on continents. Science 243, 1145–1150.

Bunn, A.G., Urban, D.L., Keitt, T.H., 2000. Landscape connectivity: a conservation application of graph theory. J. Environ. Manage. 59, 265–278.

Burns, K.C., 2007. Network properties of an epiphyte metacommunity. J. Ecol. 95, 1142–1151.

Cadotte, M.W., 2006. Metacommunity influences on community richness at multiple spatial scales: a microcosm experiment. Ecology 87, 1008–1016.

Carrara, F., Altermatt, F., Rodriguez-Iturbe, I., Rinaldo, A., 2012. Dendritic connectivity controls biodiversity patterns in experimental metacommunities. Proc. Natl. Acad. Sci. USA 109, 5761–5766.

Chase, J.M., 2005. Towards a really unified theory for metacommunities. Funct. Ecol. 19, 182–186.

Chase, J.M., Bengtsson, J., 2010. Increasing spatio-temporal scales: metacommunity ecology. In: Verhoef, H.A., Morin, P.J. (Eds.), Community Ecology: Processes, Models, and Applications. Oxford University Press, Oxford.

Chase, J.M., Leibold, M.A., 2003. Ecological Niches. Linking Classical and Contemporary Approaches. University of Chicago Press, Chicago, Illinois, USA.

Chase, J.M., Shulman, R.S., 2009. Wetland isolation facilitates larval mosquito density through the reduction of predators. Ecol. Entomol. 34, 741–747.

Chesson, P., 2000. Mechanisms of maintenance of species diversity. Annu. Rev. Ecol. Syst. 31, 343–366.

Clark, J.S., 2012. The coherence problem with the Unified Neutral Theory of Biodiversity. Trends Ecol. Evol. 27, 198–202.

Clark, J.S., Silman, M., Kern, R., Macklin, E., HilleRisLambers, J., 1999. Seed dispersal near and far: patterns accros temperate and tropical forests. Ecology 80, 1475–1494.

Clarke, A., Johnston, N.M., 1999. Scaling of metabolic rate with body mass and temperature in teleost fish. J. Anim. Ecol. 68, 893–905.

Cohen, J.E., Brian, F., Newman, C.M., 1990. Community Food Web: Data and Theory. Springer. London.

Cottenie, K., 2005. Integrating environmental and spatial processes in ecological community dynamics. Ecol. Lett. 8, 1175–1182.

D'Antonio, C., Meyerson, L.A., Denslow, J., 2001. Research priorities related to invasive exotic species. In: Soulé, M.E., Orians, G.H. (Eds.), Conservation Biology Research Priorities for the Next Decade. Island, Washington, DC, pp. 59–80.

Desjardins-Proulx, P., Gravel, D., 2012. A complex speciation-richness relationship in a simple neutral model. Ecol. Evol. 2, 1781–1790.

Diamond, M., 1975. Assembly of species communities. In: Cody, M.L., Diamond, J.M. (Eds.), Ecology and Evolution of Communities. Belknap press, pp. 342–444.

Dorazio, R.M., Kéry, M., Royle, J.A., Plattner, M., 2010. Models for inference in dynamic metacommunity systems. Ecology 91, 2466–2475.

Driscoll, D.A., Lindenmayer, D.B., 2009. Empirical tests of metacommunity theory using an isolation gradient. Ecol. Monogr. 79, 485–501.

Economo, E.P., Keitt, T.H., 2008. Species diversity in neutral metacommunities: a network approach. Ecol. Lett. 11, 52–62.

Economo, E.P., Keitt, T.H., 2010. Network isolation and local diversity in neutral metacommunities. Oikos 119, 1355–1363.

Estrada, E., Bodin, Ö., 2008. Using network centrality measures to manage landscape connectivity. Ecol. Appl. 18, 1810–1825.

Estrada, E., Rodríguez-Velázquez, J.A., 2005. Subgraph centrality in complex networks. Phys. Rev. E 71, 056103.

Fortuna, M.A., Albaladejob, R.G., Fernández, L., Apariciob, A., Bascompte, J., 2009. Networks of spatial genetic variation across species. Proc. Natl. Acad. Sci. USA 106, 19044–19049.

Freeman, C., 1979. Centrality in social networks conceptual clarification. Soc. Networks 1, 215–239.

Gilarranz, L.J., Bascompte, J., 2012. Spatial network structure and metapopulation persistence. J. Theor. Biol. 297, 11–16.

Gonzalez, A., Rayfield, B., Lindo, Z., 2011. The disentangled bank: how loss of habitat fragments and disassembles ecological networks. Am. J. Bot. 98, 503–516.

Gotelli, N.J., 1991. Metapopulation models: the rescue effect, the propagule rain, and the coresatellite hypotesis. Am. Nat. 138, 768–776.

Gotelli, N.J., 2008. A Primer of Ecology, fourth ed. Sinauer Associates, Sunderland.

Gotelli, N.J., Buckley, N.J., Wiens, J.A., 1997. Co-occurrence of Australian land birds: Diamond's assembly rules revisited. Oikos 80, 311–324.

Gravel, D., Canham, C.D., Beaudet, M., Messier, C., 2006. Reconciling niche and neutrality: the continuum hypothesis. Ecol. Lett. 9, 399–409.

Guimerà, R., Amaral, L.A.N., 2005. Functional cartography of complex metabolic networks. Nature 433, 895–900.

Hanski, I., 1999. Metapopulation Ecology. Oxford University Press, New York.

Hanski, I., 2010. The theories of island biogeography and metapopulation dynamics. In: Losos, J.B., Ricklefs, R.E. (Eds.), The Theory of Island Biogeography Revisted. Princeton University Press, Princeton and Oxford, pp. 186–213.

Hanski, I., Gilpin, M., 1991. Metapopulation dynamics: brief history and conceptual domain. Biol. J. Linn. Soc. 42, 3–16.

Harte, J., 2011. Maximum Entropy and Ecology: A Theory of Abundance, Distribution, and Energetics. Oxford University Press, Oxford.

Holt, R.D., 1993. Ecology at the mesoscale: the influence of regional processes on local communities. In: Ricklefs, R.E., Schluter, D. (Eds.), Species Diversity in Ecological Communities: Historical and Geographical Perspectives. University of Chicago Press, Ilinois, pp. 77–88.

Holt, R.D., 1997. From metapopulation dynamics to community structure: some consequences of spatial heterogeneity. In: Hanski, I., Gilpin, M. (Eds.), Metapopulation Biology. Academic Press, New York, pp. 149–164.

Holyoak, M., Leibold, M.A., Holt, R.D., 2005a. Metacommunities: Spatial Dynamics and Ecological Communities. University of Chicago Press, Chicago.

Holyoak, M., Leibold, M.A., Mouquet, N., Holt, R.D., Hoopes, M.F., 2005b. Metacommunities: a framework for large-scale community ecology. In: Holyoak, M., Leibold, M.A., Holt, R.D. (Eds.), Metacommunities Spatial Dynamics and Ecological Communities. The University of Chicago Press, Chicago, pp. 1–31.

Hubbell, S.P., 2001. The Unified Neutral Theory of Biodiversity and Biogeography (MPB-32). Princeton University Press, Princeton, New Jersey, USA.

Hutchinson, G.E., 1959. Homage to Santa Rosalia or why are there so many kinds of animals? Am. Nat. 93, 145–159.

Jacobson, B., Peres-Neto, P.R., 2010. Quantifying and disentangling dispersal in metacommunities: how close have we come? How far is there to go? Landscape Ecol. 25, 495–507.

Kadmon, R., Benjamini, Y., 2006. Effects of productivity and disturbance on species richness: a neutral model. Am. Nat. 167, 939–946.

Keddy, P.A., 1992. Assembly and response rules: two goals for predictive community ecology. J. Veg. Sci. 3, 157–164.

Keitt, T.H., Urban, D.L., Milne, B.T., 1997. Detecting critical scales in fragmented landscapes. Conserv. Ecol. 1, 4.

Kneitel, J.M., Chase, J.M., 2004. Trade-offs in community ecology: linking spatial scales and species coexistence. Ecol. Lett. 7, 69–80.

Kneitel, J.M., Miller, T.E., 2003. Dispersal rates affect species composition in metacommunities of *Sarracenia purpurea* inquilines. Am. Nat. 162, 165–171.

Krapivsky, P.L., Redner, S., 2001. Organization of growing random networks. Phys. Rev. E 63, 066123.

Laliberté, E., Shipley, B., 2011. FD: Measuring Functional Diversity from Multiple Traits, and Other Tools for Functional Ecology. R package version 1.0-11.

Leibold, M.A., Holyoak, M., Mouquet, N., Amarasekare, P., Chase, J.M., Hoopes, M.F., Holt, R.D., Shurin, J.B., Law, R., Tilman, D., Loreau, M., Gonzalez, A., 2004. The metacommunity concept: a framework for multi-scale community ecology. Ecol. Lett. 7, 601–613.

Leibold, M.A., Mikkelson, G.M., 2002. Coherence, species turnover, and boundary clumping: elements of meta-community structure. Oikos 97, 237–250.

Levins, R., 1969. Some demographic and genetic consequences of environmental heterogeneity for biological control. Bull. Entomol. Soc. Am. 15, 237—240.

Liebhold, A., Koenig, W.D., Bjørnstad, O.N., 2004. Spatial synchrony in population dynamics. Annu. Rev. Ecol. Syst. 35, 467—490.

Logue, J.B., Mouquet, N., Peter, H., Hillebrand, H., Group, T.M.W., 2011. Empirical approaches to metacommunities: a review and comparison with theory. Trends Ecol. Evol. 26, 482—491.

Loreau, M., 2010. From Population to Ecosystems: Theoretical Foundations for a New Ecological Synthesis. Princeton University Press, Oxford and Princeton.

Loreau, M., Mouquet, N., 1999. Immigration and the maintenance of local species diversity. Am. Nat. 154, 427—440.

Loreau, M., Mouquet, N., Holt, R.D., 2003. Meta-ecosystems: a theoretical framework for a spatial ecosystem ecology. Ecol. Lett. 6, 673—679.

Loreau, M., Naeem, S., Inchausti, P., Bengtsson, J., Grime, J.P., Hector, A., Hooper, D.U., Huston, M.A., Raffaelli, D., Schmid, B., Tilman, D., Wardle, D.A., 2001. Biodiversity and ecosystem functioning: current knowledge and future challenges. Science 294, 804—808.

MacArthur, R.H., 1970. Species packing and competitive equilibrium for many species. Theor. Popul. Biol. 1, 1—11.

MacArthur, R.H., Levins, R., 1964. Competition, habitat selection and character displacement in a patchy environment. Proc. Natl. Acad. Sci. USA 51, 1207—1210.

MacArthur, R.H., Levins, R., 1967. The limiting similarity, convergence and divergence of coexisting species. Am. Nat. 101, 377—385.

MacArthur, R.H., Wilson, E.O., 1967. The Theory of Island Biogeography. Princeton University Press, Princeton, New Jersey.

Maslov, S., Snepen, K., 2002. Specificity and stability in topology of protein networks. Science 296, 910—913.

Matias, M.G., Mouquet, N., Chase, J.M., 2013. Dispersal stochasticity mediates species richness in source—sink metacommunities. Oikos 122, 395—402.

Maurer, B.A., 1999. Untangling Ecological Complexity: The Macroscopic Perspective. University of Chicago Press, Chicago & London.

May, R.M., 1972. Will a large complex system be stable? Nature 238, 413—414.

May, R.M., 2006. Network structure and the biology of populations. Trends Ecol. Evol. 21, 394—399.

May, R.M., Lawton, J.H., Stork, N.E., 1995. Assessing extinction rates. In: Lawton, J.H., May, R.M. (Eds.), Extinction Rates. Oxford University Press, Oxford, pp. 1—24.

Mayr, E., 1997. This Is Biology: The Science of the Living World. Harvard University Press.

McCann, K.S., 2012. Food Webs. Princeton University Press, Oxford and Princeton.

McGill, B., 2003. Strong and weak tests of macroecological theory. Oikos 102, 679—685.

McGill, B.J., 2006. A renaissance in the study of abundance. Nature 314, 770—772.

McGill, B.J., 2010. Towards a unification of unified theories of biodiversity. Ecol. Lett. 2010, 627—642.

Melian, C.J., Bascompte, J., 2002. Food web structure and habitat loss. Ecol. Lett. 5, 37—46.

Melián, C.J., Bascompte, J., 2002. Complex networks: two ways to be robust? Ecol. Lett. 5, 705—708.

Meynard, C.N., Lavergne, S., Boulangeat, I., Garraud, L., Van Es, J., Mouquet, N., Thuiller, W., 2013. Disentangling the drivers of metacommunity structure across spatial scales. J. Biogeogr. 40, 1560—1571.

Miyazono, S., Taylor, C.M., 2013. Effects of habitat size and isolation on species immigration—extinction dynamics and community nestedness in a desert river system. Freshwater Biol. 58, 1303—1312.

Morin, P.J., 2010. Emerging frontiers of community ecology. In: Verhoef, H.A., Morin, P.J. (Eds.), Community Ecology Processes, Models, and Applications. Oxford University Press, New York.

Moritz, C., Meynard, C.N., Devictor, V., Guizien, K., Labrune, C., Guarini, J.-M., Mouquet, N., 2013. Disentangling the role of connectivity, environmental filtering, and spatial structure on metacommunity dynamics. Oikos 122, 1401—1410.

Mouquet, N., Loreau, M., 2002. Coexistence in metacommunities: the regional similarity hypothesis. Am. Nat. 159, 420—426.

Mouquet, N., Loreau, M., 2003. Community patterns in source-sink metacommunities. Am. Nat. 162, 544—557.

Munoz, F., Couteron, P., 2011. Estimating immigration in neutral communities: theoretical and practical insights into the sampling properties. Methods Ecol. Evol. 3, 152—161.

Newman, C.M., 2010. Networks an Introduction. Oxford University Press, Oxford.

Newman, M.E.J., 2003. The structure and function of complex networks. SIAM Rev. 45, 167—256.

Newman, M.E.J., 2005. Power laws, Pareto distributions and Zipf's law. Contemp. Phys. 46, 323—351.

Newman, M.E.J., Girvan, M., 2004. Finding and evaluating community structure in networks. Phys. Rev. E 69, 026113.

Nicholson, B., Mace, M.M., Armsworth, P.R., Atkinson, G., Buckle, S., Clements, T., Ewers, R.M., Fa, J.E., Gardner, T.A., Gibbons, J., Grenyer, R., Metcalfe, R., Mourato, S., Muûls, M., Osborn, D., Reuman, D.C., Watson, C., Milner-Gulland, E.J., 2009. Priority research areas for ecosystem services in a changing world. J. Appl. Ecol. 46, 1139—1144.

Olesen, J.M., Bascompte, J., Dupont, Y.L., Jordano, P., 2007. The modularity of pollination networks. Proc. Natl. Acad. Sci. USA 104, 19891—19896.

Opsahl, T., Agneessens, F., Skvoretz, J., 2010. Node centrality in weighted networks: Generalizing degree and shortest paths. Soc. Networks 32, 245—251.

Patterson, B.D., Atmar, W., 1986. Nested subsets and the structure of insular mammalian faunas and archipelagos. Biol. J. Linn. Soc. 28, 65—82.

Petchey, O.L., 2010. Maximum entropy in ecology. Oikos 119, 577.

Peterson, E.E., Ver Hoef, J.M., Isaak, D.J., Falke, J.A., Fortin, M.-J., Jordan, C.E., McNyset, K., Monestiez, P., Ruesch, A.S., Sengupta, A., Som, N., Steel, E.A., Theobald, D.M., Torgersen, C.E., Wenger, S.J., 2013. Modelling dendritic ecological networks in space: an integrated network perspective. Ecol. Lett. 16, 707—719.

Phillips, S.J., Anderson, R.P., Schapire, R.E., 2006. Maximum entropy modeling of species geographic distributions. Ecol. Model. 190, 231—259.

Pianka, E.R., 1974. Niche overlap and diffuse competition. Proc. Natl. Acad. Sci. USA 71, 2141—2145.

Pillai, P., Gonzalez, A., Loreau, M., 2012. Metacommunity theory explains the emergence of food web complexity. Proc. Natl. Acad. Sci. USA 108, 19293—19298.

Pimm, S.L., 1982. Food Webs. Chapman & Hall, London.

Platt, J.R., 1964. Strong inference. Science 146, 347—353.

Proulx, S.R., Promislow, D.E.L., Phillips, P.C., 2005. Network thinking in ecology and evolution. Trends Ecol. Evol. 26, 345—353.

R Development Core Team, 2013. R: A Language and Environment for Statistical Computing. R Foundation for Statistical Computing, Vienna, Austria. http://www.R-project.org. ISBN 3-900051-07-0.

Renner, I.W., Warton, D.I., 2013. Equivalence of MAXENT and Poisson point process models for species distribution modeling in ecology. Biometrics 69, 274–281.

Ricklefs, R.E., Schluter, D., 1993. Species Diversity in Ecological Communities: Historical and Geographical Perspectives. University of Chicago Press, Chicago, Illinois, USA.

Ritchie, M.E., 2010. Scale, Heterogeneity, and the Structure and Diversity of Ecological Communities. Princeton University Press Princeton.

Rosindell, J., Hubbell, S.P., Etienne, R.S., 2011. The unified neutral theory of biodiversity and biogeography at age ten. Trends Ecol. Evol. 26, 340–348.

Rozenfeld, A.F., Arnaud-Haond, S., Hernández-García, E., Eguíluz, V.M., Serrão, E.A., Duarte, C.M., 2008. Network analysis identifies weak and strong links in a metapopulation system. Proc. Natl. Acad. Sci. USA 105, 18824–18829.

Scheffer, M., Bascompte, J., Brock, W.A., Brovkin, V., Carpenter, S.R., Dakos, V., Held, H., van Nes, E.H., Rietkerk, M., Sugihara, G., 2009. Early-warning signals for critical transitions. Nature 461, 53–59.

Shipley, B., 2010a. Community assembly, natural selection and maximum entropy models. Oikos 119, 604–609.

Shipley, B., 2010b. From Plant Traits to Vegetation Structure: Chance and Selection in the Assembly of Ecological Communities. Cambridge University Press, Cambridge.

Shipley, B., Laughlin, D.C., Sonnier, G., Otfinowski, R., 2011. A strong test of a maximum entropy model of trait-based community assembly. Ecology 92, 507–517.

Shipley, B., Vile, D., Garnier, É., 2006. From plant traits to plant communities: a statistical mechanistic approach to biodiversity. Science 314, 812–814.

Shmida, A., Wilson, M.V., 1985. Biological determinants of species diversity. J. Biogeogr. 12, 1–20.

Shurin, J.B., Cottenie, K., Hillebrand, H., 2009. Spatial autocorrelation and dispersal limitation in freshwater organisms. Oecologia 159, 151–159.

Soininen, J., McDonald, R., Hillebrand, H., 2007. The distance decay of similarity in ecological communities. Ecography 30, 3–12.

Stouffer, D.B., Bascompte, J., 2011. Compartmentalization increases food-web persistence. Proc. Natl. Acad. Sci. USA 108, 3648–3652.

Tilman, D., 1994. Competition and biodiversity in spatially structured habitats. Ecology 75, 2–16.

Ulrich, W., Gotelli, N.J., 2007. Disentangling community patterns of nestedness and species co-occurrence. Oikos 116, 2053–2061.

Urban, D., Keitt, T.H., 2001. Landscape connectivity: a graph-theoretic perspective. Ecology 82, 1205–1218.

Urban, D.L., Minor, E.S., Treml, E.A., Chick, R.S., 2009. Graph models of habitat mosaics. Ecol. Lett. 12, 260–273.

Urban, M.C., Phillips, B.L., Skelly, D.K., Shine, R., 2008. A toad more traveled: the heterogeneous invasion dynamics of cane toads in Australia. Am. Nat. 171, E134–E148.

Vanschoenwinkel, B., Gielen, S., Seaman, M., Brendonck, L., 2008a. Any way the wind blows - frequent wind dispersal drives species sorting in ephemeral aquatic communities. Oikos 117, 125–134.

Vanschoenwinkel, B., Gielen, S., Vandewaerde, H., Seaman, M., Brendonck, L., 2008b. Relative importance of different dispersal vectors for small aquatic invertebrates in a rock pool metacommunity. Ecography 31, 567–577.

Warton, D.I., Shipley, B., Hastie, T., 2015. CATS regression – a model-based approach to studying trait-based community assembly. Methods Ecol. Evol., in press.

Wasserman, S., Faust, K., 1994. Social Network Analysis. Cambridge Univeristy Press, Cambridge.

Weiher, E., Keddy, P.A., 1999. Ecological Assembly Rules: Perspectives, Advances, Retreats. Cambridge University Press, Cambridge, UK.

Welsh Jr., H.H., Hodgson, G.R., 2011. Spatial relationship in a dendritic network: the herpetofaunal metacommunity of the Mattole River catchment of northwest California. Ecography 34, 49–66.

Williams, D.D., Berlow, E.L., Dunne, J.A., Barabási, A.-L., Martinez, N., 2002. Two degrees of separation in complex food webs. Proc. Natl. Acad. Sci. USA 99, 12913–12916.

With, K.A., Crist, T.O., 1995. Critical thresholds in species' responses landscape structure. Ecology 76, 2446–2459.

Section II

Across Aquatic Ecosystems

Chapter 5

Limited Functional Redundancy and Lack of Resilience in Coral Reefs to Human Stressors

Camilo Mora
Department of Geography, University of Hawaii, Honolulu, HI, USA

INTRODUCTION

Coral reefs are some of the most diverse, socioeconomically important, and threatened ecosystems in the world. Worldwide there are ∼835 species of reef-building corals (Veron, 1995), which in turn provide habitats for some 1–9 million other species (Reaka-Kudla, 1997). Unfortunately, a combination of co-occurring and interacting human stressors is leading to massive losses of coral reefs worldwide (Hughes, 1994; Knowlton, 2001; Gardner et al., 2003; Hughes et al., 2003a; Pandolfi et al., 2003, 2005; Bruno et al., 2007a; Wilkinson, 2008; Burrows et al., 2011). This ongoing decline of coral reefs may not only lead to the loss of species but also impair the capacity of a "multi-resource" ecosystem to deliver goods and services critical for social and economic development in many tropical countries (Moberg et al., 1999; Brander et al., 2007). Some 655 million people live within 100 km of coral reefs (Donner et al., 2007) and many rely on reefs for the delivery of food, jobs, and revenue (Costanza et al., 1997; Newton et al., 2007; Burke et al., 2011; Graham et al., 2011; Mora et al., 2013; Teh et al., 2013). The modern transformation of coral reefs is worrisome because changes could become difficult (or impossible) and expensive to revert (Knowlton, 2001; Scheffer et al., 2001, 2003; Mumby et al., 2007b; Rogers, 2013).

The role of biodiversity on ecosystem resilience has been at the core of much ecological research (Chapin et al., 1998, 2000; Sala et al., 2000; Loreau et al., 2001; Loreau, 2004; Hooper et al., 2005). Resilience is used here as the capacity of a system to return to its original functionality after being disturbed; the constituent species at the recovery state may or may not be the same.

Experimental and theoretical studies have demonstrated that the more species in a system, the more resilient the system can be to disturbances (e.g., Chapin et al., 2000; McCann, 2000, 2007; Loreau et al., 2001). This may occur as numerous species are likely to play similar functional roles but have different sensitivities to the stressor; thus, the functional role of a lost species could be compensated by that of another (Loreau, 1998; McCann, 2000; Loreau et al., 2001; Bellwood et al., 2004). In practice, however, coral reefs worldwide, despite their great diversity, are increasingly failing to recover after human disturbances (Nyström et al., 2000; Hughes et al., 2003a; Bellwood et al., 2004). Ecologists have shifted their views from seeing coral reefs as highly diverse, temporarily stable, and robust, instead seeing them as fragile and globally stressed (Mumby et al., 2008). The ability of coral reefs to recover and return to pristine states can no longer be taken for granted (Connell, 1997; Aronson et al., 2004). The standing question is then: why are coral reefs not resilient to stressors despite hosting a great diversity of species? Here I review the possibility that coral reefs, despite being diverse, are not as functionally redundant as one may expect.

DATA QUALITY

The ability to predict changes in coral reefs (e.g., an ecosystem's response and recovery to a disturbance) depends on the availability of data, which in the case of coral reefs is limited for various reasons.

First is the lack of historical baselines; without knowing what pristine coral reefs look like, it is hard to assess the relative change of coral reefs in response to a stressor or whether recovery has truly occurred (Knowlton et al., 2008). This situation is evidenced by a handful of opportunistic studies that have been able to reconstruct the "expected" status of pristine reefs (Jackson et al., 2001; Pandolfi et al., 2003, 2005, 2006; Jackson, 2008; McClenachan et al., 2008, Ward-Paige et al., 2010). In one such study, McClenachan et al. (2008) documented that the energy required to sustain monk seals in Caribbean reefs should require biomass levels of $732-1018$ g m^2 of reef, which is three to five times more than those found nowadays on typical Caribbean coral reefs and more than twice that measured on the most pristine reefs in the world.

Second, for the most part we lack data at the species level. Inherent in the concept of resilience is the idea that the functional role of one species could be replaced by another, and as such the basic functionality of the ecosystem could be sustained in spite of considerable changes in community structure. In coral reefs, the most available long-term source of data is live coral cover of the overall assemblage, which unfortunately has a high risk of failing to depict changes in community composition (Gardner et al., 2003) and even of delivering a false sense of security, as it may obscure negative change in community structure. For instance, variability in the sensitivity of coral reef species to human stressors is leading to changes in community structure from branching (highly complex) to massive (not so complex) coral species, which in turn is reducing habitat complexity and the diversity of associated fauna. This change in community structure is occurring in spite of a high live coral cover, which is now dominated by massive less complex corals (e.g., Darling et al., 2010, 2013).

Third, until very recently there was a strong geographical bias for the long-term study of coral reefs in the Caribbean, which has raised concerns over the transferability of theory to other regions. Roff et al. (2012) elucidated, for instance, that the Caribbean may have lower resilience as macroalgae can grow faster there, the region is prone to iron enrichment from Aeolian dust, it has almost completely lost the Acroporid species, and in addition it has a low abundance of and lacks entire groups of herbivore fishes. While these differences are real, and reefs on other parts of the world do indeed lack long-track research, emerging studies show that coral reefs worldwide are depicting similar declining trends (Connell, 1997; Bruno et al., 2007a).

Fourth, the patchiness of available data often leads to consolidating data from different places and times (the so-called meta-analyses), often giving rise to inconsistencies from one compilation to the next (Hughes et al., 2010). Hughes et al. (2010) noted, for instance, how different meta-analyses of trends in coral cover for the same region differ by up to 17% in absolute terms for the same years. Due to their integrative nature, meta-analyses often rely on simple metrics such as coral cover, which as noted earlier can also fail to identify underlying community changes (Gardner et al., 2003).

PATTERN OF CHANGE

In spite of data limitations, there is large agreement that the world's coral reefs are in decline, moving from a dominance by live corals to a dominance by macroalgae and at times invertebrates such as bivalves, sponges, tunicates, and zoanthids (Schutte et al., 2010; Hughes, 1994; Gardner et al., 2003; Hughes et al., 2003a; Pandolfi et al., 2003, 2005; Bellwood et al., 2004; Cote et al., 2005; Bruno et al., 2007a, Wilkinson, 2008; Norström et al., 2009; Burrows et al., 2011). This pattern has been coined as the phrases "phase shifts" (Done, 1992; Hughes, 1994), "alternative states" (Knowlton, 1992; Bellwood et al., 2004), "the slippery slope to slime" (Pandolfi et al., 2005), "trophic cascades" (Dulvy et al., 2004; Mumby et al., 2006, 2007a), and "the straw that broke the camel's back" (Knowlton, 2001) (note: Dudgeon et al. (2010) cautions on the need to differentiate terms such as phase shifts and stable states for describing different patterns of ongoing coral reef transformations).

Perhaps the best documented example of coral reef "transformation" has been reported for coral reefs in the Caribbean. By the end of 1970s, live coral cover in this region was commonly above 75%, with great dominance by highly 3-D complex Acroporid species. During the 1980s, a mixture of hurricane Allen, lime coral disease, extreme overfishing of herbivorous fishes, and an unfortunate massive die-off of the urchin grazer *Diadema antillarum* reduced coral cover to just ~5% by the 1990s, and it has remained so ever since (Schutte et al., 2010; Hughes, 1994; Nyström et al., 2000; Gardner et al., 2003; Mumby et al., 2007b). The mortality of corals has been followed by a demographic dominance of fast-growing algae (Roff et al., 2012) but see Bruno et al. (2009), reductions in coral reef complexity (Alvarez-Filip et al., 2009) and in the diversity and abundance of associated fishes (Paddack et al., 2009). Despite data limitations, similar patterns of coral reef transformation have been documented in other regions (Connell, 1997; Bruno et al., 2007a; Baker et al., 2008).

With the exception of a few reefs (e.g., Kaneohe Bay—Hawaii (Hunter et al., 1995)), there have been very few documented examples of significant phase-shift reversal to coral dominance (Nyström et al., 2000; Bellwood et al., 2004). Significant recovery from bleaching events, not necessarily coral mortality or algae domination, has been documented in the Indian Ocean but not on western Atlantic reefs (Baker et al., 2008; Sheppard et al., 2008). No clear trends of recovery are apparent in the eastern Pacific, the central—southern—western Pacific or the Arabian Gulf, where some reefs are recovering while others are not (Connell, 1997; Baker et al., 2008; Roff et al., 2012). These regional differences are commonly associated with spatial variations in the diversity of ecosystems and thus resilience (Bellwood et al., 2004; Roff et al., 2012), although spatial variations in the intensity of human stressors cannot be completely ruled out (i.e., even though resilience or lack thereof may be similar, reefs around the world are located at different stages along the gradients of human disturbance). Reefs in the Caribbean, for instance, have been exposed to a longer history of insults (Jackson, 1997) and are considerably more populated than other reefs around the world (Mora et al., 2011a), which may explain their greater degradation and failure to recover.

The magnitude of coral reef transformation is clearly indicated by the fact that ~19% of coral reefs globally are effectively lost, 15% are critically impacted by humans and likely to be lost within 10—20 years (Wilkinson, 2008; Burke et al., 2011), and the remaining reefs distant from direct human stress will become increasingly vulnerable to global changes in temperature (Mora et al., 2013; Van Hooidonk et al., 2013; van Hooidonk et al., 2013) and acidification (Mora et al., 2013; Ricke et al., 2013). Very few reefs in the world, if any, are currently considered to be in a pristine state (Burke et al., 2011), and some 32.8% of coral reef species worldwide are currently facing elevated risk of extinction (Carpenter et al., 2008).

DRIVERS OF CHANGE

The causes of coral reef loss are diverse and intricate. First, it needs to be recognized that human activities have been detrimental to coral reefs over millennia (e.g., several mega-fauna have been already completely extirpated and many ongoing declines may have started before the onset of scientific research (Jackson, 1997; Jackson et al., 2001; Pandolfi et al., 2003)). Historical reconstructions, however, have illustrated the accelerated impact of human activities over the past 50 years (Jackson et al., 2001; Pandolfi et al., 2003; Wilkinson, 2008). Modern human disturbances (not in order of importance to avoid debate over their relative effect and interactive nature) include unprecedented extreme climates, overfishing, pollution, invasive species, habitat loss, destabilized ecological interactions (e.g., trophic cascades and coral−algal and coral−pathogen interactions), increasing spread and susceptibility to diseases, and impaired metapopulation dynamics (Jackson et al., 2001, 2005; Knowlton, 2001; Harvell et al., 2002; Aronson et al., 2003; Bruno et al., 2003, 2007b; Hughes et al., 2003a,b; Bellwood et al., 2004; Pandolfi et al., 2005; Hoegh-Guldberg et al., 2007; Jackson, 2008; Knowlton et al., 2008; Mora, 2008, 2009; Wilkinson, 2008; Sale, 2011; Ateweberhan et al., 2013). These disturbances vary from local to global scales (Knowlton, 2001; Knowlton et al., 2008, Mora, 2008; Mora et al., 2011a), in their effects from additive to synergistic (Knowlton, 2001; Darling et al., 2008, 2010; Knowlton et al., 2008; Ateweberhan et al., 2013), and from having direct to indirect effects over species (Hughes et al., 2003a,b; Bellwood et al., 2004; Cote et al., 2010; Ateweberhan et al., 2013). Stressors also trigger ecological feedbacks, as weakening of corals increases sensitivity to coral diseases, facilitates the breakdown of reef framework by bioeroders, and leads to the loss of critical habitats and food for associated biota.

ARE CORAL REEFS FUNCTIONALLY REDUNDANT?

It is intuitive that in a large pool of species, there should be species with redundant functional roles. However, functional redundancy is often questioned, as this could drive species to extinction mediated by competition (Loreau, 2004). One possibility to avoid such a scenario is via spatial and/or temporal segregation, or via niche specialization. For coral reefs, the idea of functional redundancy is supported by the fact that the number of functional groups often saturates as the number of species increases (Halpern et al., 2008; Mora et al., 2011a). These studies, however, are based on gross classifications of functional groups that can overestimate the extent of functional redundancy. For instance, deeper exploration of morphological (Price et al., 2011) and dietary (Robertson et al., 1986) characteristics of species generally classified as herbivorous has revealed the existence of considerable differences among species. Such small differences may result from niche specialization and have nontrivial effects on coral reefs. For instance, variations in the palatability of algae (e.g., Littler et al., 1983; Ledlie et al., 2007) suggest that a broad portfolio of "herbivores" is required to keep algae cover at check. If specialization is dominant on coral reefs, then redundancy is not as extensive as originally thought, as different species are likely to have different functional roles.

If specialization is dominant and functional redundancy is limited, this could lead to each species adding rather than subtracting (due to the cost of ecological interactions) to the functioning (productivity) of ecosystems. This should cause exponential/concave-up biodiversity—ecosystem functioning (B-EF) relationships, which have been recently documented for coral reef fishes (Mora et al., 2011a). Concave-up B-EF relationships are supported by experimental studies showing that the effect of biodiversity on ecosystem functioning increases over time as species "become more functionally unique through time" (Reich et al., 2012). Limited redundancy may also occur if the maintenance of a given function requires the same functional groups but represented by different species at different times and places, and for different environmental conditions (Peterson et al., 1998; Isbell et al., 2011). In other words, apparently redundant species may actually operate at different scales and conditions, supporting the idea of limited functional redundancy and highlighting the high vulnerability of ecosystems to the loss of even a few species.

A saturating relationship between richness and functional diversity is also deceiving of functional redundancy, as it fails to indicate the distribution of species within functional groups. For instance, exploration of the frequency distribution of species per functional group has revealed strong right-skew frequencies, with few functional groups having numerous species and most being constituted by a handful of and at times a single species (Bellwood et al., 2004). Indeed, key functional groups such as bioeroders in the Indo-Pacific are almost entirely represented by a single species (the giant humphead parrotfish, *Bolbometopon muricatum*), which given its large body size is unfortunately prone to extensive fishing and thus is missing on most reefs near human settlements (Bellwood et al., 2003). Likewise, branching corals in the Caribbean are solely represented by Acroporid species that have undergone extensive mortality by a combination of severe storms and diseases (Nyström et al., 2000; Roff et al., 2012; Rogers, 2013). Further, Mouillot et al. (2013) recently showed how even rare species often lack functional analogs and carry out some key vulnerable functions in ecosystems, suggesting that even the loss of rare species could have severe consequences for coral reef resilience (see also Jain et al., 2014). This pattern of limited redundancy may be common on coral reefs (Micheli et al., 2005).

Inherent in the idea that biodiversity confers higher resilience is also the idea that similar functional species will have differential sensitivity to stressors in order to ensure ecosystem recovery by the more resistant species (Elmqvist et al., 2003). However, high diversity may offer limited resilience, if all species within a functional group respond the same way to the same stressors, which may also be the case on coral reefs (Nyström et al., 2000, 2008; Bellwood et al., 2004). Fishing, for instance, can impose a similar detrimental impact over most species of predators and herbivores on coral reefs (Nyström et al., 2000; Bellwood et al., 2004; Micheli et al., 2005). Likewise, warming and diseases appear to have similar deterring effects on the branching corals that provide most of the complexity to coral reefs (Darling et al., 2013; Rogers, 2013).

SOLUTIONS TO ENSURE RESILIENCE

Clearly, coral reefs are not as functionally redundant as would be expected from their great number of species. This can explain their reduced resilience to human impact and highlights the urgent need for effective conservation actions. Such actions are urgently needed, because the longer coral reefs move along this declining slippery slope, the harder it will be to reverse such movement, and the larger the risk that it cannot be reversed at all. Mumby et al. (2007b) documented, for instance, that reverting coral—algae phase shifts through the restoration of herbivore fishes would require a fourfold increase of herbivores at coral covers of ~5% but only a twofold increase at coral covers of ~30%. There are also chances that human stressors could be so extensive as to impose selective pressures, thus inducing genetic loss that will reduce environmental adaptability and increase chances of random genetic drift, the loss of important genes, and inbreeding depression—not to mention Allee effects of increasingly smaller populations. Human impacts pose real risks of irreversibility of phase shifts and even risk of extinction of coral reefs, which in turn could be detrimental to the welfare of millions of people whose livelihoods depend on coral reefs. This stresses the urgent call for effective conservation actions.

Are there other Solutions Available?

There are many proposed solutions to reverse ongoing declines in biodiversity (Butchart et al., 2010), but for coral reefs they are ineffective as clearly demonstrated by the ongoing losses of such reefs (Mora et al., 2011b), and several reasons may explain this. One possibility is that the large scales over which coral reef populations operate, and the global pensiveness of human stressors and their feedback social loops, call for multinational strategies that are largely lacking (Rockstrom et al., 2009). One potential feedback is that stopping fishing could lead to an increasing demand for agriculture or livestock, which in turn could lead to excessive use of fertilizers and pesticides and land conversion, which in turn could impact coral reefs through runoff. Another explanation for the failure to revert biodiversity loss is that most solutions inherently try to restrict different human uses (e.g., by reductions in fishing, CO_2, development), often causing a disparity in the costs and benefits of solving problems (Nyström et al., 2000; Donner et al., 2007); these conditions cause considerable conflicts and resistance to proposed solutions

(West et al., 2006; Mora et al., 2011b). Another explanation is the issue of selfishness. On one hand, if the problem is global and many countries are concerned, there is always a temptation to just get a free ride (Nyström et al., 2000). On the other hand, for many countries it is beneficial to gain revenues at the expense of polluting the world (Nyström et al., 2000). Finally, it is possible that current conservation strategies have had a very narrow focus on proximal rather than ultimate causes of coral reef damage. Proximal drivers include overfishing, habitat loss, invasive species, the spread of diseases, climate change, pollution, sedimentation, and eutrophication, whereas the ultimate driver is an ever-growing and expanding human population. This narrow focus leads to adaptive strategies that will struggle in a perpetual effort to maintain or restore biodiversity (Mora et al., 2011b).

CONCLUDING REMARKS

The capacity of coral reefs to withstand human assaults can no longer be assumed. Over the next 50 years, it is projected that over 80% of countries with coral reefs will double their human populations (Mora et al., 2011a), not to mention that 1 billion to 2 billion people will be added to the total global population. Since virtually all factors associated with the ongoing decline of coral reefs have human origins, it is easy to predict that stressors on coral reefs will continue and likely exacerbate each other, and that in the next century we will likely witness massive transformation of coral reefs. At the same time, if the evidence suggests a linkage between stressors and people, one has to wonder if longer term, more permanent and perhaps cheaper strategies to mitigate stressors on coral reefs should also include humane solutions to avert population growth (e.g., empowering women, sex education, free or affordable family planning, revisiting subsidies that promote natality, and allowing educated choices on whether or when to have children by better divulgation on the different childbearing costs to the environment, climate, state, family, and individuals). Coral reefs have endured the force of time but are succumbing to that of humans. The ongoing decline of coral reefs despite increasing conservation efforts suggests that we need an urgent paradigm shift that puts the focus on the ultimate drivers of environmental damage rather than expecting that this ecosystem will be resilient to our stressors.

REFERENCES

Alvarez-Filip, L., et al., 2009. Flattening of Caribbean coral reefs: region-wide declines in architectural complexity. Proc. R. Soc. B: Biol. Sci. 276, 3019–3025.

Aronson, R.B., et al., 2003. Causes of coral reef degradation. Science 302, 1502.

Aronson, R.B., et al., 2004. Phase shifts, alternative states, and the unprecedented convergence of two reef systems. Ecology 85, 1876–1891.

Ateweberhan, M., et al., 2013. Climate change impacts on coral reefs: synergies with local effects, possibilities for acclimation, and management implications. Mar. Pollut. Bull. 74, 526–539.

Baker, A.C., et al., 2008. Climate change and coral reef bleaching: an ecological assessment of long-term impacts, recovery trends and future outlook. Estuarine, Coastal Shelf Sci. 80, 435–471.

Bellwood, D.R., et al., 2003. Limited functional redundancy in high diversity systems: resilience and ecosystem function on coral reefs. Ecol. Lett. 6, 281–285.

Bellwood, D.R., et al., 2004. Confronting the coral reef crisis. Nature 429, 827–833.

Brander, L.M., et al., 2007. The recreational value of coral reefs: a meta-analysis. Ecol. Econ. 63, 209–218.

Bruno, J.F., et al., 2003. Nutrient enrichment can increase the severity of coral diseases. Ecol. Lett. 6, 1056–1061.

Bruno, J.F., et al., 2007a. Regional decline of coral cover in the Indo-Pacific: timing, extent, and subregional comparisons. PLoS One 2, e711, 711–718.

Bruno, J.F., et al., 2007b. Thermal stress and coral cover as drivers of coral disease outbreaks. PLoS Biol. 5, 1220–1227.

Bruno, J.F., et al., 2009. Assessing evidence of phase shifts from coral to macroalgal dominance on coral reefs. Ecology 90, 1478–1484.

Burke, L.M., et al., 2011. Reefs at risk revisited. World Resources Institute, Washington, DC.

Burrows, M.T., et al., 2011. The pace of shifting climate in marine and terrestrial ecosystems. Science 334, 652–655.

Butchart, S.H.M., et al., 2010. Global biodiversity: indicators of recent declines. Science 328, 1164–1168.

Carpenter, K.E., et al., 2008. One-third of reef-building corals face elevated extinction risk from climate change and local impacts. Science 321, 560–563.

Chapin III, F.S., et al., 1998. Ecosystem consequences of changing biodiversity. BioScience 48, 45–52.

Chapin, F.S., et al., 2000. Consequences of changing biodiversity. Nature 405, 234–242.

Connell, J., 1997. Disturbance and recovery of coral assemblages. Coral Reefs 16, S101–S113.

Costanza, R., et al., 1997. The value of the world's ecosystem services and natural capital. Nature 387, 253–260.

Cote, I.M., et al., 2010. Rethinking ecosystem resilience in the face of climate change. PLoS Biol. 8, e1000438.

Cote, I.M., et al., 2005. Measuring coral reef decline through meta-analyses. Philos. Trans. R. Soc. B: Biol. Sci. 360, 385–395.

Darling, E.S., et al., 2008. Quantifying the evidence for ecological synergies. Ecol. Lett. 11, 1278–1286.

Darling, E.S., et al., 2010. Combined effects of two stressors on Kenyan coral reefs are additive or antagonistic, not synergistic. Conserv. Lett. 3, 122–130.

Darling, E.S., et al., 2013. Life histories predict coral community disassembly under multiple stressors. Glob. Change Biol. 19, 1930–1940.

Done, T.J., 1992. Phase-shifts in coral reef communities and their ecological significance. Hydrobiologia 247, 121–132.

Donner, S.D., et al., 2007. The inequity of the global threat to coral reefs. Bioscience 57, 214–215.

Dudgeon, S.R., et al., 2010. Phase shifts and stable states on coral reefs. Mar. Ecol. Prog. Ser. 413, 201–216.

Dulvy, N.K., et al., 2004. Coral reef cascades and the indirect effects of predator removal by exploitation. Ecol. Lett. 7, 410–416.

Elmqvist, T., et al., 2003. Response diversity, ecosystem change, and resilience. Front. Ecol. Environ. 1, 488–494.

Gardner, T.A., et al., 2003. Long-term region-wide declines in Caribbean corals. Science 301, 958–960.

Graham, N.A., et al., 2011. From microbes to people: tractable benefits of no-take areas for coral reefs. Oceanogr. Mar. Biol. Annu. Rev. 49, 105–136.

Halpern, B.S., et al., 2008. Functional diversity responses to changing species richness in reef fish communities. Mar. Ecol. Prog. Ser. 364, 147–156.

Harvell, C.D., et al., 2002. Climate warming and disease risks for terrestrial and marine biota. Science 296, 2158–2162.

van Hooidonk, R., et al., 2013. Opposite latitudinal gradients in projected ocean acidification and bleaching impacts on coral reefs. Global Change Biol. 20, 103–112.

Hoegh-Guldberg, O., et al., 2007. Coral reefs under rapid climate change and ocean acidification. Science 318, 1737–1742.

Hooper, D.U., et al., 2005. Effects of biodiversity on ecosystem functioning: a consensus of current knowledge. Ecol. Monogr. 75, 3–35.

Hughes, T.P., 1994. Catastrophes, phase shifts, and large-scale degradation of a Caribbean coral reef. Science Ed. 265, 1547–1551.

Hughes, T.P., et al., 2003a. Climate change, human impacts, and the resilience of coral reefs. Science 301, 929–933.

Hughes, T.P., et al., 2003b. Causes of coral reef degradation - response. Science 302, 1503–1504.

Hughes, T.P., et al., 2010. Rising to the challenge of sustaining coral reef resilience. Trends Ecol. Evol. 25, 633–642.

Hunter, C.L., et al., 1995. Coral reefs in Kaneohe Bay, Hawaii: two centuries of western influence and two decades of data. Bull. Mar. Sci. 57, 501–515.

Isbell, F., et al., 2011. High plant diversity is needed to maintain ecosystem services. Nature 477, 199–202.

Jackson, J.B., 1997. Reefs since Columbus. Coral Reefs 16, S23–S32.

Jackson, J.B.C., 2008. Ecological extinction and evolution in the brave new ocean. Proc. Natl. Acad. Sci. USA 105, 11458–11465.

Jackson, J.B.C., et al., 2001. Historical overfishing and the recent collapse of coastal ecosystems. Science 293, 629–638.

Jackson, J.B.C., et al., 2005. Reassessing US coral reefs - response. Science 308, 1741–1742.

Jain, M., et al., 2014. The importance of rare species: a trait-based assessment of rare species contributions to functional diversity and possible ecosystem function in tall-grass prairies. Ecol. Evol. 4, 104–112.

Knowlton, N., 1992. Thresholds and multiple stable states in coral reef community dynamics. Am. Zool. 32, 674–682.

Knowlton, N., 2001. The future of coral reefs. Proc. Natl. Acad. Sci. USA 98, 5419–5425.

Knowlton, N., et al., 2008. Shifting baselines, local impacts, and global change on coral reefs. PLoS Biol. 6, 215–220.

Ledlie, M., et al., 2007. Phase shifts and the role of herbivory in the resilience of coral reefs. Coral Reefs 26, 641–653.

Littler, M.M., et al., 1983. Algal resistance to herbivory on a Caribbean barrier reef. Coral Reefs 2, 111–118.

Loreau, M., 1998. Biodiversity and ecosystem functioning: a mechanistic model. Proc. Natl. Acad. Sci. USA 95, 5632–5636.

Loreau, M., 2004. Does functional redundancy exist? Oikos 104, 606–611.

Loreau, M., et al., 2001. Ecology - biodiversity and ecosystem functioning: current knowledge and future challenges. Science 294, 804–808.

McCann, K., 2007. Protecting biostructure. Nature 446, 29.
McCann, K.S., 2000. The diversity-stability debate. Nature 405, 228–233.
McClenachan, L., et al., 2008. Extinction rate, historical population structure and ecological role of the Caribbean monk seal. Proc. R. Soc. B: Biol. Sci. 275, 1351–1358.
Micheli, F., et al., 2005. Low functional redundancy in coastal marine assemblages. Ecol. Lett. 8, 391–400.
Moberg, F., et al., 1999. Ecological goods and services of coral reef ecosystems. Ecol. Econ. 29, 215–233.
Mora, C., 2008. A clear human footprint in the coral reefs of the Caribbean. Proc. R. Soc. B: Biol. Sci. 275, 767–773.
Mora, C., 2009. Degradation of Caribbean coral reefs: focusing on proximal rather than ultimate drivers. Reply to Rogers. Proc. R. Soc. B: Biol. Sci. 276, 199–200.
Mora, C., et al., 2011a. Global human footprint on the linkage between biodiversity and ecosystem functioning in reef fishes. PLoS Biol. 9, e1000606.
Mora, C., et al., 2011b. Ongoing global biodiversity loss and the need to move beyond protected areas: a review of the technical and practical shortcomings of protected areas on land and sea. Mar. Ecol. Prog. Ser. 434, 251–266.
Mora, C., et al., 2013. Biotic and human vulnerability to projected changes in ocean biogeochemistry over the 21st century. PLoS Biol. 11, e1001682.
Mouillot, D., et al., 2013. Rare species Support vulnerable functions in high-diversity ecosystems. PLoS Biol. 11, e1001569.
Mumby, P.J., et al., 2006. Fishing, trophic cascades, and the process of grazing on coral reefs. Science 311, 98–101.
Mumby, P.J., et al., 2007a. Trophic cascade facilitates coral recruitment in a marine reserve. Proc. Natl. Acad. Sci. USA 104, 8362–8367.
Mumby, P.J., et al., 2007b. Thresholds and the resilience of Caribbean coral reefs. Nature 450, 98–101.
Mumby, P.J., et al., 2008. Coral reef management and conservation in light of rapidly evolving ecological paradigms. Trends Ecol. Evol. 23, 555–563.
Newton, K., et al., 2007. Current and future sustainability of island coral reef fisheries. Curr. Biol. 17, 655–658.
Norström, A.V., et al., 2009. Alternative states on coral reefs: beyond coral-macroalgal phase shifts. Mar. Ecol. Prog. Ser. 376, 295–306.
Nyström, M., et al., 2000. Coral reef disturbance and resilience in a human-dominated environment. Trends Ecol. Evol. 15, 413–417.
Nyström, M., et al., 2008. Capturing the cornerstones of coral reef resilience: linking theory to practice. Coral Reefs 27, 795–809.
Paddack, M.J., et al., 2009. Recent region-wide declines in Caribbean reef fish abundance. Curr. Biol. 19, 590–595.
Pandolfi, J.M., et al., 2003. Global trajectories of the long-term decline of coral reef ecosystems. Science 301, 955–958.
Pandolfi, J.M., et al., 2006. Ecological persistence interrupted in Caribbean coral reefs. Ecol. Lett. 9, 818–826.
Pandolfi, J.M., et al., 2005. Are US coral reefs on the slippery slope to slime? Science 307, 1725–1726.
Peterson, G., et al., 1998. Ecological resilience, biodiversity, and scale. Ecosystems 1, 6–18.
Price, S.A., et al., 2011. Coral reefs promote the evolution of morphological diversity and ecological novelty in labrid fishes. Ecol. Lett. 14, 462–469.

Reaka-Kudla, M.L., 1997. The global biodiversity of coral reefs: a comparison with rainforests. In: Reaka-Kudla, M.L., Wilson, D.E., Wilson, E.O. (Eds.), Biodiversity II: Understanding and Protecting Our Biological Resources. Joseph Henry Press, pp. 83–108.

Reich, P.B., et al., 2012. Impacts of biodiversity loss escalate through time as redundancy fades. Science 336, 589–592.

Ricke, K.L., et al., 2013. Risks to coral reefs from ocean carbonate chemistry changes in recent earth system model projections. Environ. Res. Lett. 8, 034003.

Robertson, D.R., et al., 1986. Interference compoetition structures habitat use in a local assemblage of coral reef surgeonfishes. Ecology 1372–1383.

Rockstrom, J., et al., 2009. A safe operating space for humanity. Nature 461, 472–475.

Roff, G., et al., 2012. Global disparity in the resilience of coral reefs. Trends Ecol. Evol. 27, 404–413.

Rogers, C.S., 2013. Coral reef resilience through biodiversity. ISRN Oceanogr. 2013.

Sala, O.E., et al., 2000. Global biodiversity scenarios for the year 2100. Science 287, 1770–1774.

Sale, P., 2011. Our Dying Planet: An Ecologist's View of the Crisis We Face. Univ of California Press.

Scheffer, M., et al., 2001. Catastrophic shifts in ecosystems. Nature 413, 591–596.

Scheffer, M., et al., 2003. Catastrophic regime shifts in ecosystems: linking theory to observation. Trends Ecol. Evol. 18, 648–656.

Schutte, V.G.W., et al., 2010. Regional spatio-temporal trends in Caribbean coral reef benthic communities. Mar. Ecol. Prog. Ser. 402, 115–122.

Sheppard, C., et al., 2008. Archipelago-wide coral recovery patterns since 1998 in the Chagos Archipelago, central Indian Ocean. Mar. Ecol. Prog. Ser. 362, 109–117.

Teh, L.S.L., et al., 2013. A global estimate of the number of coral reef fishers. PLoS One 8, e65397.

Van Hooidonk, R., et al., 2013. Temporary refugia for coral reefs in a warming world. Nat. Clim. Change 3, 508–511.

Veron, J.E.N., 1995. Corals in Space and Time: The Biogeography and Evolution of the Scleractinia (Comstock Book) by May -Cornell Univiversity Press.

Ward-Paige, C.A., et al., 2010. Large-scale absence of sharks on reefs in the greater-Caribbean: a footprint of human pressures. PloS One 5, e11968.

West, P., et al., 2006. Parks and peoples: the social impact of protected areas. Annu. Rev. Anthropol. 35, 251–277.

Wilkinson, C., 2008. Status of Coral Reefs of the World. Australian Institute of Marine Science.

Chapter 6

Biodiversity, Ecosystem Functioning, and Services in Fresh Waters: Ecological and Evolutionary Implications of Climate Change

Guy Woodward and Daniel M. Perkins
Department of Life Sciences, Imperial College London, Ascot, Berkshire, UK

INTRODUCTION

Fresh waters face a cocktail of environmental stressors in the twenty-first century, with climate change widely predicted to usurp many of the other, more localized, impacts of human activity that typified the twentieth century, such as habitat loss and fragmentation, eutrophication, and acidification (Vörösmarty et al., 2010). Climate change thus poses one of the most pervasive, yet also poorly understood, emerging threats to biodiversity and ecosystem functioning, as well as to the provision of socioeconomically valuable goods and services (Parmesan and Yohe, 2003; Daily and Matson, 2008). It also has the potential to interact with, and modulate, other stressors (both directly and indirectly) that could generate potentially dangerous synergies, especially in the many systems that are already exposed to other stressors. These risks are likely to be most pronounced at the higher trophic levels, where stressors can have disproportionately strong impacts on large, rare organisms (Raffaelli, 2004). We also need to understand how these effects ripple through the network of direct and indirect interactions in the food web, and how this translates to responses at organizational levels both below (e.g., species populations) and above (e.g., ecosystem processes) those immediately connected to the trophic network (Loreau, 2010).

At present we still tend to study stressors in a piecemeal fashion, an approach that even applies to climate change, itself an amalgam of different abiotic and biotic drivers (e.g., temperature change, altered hydrology, and species range shifts) that are rarely considered together (but see Berger et al., 2010). Still, before we can consider synergies among these climate-change components or with other orthogonal stressors, we first need to understand the simplest causal relationships and build from there. Ultimately, we will need to achieve sufficient understanding to be able to predict how multiple drivers affect multiple responses at different organizational levels, and across multiple scales in time and space (Figure 1), work that is still very much in an embryonic state. This will extend beyond traditional ecological or evolutionary responses, and embraces the emerging field of eco-evolutionary dynamics in which complex reciprocal relationships can exist between environmental and

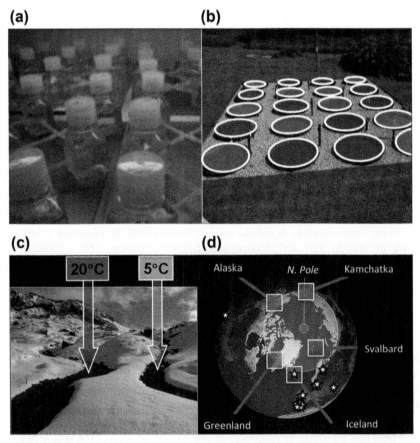

FIGURE 1 Scaling up in attempts to glimpse the bigger picture: (a) laboratory microcosms in controlled-temperature rooms (e.g., Petchey et al., 1999); (b) warmed pond mesocosms (e.g., Yvon-Durocher et al., 2010b); (c) Icelandic geothermal "natural experiment" model system (e.g., Woodward et al., 2010b); and (d) freshwater geothermal sites (squares) and examples of locations of >200 mesocosms used to date (or currently under construction) for warming experiments that could be used for future meta-analysis (stars).

biotic systems (Moya-Laraño et al., 2014), an area we will return to later with a few pertinent case studies.

After decades of focusing on individuals or species, there is now a growing recognition of the need to develop more holistic approaches that consider not just the nodes in ecological networks but also the interactions between them and that can also help connect different organizational levels (Bersier, 2007; Woodward et al., 2010a). Network approaches can also be used to form a bridge from autecological to synecological functional diversity that can account for both direct and indirect species interactions, moving beyond the current focus on organismal traits. This change of direction in general ecology is increasingly being manifested in the practical aspects of biomonitoring and assessment (see Chapter 10; Gray et al., 2014), and given the huge recent increases in the gathering of "big data" across large spatial environmental gradients, there is considerable potential for applied disciplines to develop the new methodologies that will be needed. These include remote sensing and GIS technologies, local-to-global surveying activities of citizen scientists and the rapidly increasing capacity of next generation sequencing (NGS) and eDNA in molecular ecology, plus huge advances in processing power for modeling complex driver—response systems. In this new age of big data we are at the cusp of being able to join many of the previously disparate fields in ecology and other disciplines. This promises to provide us with important new insights that will improve our currently very limited ability to predict the responses of the different levels of biodiversity—from genes to ecological networks to ecosystems—to climate change. Our principal aim here is to provide a brief overview of progress to date and assess prospects for future advances in the field.

Climate Change: An Environmental Stressor That Is More Than Just the Sum of Its Parts?

Throughout the twentieth century, the focus of applied ecology of fresh waters in general was on eutrophication and organic pollution and, to a lesser degree, acidification resulting from industrial emissions (Friberg et al., 2011). As these impacts have ameliorated in many areas in the early years of the twenty-first century, at least in the developed world, the focus has shifted to dealing with other stressors (e.g., climate change and habitat degradation) that are emerging as the new principal axes of community change (Gray et al., 2014). The assessment and monitoring tools needed to track and predict the relevant biotic responses remain relatively understudied compared with the major twentieth-century stressors, however, and we still lack an unequivocal mechanistic, phenomenological, or statistical framework that can deal with these emerging stressors. We are therefore still forced to resort to piecemeal case studies, experiments, and models, often based on extrapolations from first principles, in the interim. These emerging empirical data are bolstered by theoretical advances in our understanding of the fundamental controlling variables, such as the body size—temperature—metabolism relationships that seemingly underpin all levels of biological organization (Brown et al., 2004), that can provide a powerful heuristic framework to push our understanding forward.

Climate change is an especially challenging topic to study because it is composed of multiple drivers, all of which can affect freshwater ecosystems, both when acting alone and in combination. These include temperature, hydrological, and hydrological and atmospheric changes, as well as the direct and indirect effects associated with species range shifts (e.g., poleward migration of warmer-water species) (Bonada et al., 2007; Brown et al., 2007). The connections between these biotic and abiotic components are often not straightforward or obvious: for instance, not all periods of high rainfall lead to floods, not all droughts are due to extended hot, dry periods (Lake, 2003), and species range shifts rarely map perfectly onto their projected bioclimatic envelopes. Layered on top of this is the potential for climate change to have complex synergies with other stressors that may be only partly related, or orthogonal, to climate change per se, including the long-standing drivers of water quality, water loss, and land-use change (Woodward et al., 2002; Durance and Ormerod, 2007, 2009; Friberg et al., 2011). Given this complex mix of drivers it may seem impossible to make any meaningful predictions about the likely state of freshwater ecosystems under any future climate scenarios. However, complex systems are not necessarily inherently unpredictable—e.g., the cell cycle is complex, yet well understood—and there are some simplifying rules that we can apply to help us understand the ecological and evolutionary consequences of climate change. This does require drawing together different disciplines and theoretical frameworks, however, as well as connecting the physical, social, and natural sciences more intimately, and there is still much work to do in this regard. Nonetheless, by breaking down the overarching driver and its impacts into smaller, more manageable components, we can at least start to progress toward the more integrated approach that will ultimately be needed: at present research effort is still very skewed in terms of the ecosystem type under study and the component(s) of climate change being investigated (Figure 2), even though total research effort is increasing exponentially (Figure 3).

To predict climate change impacts on functional diversity and the delivery of goods and services in freshwater ecosystems, we need to know how, and to what extent, normal functioning might be impaired and what the likely consequences of biodiversity loss will be—the first step is to identify a few key drivers and mechanisms and to build from there.

Temperature and Metabolism: The Master Variables in Biological Responses to Global Warming

Temperature sets the pace of life, and insights into its ubiquitous effects on all biological processes, from enzyme kinetics and individual metabolism to food

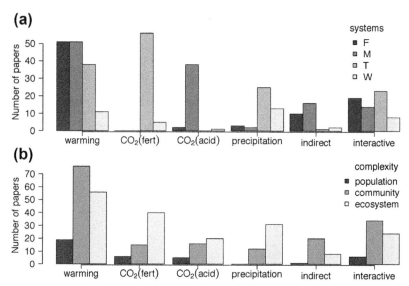

FIGURE 2 Biases in coverage of the different components of climate change in different systems types (a) and across organizational levels (b) in 276 mesocosm studies between 1990 and 2013. *(After Stewart et al. (2013).)* Components of climate change are divided into direct effects: warming, CO_2 fertilization, CO_2 acidification, and changes in precipitation patterns. Indirect components of climate change are, for example: increased UV radiation, changes in cloudiness, changes in the quality and quantity of surface water runoff (and consequences for cross-ecosystem subsidies), changes in salinity, and changes in seawater level. Interactive components refer to the interaction between components of climate change on other stressors—for example, nutrient enrichment, pollution and alteration of habitat, and diversity. In (a) F,M,T and W correspond to freshwater, marine, terrestrial and wetland ecosystems, respectively.

web dynamics and ecosystem functioning, have helped to overcome some of the challenges faced in understanding and predicting the ecological consequences of climate change (Rall et al., 2010; Reuman et al., 2014). Many biological thermal responses are both strong and relatively predictable (Friberg et al., 2009; Demars et al., 2011a), at least compared with the other major stressors in fresh waters. Also, since essentially every single ecosystem process that contributes to overall functioning is (ultimately) temperature dependent, with a characteristic thermal sensitivity (i.e., activation energy), we can start to make some reasonable predictions about the effects of global warming as one of the key components of climate change. One seemingly common biotic response to warming, which is predicted by the so-called metabolic theory of ecology (MTE: Brown et al., 2004), and various other temperature–size rules, is that smaller organisms will be favored both among and within species (Daufresne et al., 2009; Sheridan and Bickford, 2011). These responses appear to be especially pronounced in aquatic systems (Forster et al., 2012), although this apparent orthodoxy has been questioned in some more recent papers (O'Gorman et al., 2012; Adams et al., 2013).

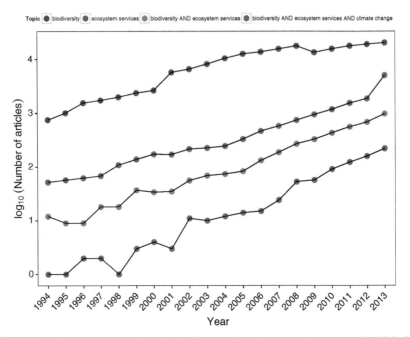

FIGURE 3 Emerging research themes in the literature. Results represent the number of published articles (on logarithmic axis) that included the terms "biodiversity," "ecosystem services," "climate change," and combinations of these three, over the past two decades. *Data were obtained from* Web of Science, *entering search terms as topics (i.e., terms occurring in the title, in the abstract, or as keywords in published articles)*.

Several studies have shown a close fit between the predictions of the MTE and empirical measures of process rates, including for three key components of the carbon cycle that feed back into climate change: primary production, ecosystem respiration, and methane efflux (Yvon-Durocher et al., 2010a, 2010b, 2011b). These processes translate both directly and indirectly into important ecosystem services when placed in an anthropocentric context: e.g., carbon sequestration and regulatory services ultimately support socioeconomically valuable fisheries. As such, there appears to be a hierarchical structure to the delivery of services that is related to the network of species, their functional roles, and their interactions within the food web, from basal processes to apex predators (Figure 4), that is in turn influenced by environmental drivers. A range of carbon-flux models have been tested and validated with experimental and empirical survey data, and one clear general pattern emerges: as temperatures rise, rates increase in a (broadly) predictable fashion. However, the key point to note is that not all processes respond with the *same* thermal sensitivity (Yvon-Durocher et al., 2010b; Demars et al., 2011b; Veraart et al., 2011), and this can lead to imbalances in biogeochemical cycles, especially in fresh waters, which are relatively open rather than closed systems

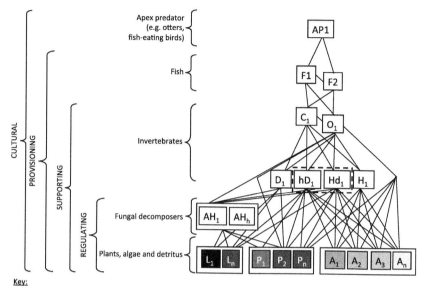

Key:
AP: apex vertebrate predator; F = fish; C: carnivorous invertebrate predator; O: omnivorous invertebrate predator; D: detritivore; hD / Hd: herbivore-detritivore; H: herbivore; AH: aquatic hyphomycete; L: leaf-litter; P: plant; A: algae. Rotated text highlights main areas of ecosystem service delivery within the components of the food web.

FIGURE 4 The food web context of biodiversity, ecosystem functioning, and associated services in fresh waters. In this schematic of a "generic" stream food web, certain services are generally linked to particular trophic levels, whereas others (e.g., community stability) represent an amalgam across levels up to the full trophic network.

reliant on energy inputs from terrestrial carbon subsidies at the base of the food web. For instance, if respiration starts to outstrip primary production, more carbon and methane may be released than is fixed, with greenhouse gases being released from long-term sediment stores. Activation energy, which describes the strength of these temperature dependencies, increases progressively for photosynthesis, followed by respiration, and finally methane efflux (Yvon-Durocher et al., 2010a), and this differential response could trigger potentially dangerous positive feedbacks between the biota and a warming climate.

Theoretical Frameworks: The Metabolic Theory of Ecology and Beyond

Despite its apparent success in predicting some of these ecological responses at the higher-ecosystem-level tiers of biodiversity or community size structure (Yvon-Durocher et al., 2011a; Dossena et al., 2012), the MTE does not fully capture all the important aspects of biotic responses to temperature change (Forster and Hirst, 2012). Indeed, in some aquatic organisms the supposedly fundamental scaling "laws" are not always close matches to the quarter-power scaling widely advocated in the earlier MTE papers, or at least not in an obvious or simple manner (Hirst et al., 2014). This suggests the underlying mechanism for individual-level responses, on which high-level responses are built, is not perfectly defined by the simple fractal properties of branching circulatory systems that were proposed initially by the advocates of the MTE. Indeed, Hirst and colleagues have suggested that surface area-to-volume ratios and the shapes of organisms themselves, in addition to their sizes, provide better fits to empirical data related to individual metabolism. These functional traits could provide an important aspect of biodiversity—beyond the supertrait of body size and taxonomic identity—that has been largely ignored in much of the MTE. In addition, the underlying allometries assumed by the MTE may be modulated by species interactions, such that their expression is not a perfect fit to temperature and/or body size—this suggests that there is still scope for refining the theory still further once these additional aspects have been taken on board.

Many of the responses that are best predicted by warming are manifested most clearly at the individual (e.g., temperature–size rules) or ecosystem level, whereas those at the assemblage or community level appear to be much more noisy and unpredictable. This is not entirely unexpected, and part of the explanation may be because species redundancy is rife, such that community composition can undergo huge turnover in response to warming, while fundamental enzymes and genes that drive individual metabolism are conserved across very different taxa. Even so, there is still clear evidence of strong responses by multispecies systems to temperature change, even though the particular species responses may be contingent on local and regional biogeographical contexts (Walther, 2010; Woodward et al., 2010a, 2010c). If certain properties are indeed highly conserved at the molecular or biochemical level across widely separated branches of the phylogenetic tree, as these results hint at, there might be only limited scope for evolutionary responses to warming for key ecosystem processes and the services they provide.

As in the differential responses to warming seen among different ecosystem processes, other forms of biological "mismatches" can arise at different organizational levels if temperature effects are not uniform across the food web. For instance, in food webs consumer and resource populations may shift out of phase with one another due to differentially altered phenology, or interaction strengths may be altered due to changes in encounter rates arising from differential effects of temperature on prey and predator mobility (Thackeray et al., 2010). Some of these changes may be manifested in transient changes in the slope of individual size distributions, with smaller organisms being favored over larger organisms as a system warms (Daufresne et al., 2009), and this structural change in the community alters the transfer efficiency of biomass between trophic levels (Yvon-Durocher et al., 2011a). In some of these instances, the MTE has proved surprisingly effective in accounting for individual-to-ecosystem-level changes. In others it has been less successful, especially when attempting to predict which species or groups of species will be losers or winners (e.g., among diatoms and macroinvertebrates (O'Gorman et al., 2012; Adams et al., 2013)), and here a more phenomenological or trait-based approach may be needed to cope with the vagaries of taxonomy.

Whether the exceptions or caveats to the MTE that have been reported recently eventually amount to footnotes that need to be appended to the current general theory, or manage to provide a more compelling case to replace it with a better alternative, remains to be seen. Either way, it is clear that metabolism scales with both body size and temperature in a (largely) predictable manner, and that what happens at the individual level also appears to have ecosystem-level consequences. Understanding these issues, and how individual performance varies across thermal gradients, can help to provide the first steps that will ultimately lead to formal consideration of the role of not just individuals, but their traits and interactions within ecological networks in response to temperature change.

An important development in this area comes from the data-driven collation of empirical information on species traits from the literature and ongoing experiments described in Chapter 1. This will ultimately enable us to discern functional biodiversity for a wide range of organisms spanning a broad phylogenetic spectrum, by describing empirically based performance curves for taxa and traits. It will also allow us to assign probabilities that can be used to parameterize a new generation of models that can include key functional traits, such as, ingestion and movement rates among consumers and resources (Rall et al., 2010). Thus, we are moving closer to being able to predict whether certain entities (species or traits) will encounter one another and how they will subsequently interact (e.g., who eats whom and how quickly?) under different environmental contexts and in different, even novel, communities under climate change.

Such trait-based approaches are especially appealing because they should ultimately have far more universal applicability than purely taxonomic approaches because they may provide a means of (eventually) removing the contingencies of biogeographical species pools by focusing on functional attributes (Bonada et al., 2007). They can also be used to formulate and test more mechanistic-based theory, which can then be validated experimentally. Some of these are already underway in different parts of the world, and this will help us to move from phenomenological to more predictive and functional-based frameworks, and combining the MTE (or its refined derivatives) with trait-based "big data" ecoinformatics is a promising avenue of future research.

We should bear in mind, however, that although most ecological models and data focused on warming have tended to focus on gradual warming trends, more extreme spikes and even cooling are predicted in certain areas within this general trend (Thomson et al., 2012; Thompson et al., 2013).

Notwithstanding these caveats, it is clear we have already made significant progress over the last decade or so, and we are now at the stage where we can start to make some broad generalizations based on advances made in both theory and data.

Biodiversity—Ecosystem Functioning Relationships: How Many Species Do We Need to Maintain Functioning and Services in a Changing Climate?

A logical next step from understanding the traits of individuals and species in isolation, or in simple pairwise interactions, is to move into the realm of entire assemblages or communities of multiple interacting entities (e.g., Dossena et al., 2012; Ledger et al., 2012). Over the past two decades, huge effort has been devoted to unraveling biodiversity—ecosystem functioning (B—EF), and more latterly, but to a far lesser extent, biodiversity—ecosystem services (B—ES) relationships (Balvanera et al., 2006; Cardinale et al., 2006, 2012; Reiss et al., 2009; Figure 3). These studies have been conducted mostly in the context of random species loss rather than in connection to explicit environmental drivers per se, especially those related to climate change (Figures 2 and 3). This work has been reviewed extensively elsewhere (e.g., see Cardinale et al., 2012), so we will not cover it in detail again here, but the main salient points that have emerged from freshwater studies to date can be summarized as:

1. For single processes, rates saturate at low levels of species richness—far below those of most natural systems.
2. Redundancy therefore appears to be prevalent, although some key species may have disproportionate effects within this general schema.
3. Functional diversity is far lower than taxonomic diversity, but when entire functional groups are lost, process rates and service provision can decline rapidly.
4. Body size is a key functional trait that can outweigh taxonomic identity or species richness effects, largely due to its close correlation with the metabolic capacity of the individuals within an assemblage, and it is especially important in the context of temperature change.
5. Environmental stressors can modulate B—EF and B—ES relationships, and often nonlinearly, such that tipping points may be passed that lead to regime shifts.
6. B—EF relationships are far better understood than B—ES relationships, largely due to the logistical and philosophical constraints of detecting mechanistic relationships at large scales in complex socioeconomic systems.

A crucial question that arises from the emerging body of data, models, and meta-analyses summarized in these six points is: what level of biodiversity turnover is occurring beneath ecosystem-level responses in natural systems, rather than in the artificially assembled communities commonly used in such experiments? An understanding of this would provide important insight into the true extent of functional redundancy and its scope to preserve functioning and services in a changing environment. For instance, recent work that involved experimental translocations of very different microbial communities from geothermally warmed streams of different temperature but from a single catchment in Iceland revealed no evidence of a "biodiversity effect" on activation energies or absolute rates of ecosystem respiration, once biofilm biomass had been taken into account (Perkins et al., 2012). Essentially, each stream, despite having negligible species overlap at the extremes of the temperature gradient, responded in exactly the same way to temperature change (Figure 5). This suggests that for at least this functional measure, community-level species richness and identity differences had negligible effects on ecosystem-level responses in both the field and the lab (O'Gorman et al., 2012; Perkins et al., 2012): different microbes were performing equivalent functional roles.

At present, much of the evidence suggests that ecosystem process rates tend to saturate at single-figure numbers of species richness in fresh waters (Cardinale et al., 2006): this seems to hold for most macrofaunal taxa (Jonsson and Malmqvist, 2003; McKie et al., 2008), but appears to be higher for microbes (Bell et al., 2005; but see Reiss et al., 2010). Nonetheless, the simple bivariate B—EF redundancy model seems to be the most appropriate most of the time, although idiosyncrasy can arise where keystone species, ecosystem engineers, or foundation species that have disproportionately strong effects are present (Schmid et al., 2009). Identifying these particularly influential nodes would improve our predictive power, although this will be context-dependent, as the traits that make one species dominant under certain conditions are likely to disfavor it at the other end of the climatic gradient.

It seems likely too that there are strong phylogenetic and food web contexts to redundancy and where it is distributed among the biota in local communities. Although many invertebrates and fishes have clear thermal optima that differ in their peaks, spans, and shapes, there are suggestions that some microbial taxa, such as ciliates, may be relatively insensitive to environmental change and have (seemingly) far wider tolerance limits than for the macrofauna on which most freshwater B—EF work has been based (O'Gorman et al., 2012). This could indicate that these microscopic species at the base of the food web are far more plastic in their functional roles, and/or that biodiversity needs to be viewed in a more sophisticated manner than the traditional reliance on species richness. Indeed, our ability to differentiate "types" of organisms—and even the entire notion of the species concept—is inadequate for many groups, so the x-axis of traditional B—EF/S studies may be fundamentally different from those applied to taxa higher in the web (Leitch et al., 2014).

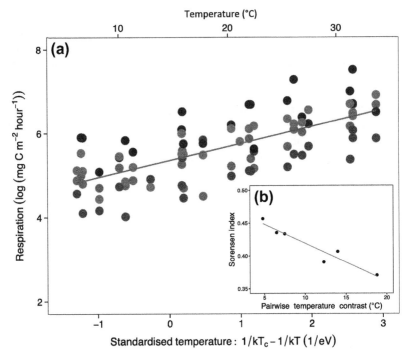

FIGURE 5 Consistent metabolic responses to warming for stream biofilms contrasting in ambient temperature and species composition. *(Redrawn from Perkins et al. (2012).)* (a) Arrhenius plot of (logarithmically transformed) biofilm respiration rates as a function of standardized experimental temperature ($1/kT_c - 1/kT$), where k is Boltzmann's constant (8.62×10^{-5} eV K^{-1}) and T is temperature (in Kelvin); experimental temperature gradient in °C is given for comparison; the slope of this relationship yields the average activation energy (in electron volts), which is the fundamental parameter that determines the temperature sensitivity of metabolism and was found to be indistinguishable among biofilm assemblages from different streams (color symbols; scaled to stream temperature with blue (dark gray in print versions) the coldest (6 °C) and red (light gray in print versions) the warmest (25 °C) stream). (b) Sørensen similarity index for pairwise comparisons of taxonomic biofilm assemblages (diatoms, ciliates, and micrometazoans) from benthic biofilm samples used in laboratory incubations, as a function of pairwise temperature contrasts ($r^2 = 0.92$, $n = 6$, $P = 0.003$).

Trait-based approaches are sure to be used far more widely in future B–EF and B–ES studies, especially in cases where classical Latin binomial taxonomy breaks down, although we also need to bear in mind that traits tend to come in multivariate suites rather than as simple free univariate descriptors. Thus, when we think of the x-axis in B–EF or B–ES studies, we need to move beyond the species-richness emphasis that has permeated the field to date, no doubt reflecting the roots of classical entomology that underpins much of (benthic) freshwater community ecology. Functional diversity is thus shaped by both the current environmental filters as well as past evolutionary and phylogenetic history. By embracing other measures of biodiversity—whether it be operational taxonomic units among microbial groups, functional feeding groups, size classes, other trait-based classifications, or more network-based views—will help us gain a better view of where redundancy and insurance might be nested within the system, as well as what drives the functions or services on the y-axis. The notion of a gradient that moves from classical taxonomy at one end, to trait-based autecology, and finally to the functional synecology of organisms in the context of the food web at the other, highlights how B—EF research can ultimately develop into something closer to understanding B—ES relationships in the real world. That is, the functional diversity that drives services is likely to be plastic and reactive to both biotic and abiotic contexts, rather than a simple agglomeration of fixed "traits" that map passively onto an environmental template. This will help us to better understand many of the seemingly subtle effects of climate change on multispecies systems.

Are We Measuring the Relevant Drivers and Responses in Biodiversity—Ecosystem Functioning Studies?

More effort clearly needs to be directed into resolving both the (multiple) x- and y-axes of B—EF studies, as the idea that species can be lost or replaced without any notable loss of functioning is widespread. However, this notion might prove to be erroneous if multifunctionality (Hector and Bagchi, 2007) is the rule—i.e., if higher levels of diversity are always needed to drive high levels of multiple processes simultaneously. Indeed, a recent study from fresh waters provides evidence for the increased importance of biodiversity in driving ecosystem multifunctionality across thermal regimes (Perkins et al., 2014). Perkins and coauthors manipulated species diversity and environmental temperature simultaneously in a simple but tightly controlled laboratory experiment, which allowed for links between the contribution of different macrofaunal assemblages and temperature regimes to a range of single processes and ecosystem multifunctionality. Their results clearly demonstrate the context dependency of biodiversity effects—although species richness had negligible effects on individual processes, it influenced multifunctionality, but only at the coldest and the warmest temperatures (Figure 6). It seems intuitive that species *must* differ in their contributions to different processes, just as they must differ in niche use if they are to avoid competitive exclusion in the face of limiting resources. Nonetheless, this is still a very embryonic area of B—EF research, especially in the context of climate change, and one that has barely been touched on in laboratory studies, let alone properly controlled multifactorial field experiments, so it is hard to make meaningful generalizations at this stage. Nonetheless, it is clear that we may be in danger of significantly overestimating the true levels of redundancy and thus the potential robustness of freshwater ecosystems to climate change. Future studies will need to measure multifunctionality and to do so across a range of environmental conditions, to bring greater realism and predictive power to future research in the area.

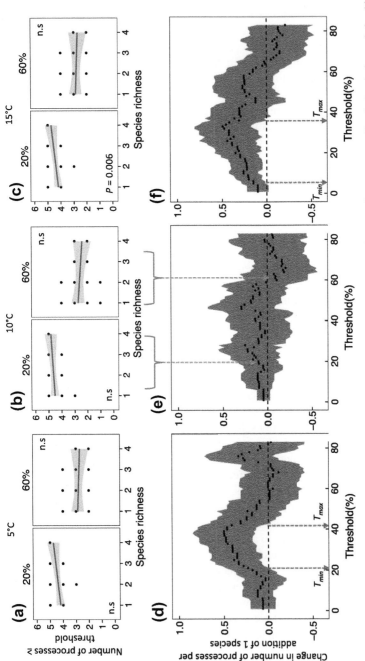

FIGURE 6 Context dependency of biodiversity effects on ecosystem multifunctionality across environmental temperatures. Panels (a–c) show relationships between richness of freshwater macroinvertebrates and multifunctionality at threshold values of 20% and 60% for each of the experimental temperatures in the Perkins et al. (2014) study. Threshold values represent the percentage of the maximum observed rates for each individual process in the study; a value of 20%, for example, means all processes are "functioning" at 20% of the maximal rates. Panels (d–f) show relationships between species richness and multifunctionality across multiple thresholds (1%–83%), where each data point corresponds to the slope of the relationship between species richness and the number of ecosystem processes that surpass threshold values (indicated for thresholds of 20% and 60% by blue (dark gray in print versions) lines in the central panel). The 95% confidence intervals (indicated in grey) around the estimated slopes (filled data points) indicate whether the intervals contain zero, giving a test of the threshold values at which diversity has no effect on multifunctionality. T_{min} and T_{max} indicate the minimum and maximal thresholds where diversity has significant effects on multifunctionality, which were found only to occur at 5 °C and 15 °C.

Traits and Functional Diversity in a Changing Climate: Beyond Body Size

Fortunately, some attributes of organisms that exert disproportionate effects on functioning in local assemblages do recur, and body size once again emerges as a prime variable. Keystone species that confer especially strong idiosyncratic effects on ecosystem functioning are often, but not always, large and/or predatory organisms—traits that are also often associated with increased vulnerability to climate change. For instance, *Gammarus* spp., although primarily detritivorous, are often among the largest but also most abundant macroinvertebrates in many lowland UK rivers, and they are key conduits of energy flux in the food web and major drivers of decomposition (Woodward et al., 2008).

Decomposition processes are strongly temperature-dependent (Friberg et al., 2009; Perkins et al., 2010) and in turn underpin a range of other ecosystem services (Costanza et al., 1998; Odum and Odum, 2000; Raffaelli and White, 2013). These include water purification and pollutant processing, fishery productivity, and the support of iconic species of cultural value at the top of detrital-based food chains, such as piscivorous birds, mammals, and herptiles. If populations of keystone species such as *Gammarus* are unable to keep pace with a changing climate, the rest of the web may start to unravel. For instance, apex predators not only are especially vulnerable to climate change because they are numerically rare and require large populations of prey to support them, but also can exert strong top-down effects in freshwaters, so their loss can have profound consequences that ripple back down through the food web: such indirect effects can result in surprising counterintuitive consequences. Top-down effects have the potential to alter eco-evolutionary dynamics, especially as changes in trait (such as body size) distributions will influence network structure and the strength of interactions within the food web. Although the importance of such phenomena are still largely unknown in the context of climate change, their potential importance has been highlighted in recent theoretical work (Moya-Laraño et al., 2014).

Structure and Functioning of Freshwater Food Webs

Freshwater food webs are typically characterized by trophic generalists, with most species being only a couple of links away from every other. This means that not only are species functionally similar in their trophic autecology and synecology, as they overlap in body size and share many of the same predators and prey, but they are also intimately connected to one another through the food web (Woodward and Hildrew, 2002). Perturbations can therefore propagate quickly, but they can also dissipate rapidly through the "small-world" network when generalist nodes are affected. This may be especially true in more physically disturbed systems and those that are detrital-based such as headwater streams, as both these system traits are likely to favor generalism and hence the prevalence of small-world traits within the constituent members of the food web. In contrast, more productive autochthonous systems such as shallow lowland lakes may have less reticulate but more species-rich food webs with longer food chains that are more prone to oscillatory dynamics and catastrophic regime shifts (Scheffer et al., 2001). If these more "fragile" systems are perturbed, recovery may be much slower, and may even fail if a full-scale regime shift is triggered. A prevalence of detrital chains could therefore buffer the potentially destabilizing effects of algal–herbivore pathways on the network as a whole (Rooney et al., 2006), and many invertebrates possess functional traits that align them more closely with one pathway or mode of feeding than another—in terms of the so-called "functional feeding groups" frequently encountered in the literature, shredders tend to be detritivorous, whereas grazers tend to be herbivorous. These groupings also often contain strong phylogenetic associations and evolutionary signatures: snails and mayflies tend to graze algae, whereas stoneflies and shrimps tend to be more detritivorous, with the larger predators being predominantly fish species (and ultimately the birds, mammals, and herptiles that prey on them in turn).

Thus, freshwater food webs can be roughly compartmentalized into broad functional modules defined by body size, feeding mode, and taxonomy. For instance, in mass-abundance "trivariate" plots of food webs there are often three distinct clouds of nodes in freshwaters—one of abundant small diatoms at the base of the web that are eaten by a cloud of larger but rarer herbivorous or omnivorous invertebrates that are in turn eaten by another cloud of much larger and rarer fishes (e.g., O'Gorman et al., 2012). The size structure of these trophic networks is striking and repeated in very disparate systems, even where there are few or no shared species (Woodward et al., 2005), providing a useful and taxonomically free way to characterize systems in both reference and perturbed states. There are subtle but potentially important differences within these broadly similar system traits, however; for instance, the three "clouds" described above are more distinct in standing waters than in running waters, and as this shows the spread of a key functional trait within the food web—body size—it could also be related to the prevalence for the system to be perturbed. In headwater streams, for instance, the food web is less clearly compartmentalized into these clouds than it is in the pelagic zone of lakes, and thus there is scope for more omnivory and the weakening of size-structured cascades as a result. This could help make them less susceptible to perturbations in response to climate change, although this notion has yet to be tested experimentally.

Where individual-based data are available concurrently with trivariate information, these webs can be replotted as "individual size distributions" or size spectra (White et al., 2007). The slope, intercept, and span of these taxon-free size-based plots can then be used to gauge the link between structure (e.g., biomass and abundance) and functioning (e.g., energy transfer through the web). This last approach has been well developed in marine systems for many years, but is less familiar in freshwaters (but see Yvon-Durocher et al., 2011a; Dossena et al., 2012). It provides a relatively simple way of condensing a food web, irrespective of species identity, into a few functional size-class groups that can provide an overview of the broad patterns of energy and biomass flux through the system based on the body size of individuals. Rather than have clouds of points representing separate species, the data are simply aggregated into size bins that often contain many individuals of different species, akin to Elton's "pyramids of numbers" (Elton, 1927). Different bodies of theory have developed around trivariate and size-spectra approaches, but both can ultimately be linked to individual metabolism and how the system as a whole should respond to perturbations in general, and warming in particular (Woodward et al., 2005, 2010b).

From Averages to Individuals: The Common Currency of Freshwater Ecology

For over a century, ecologists have depicted food webs as static species-averaged networks of nodes and links that do not allow for individual variation. However, it is increasingly apparent (especially in aquatic systems) that the functional role and phenotypic plasticity of a species is often more strongly determined by its size than its taxonomic identity. Not only that, but at an even finer scale, variation across individuals can be extremely important; but again, that is largely overlooked in traditional food web ecology (Bolnick et al., 2011). Of course, individual-level variation is the engine that drives evolution via natural selection, and once again it is this organizational level that we return to as a fundamental unit for linking ecological and evolutionary processes, as well as for understanding the structure and dynamics of food webs and other higher organizational levels in a changing environment.

Adaptive phenotypic plasticity and evolutionary responses among prey are common in many species and systems, such as among zooplankton in lakes, which form a critical link in the trophic cascade between primary producers and apex predators that, in extreme cases, can trigger regime shifts and alter ecosystem functioning profoundly (Scheffer et al., 2001). Over the last decade or so, increasing efforts have been directed to considering the relative importance of warming to nutrient concentrations in freshwater mesocosms, in terms of their respective abilities to mold community structures and ecosystem processes. The results have been equivocal so far, with some revealing profound effects, whereas others have revealed no strong (or at least seemingly predictable) responses, at least at the community level (McKee et al., 2003; Moss et al., 2003, 2011; Jeppesen et al., 2010; Moss, 2010). We can make progress in this area via the development of the phenomenological databases of species traits and thermal performance described earlier, for instance, as well as through more sophisticated modeling and experimental work, but at present the bridge between community- and ecosystem-level responses to climate change is still a weak spot in our overall understanding.

Scaling Up: Cross-System Subsidies and Source—Sink Dynamics in Fresh waters

Although the activation energies of key ecosystem processes (e.g., ecosystem respiration) at short timescales (i.e., diel cycles) are remarkably similar across multiple aquatic and terrestrial systems, at longer timescales marked differences emerge among ecosystem types as the effects of resource subsidy and substrate limitation come into play (Yvon-Durocher et al., 2012). This is especially true in running waters, where the realized temperature-dependence is lower than theoretically predicted due to external carbon inputs (Yvon-Durocher et al., 2012). These cross-system subsidies that are especially prevalent in freshwater systems mean they are not closed systems, and that they need to be considered in the context of the wider landscape, particularly over longer timescales. This is particularly important if climate change is not spatially homogenous such that it alters source—sink dynamics at large temporal and spatial scales, as this has ramifications for not just ecosystem processes and services but also the community dynamics that underpin them. The roles of refugia and source—sink dynamics are therefore especially pertinent in fresh waters because they are essentially fragmented patches in a terrestrial mosaic (Hagen et al., 2012). It also becomes evident that the basin-level perspective that is often used in resource management, or for assessing ecosystem services in the landscape, is a sensible way to view these systems at those organizational levels, because human activities on the land might be just as important as what is happening in the water (Woodward and Hildrew, 2002). The interconnections between patches of freshwater and their links to the adjacent terrestrial (and marine) systems will affect their responses to climate change. However, as freshwater ecologists we still tend to focus on local, proximate drivers, rather than looking beyond the stream bank or lakeshore at the ultimate drivers that are operating at far larger scales: often this is where climate change effects will be especially evident, where the interplay between ecology and evolution will be most apparent, and where ecosystem services are manifested.

This patchy and isolated coverage of fresh waters within the terrestrial landscape is one of their defining features, but it makes them particularly prone to alterations in spatial connectivity in the "hydroscape" that arise via range shifts in response to warming and habitat loss and fragmentation under drought conditions (Dewson et al., 2007). Essentially, freshwater food webs are "networks of networks" in the landscape, and metacommunity and metanetwork properties become important for defining ecological resilience, so dispersal constraints and source—sink dynamics will need to be considered in future climate change research. There is also a food web context to the ability of different freshwater taxa to respond to fragmentation and isolation, partly reflected by body size, with dispersal abilities being high among the microbial primary producers and insect consumers. The former can form a large-scale diaspora, with some even having a truly global distribution (Finlay, 2002), whereas the latter typically have an aerial dispersal phase that can span multiple catchments over many kilometers, via either active flight or passive windblown dispersal (Hagen et al., 2012).

The predominance of insects in many fresh waters, especially headwater streams, reflects strong selection for this high dispersal ability to cope with a patchy environment: indeed, the combination of classic Lévy flight when dispersing in the adult stage and strong density-dependent mortality early in the life cycle means only a few individuals are needed to repopulate an entire system. Glimpses of this huge capacity for rapid population recovery have come from a range of "natural experiments" as well as more direct experimental manipulations of the life cycle and recruitment rates of particular taxa (Hildrew et al., 2004). The ecological data are also complemented with insights gleaned from population genetics studies that have shown that many populations can be panmictic over even very large scales, but also for strong density-dependent survival and regeneration of populations from just a handful of founders or even a single gravid female. Other taxa, such as molluscs and crustaceans, are tied to the water body for their entire life cycles. Fishes are higher in the food web, so their ability to disperse across systems is far more limited due to this lack of insects' "island-hopping" ability (Hagen et al., 2012).

Larval insects must disperse in the same way as these other taxa, via active swimming, active crawling, or passive drift in the flow, but these distances are typically far less than those that are covered in the adult stage, and they are also restricted to moving within the same water body or those directly connected to it. Dispersal is therefore a key functional trait, over and above that of body size per se, but one that is highly context specific in terms of both the system and species: it is likely to be far less important in large permanent water bodies such as lakes than in small transient systems such as intermittent headwater streams, and this helps to explain much of the taxonomic and functional composition of different freshwater communities (Hagen et al., 2012).

Gauging Ecological and Evolutionary Responses to Climate Change in the Real World: Sentinel Systems and Future Refugia for Functional Diversity

Under a changing climate, we would predict large non-insect and wholly aquatic consumers to be especially vulnerable, as these are the taxa least able to manifest range shifts. Classic examples of genetic bottlenecks and local extinctions include landlocked glacial relict populations of Arctic char in Europe, which cannot

Continued

Gauging Ecological and Evolutionary Responses to Climate Change in the Real World: Sentinel Systems and Future Refugia for Functional Diversity—cont'd

migrate and must therefore either adapt or perish as temperatures rise (Rouse et al., 1997). In contrast, insects in running waters are far more likely to be able to cope with warming, as even in the small tendrils of headwaters in the upper reaches, most taxa possess an adult aerial dispersal phase. There is another important element to this ability to cover large distances rapidly, however, that has been overlooked in the ecological climate change literature, and that is the presence of "future refugia" provided by geothermally warmed systems in otherwise cold climates (O'Gorman et al., 2014). For instance, in the many geothermal areas from across the world, there are vast numbers of water bodies that span the full biological range of thermal tolerance due to natural geothermal warming that has been ongoing since even before the last glaciation. Thus, there are already species in place that may have passed through many thousands of generations while being exposed to far higher temperatures than predicted for the next century even under the worst-case scenarios of the various IPCC models (IPCC, 2007, 2013). These are not necessarily the extreme thermophiles that often spring to mind when we think of geothermal areas—many are in fact taxa that are common and widespread in temperate regions: for instance, the detritivorous caddis *Potamophylax cingulatus*, the herbivorous snail *Radix peregra*, and the predatory brown trout *Salmo trutta* can all be found in warmed Icelandic streams (O'Gorman et al., 2012) as well as across much of temperate Europe.

Some of these taxa, including the three above, are rare or absent under ambient conditions in Iceland, but are common in warmed systems and would otherwise be limited to warmer climes at lower latitudes. Thus geothermal hotspots could provide preadapted populations and taxa that would capitalize from environmental warming as well as providing a possible means of accelerating "evolutionary rescue" (sensu Gonzalez et al., 2013). Essentially, geothermal sites provide sentinel systems for anticipating climate change as well as providing future refugia for coping with its effects, and their presence could help to buffer the effects of warming by bypassing the need for range shifts and rapid evolution that track the poleward shifting of the bioclimatic envelope of temperate species. Of course, it does mean that local cold-adapted taxa are more likely to be usurped by this ready-made pool of competitors and predators, but it could also be a way of maintaining functioning, given that there are populations that have been exposed to long-term warming over multiple generations. These geothermal systems can be found in many parts of the world: for instance, there is a circumpolar ring of sites that includes stream catchments that span comparably thermal gradients of about 5–20 °C around the Arctic and Boreal zones (Figure 1(d)) that include Svalbard, Iceland, Greenland, Alaska, and Kamchatka (O'Gorman et al., 2014). This aspect of biogeography, with preexisting warmed systems interdigitating with ambient areas merits further study, as it could enable ecosystems to respond far faster to climate change than previously thought, in both ecological and evolutionary terms.

Eco-evolutionary Dynamics: Reciprocal Feedbacks between Ecology and Evolution, and Interactions between Biotic and Abiotic Drivers in Multispecies Systems

In addition to the potential role of "future refugia" and evolutionary rescue to mitigate (or amplify) the effects of climate change, it is becoming increasingly evident that environment—biota connections are far more dynamic, closely coupled, and reactive than previously imagined. Therefore, evolutionary responses can occur surprisingly rapidly—even at what have been traditionally viewed as ecological timescales. This potential for reciprocity between ecological and evolutionary phenomena in multispecies systems has triggered the emergence of a new subdiscipline focused on "eco-evolutionary dynamics." This area of research has developed as part of a general growing shift in focus toward explicitly recognizing the role of species interactions in natural systems. Eco-evolutionary approaches in aquatic systems have a somewhat different history than those in terrestrial systems, where network-based approaches are already well underway: aquatic ecologists have only recently explicitly connected eco-evolutionary dynamics with network theory and in the context of environmental change (Melián et al., 2011). Nonetheless, some recent work has highlighted how these little-known mechanisms could be far more prevalent in freshwaters than previously thought (Travis et al., 2014), and also that they could play important roles in modulating biotic responses to climate change in general (Moya-Laraño et al., 2014).

Synergies between Multiple Stressors and the Modulation of Ecological and Evolutionary Responses in a Changing Climate

At the start of this paper, we touched on the potential synergistic roles of multiple stressors in fresh waters, a point we return to now at the close. We are still grappling with the consequences of single or multiple stressors acting in an isolated or at best additive fashion, and a huge gap in our knowledge remains as to how synergies among them could modulate their individual effects on the biota (Friberg et al., 2011). For instance, many stressors are likely to have additive or multiplicative effects, as most seem to exert the strongest effects on species that are large, rare, and high in the food web: this appears to be largely true for warming, habitat fragmentation, and drought, which tend to be autocorrelated in many climate change scenarios (Greig et al., 2012). As such, we should expect, even just based on simple conditional probabilities, that when these stressors co-occur in nature—as they often do—the impacts on the biota are magnified and in the same general direction. Many stressors not associated with climate change may also amplify its likely effects: acidification and heavy metal pollution also shorten food chains and are especially damaging for apex predators (Layer et al., 2011), for instance, whereas other stressors might operate antagonistically.

Agrochemicals in particular are an increasing concern as the long-term effects of acidification and eutrophication subside, and yet they can have very different effects depending on their original intended targets' positions in the food web (e.g., herbicides, insecticides, fungicides) and mode of action (e.g., lethal, sublethal, and biomagnifiers vs. bioaccumulators). A notable exception to the general trend of stressor impacts being concentrated in the higher trophic levels, as appears to be true for many components of climate change, is pesticides (usually herbicides, fungicides, or insecticides) designed to knock out specific food web components, particularly those at the lower or intermediate trophic levels. Even though they are rarely the target systems, fresh waters are especially vulnerable, as they act as the main conduits for pollutants applied to the land on their journey to the coast. Climate change will affect these waste-disposal ecosystem services offered by fresh waters, both directly (e.g., due to reduced ability to lower toxin concentrations where water availability is compromised during droughts) and indirectly (e.g., by compromising individual metabolism, which is also strongly dependent on temperature), but we are still only beginning to realize the full implications of this.

Ultimately, however, we might expect to see the impacts of all these chemicals manifested at the top of the food web, as that is where the different (compromised) food chains tend to integrate. Even if one route may be unaffected, another may be impaired, and overall the total flux of biomass is likely to be reduced as a result. However, in the transient phase of impact and recovery, unexpected trajectories are likely to appear: e.g., knocking out fungal-based pathways will have consequences that are immediately very different from those that arise from the loss of plants or insects. Climate change could generate a wide range of synergistic responses, including: concentration of toxins into isolated patches under drought, increasing the spatial and temporal heterogeneity of pesticide impacts; elevated metabolic demands under warming placing increased strain on the detoxifying apparatus of individuals; and altering encounter rates and handling times under warming where consumer and/or resource behavior is modified differentially by toxins.

Another emerging threat to biodiversity in running waters in particular is the effect of habitat loss, degradation, and fragmentation, which has been a feature of the European fluvial landscape for centuries, as floodplains have been drained and rivers have been dredged, straightened, and fragmented by the introduction of dams and weirs in almost every watershed across the continent. As the effects of organic pollution have started to subside in many areas, as have those of acidification in Northwestern Europe in recent decades, so the role of the physical matrix—the streambed itself—in influencing biodiversity has become an increasing focus of attention. Many ecosystems are now being actively reengineered or "rewilded" back to what is assumed to be something approximating their natural states (Bullock et al., 2011), but even so, most fresh waters have been so dramatically altered and homogenized by human activity that much of the potential functional redundancy that might have buffered the effects of climate change may have already been severely compromised. In particular, the fragmentation and isolation of many waterways could reduce the ability for species range shifts to track climate change, whereas conversely some of the large-scale engineering works on the biggest rivers have exacerbated range shifts of exotic species, as can be seen from the incessant waves of Pontocaspian invaders moving through the Danube basin, and similar invasional "meltdowns" in the River Thames (Jackson and Grey, 2013).

In summary, the full picture of climate change impacts in fresh waters at the landscape scale, at least in terms of community change, is likely to be complex and difficult to predict due to the contingencies of past and present resource use. Clearly, the historical context needs to be taken into account here, as well as the identity and traits of potential invaders as connections between water bodies are forged or broken over time and space. Even so, we need to be able to assess what are likely to be the major individual components and synergies that are likely to drive future change, and then to refine our efforts for future research accordingly.

We now know that temperature is a major controlling variable, and that it has (relatively) predictable effects on individuals and ecosystem processes, and the same appears to be true for hydrological change, where large rare taxa high in the food web are once again those that are most negatively affected by droughts, for instance (Ledger et al., 2013). We also have plenty of empirical data and theory to make reasonable predictions about responses to major axes of chemical pollution in the form of acidification and organic pollution, plus an ever-growing set of case studies with which to start to make generalizations or specific predictions about invasive species and agrochemicals. Some steps have been made to explore interactions between these stressors in attempts to gauge both their relative importance and whether they amplify or mitigate one another's impacts, as this has implications for both ecological and evolutionary processes, especially if the different stressors are selecting for similar versus contrasting traits.

Future Directions and Concluding Remarks

Freshwater ecology has come a long way in recent years, and we now have a much better understanding of the types of species we should expect to find in a given water body based on its physical and chemical properties, albeit from a phenomenological rather than mechanistic viewpoint. We also have a good idea as to how individuals and ecosystem processes are likely to respond to temperature change, based on first principles and empirical data. We have a much weaker understanding of the community-level responses to warming, and still very little idea of the likely impacts of the other components of climate change (atmospheric change, floods, etc.) and their interactions with other stressors at any level of biological organization (but see Christensen et al., 2006; Feuchtmayr et al., 2010). Much of the existing data are also still either observational or outputs of mathematical models, so the mechanistic cause-and-effect relationships that can only be validated experimentally are largely lacking. The connections from B–EF to B–ES are also still in their infancy, although considerable research effort is now being directed to plugging these gaps. Far more complex multifactorial experiments are now needed, in conjunction with larger-scale spatial and temporal surveys that include several stressor gradients as well as more sophisticated modeling approaches (Chave, 2013). The expansion of microcosm and mesocosm experiments in particular can help to address the first issue (Petchey et al., 1999; Benton et al., 2007; Liboriussen et al., 2011; Stewart et al., 2013), the use of coordinated sampling, including the use of citizen science and biomonitoring data, can help with the second (Gray et al., 2014), and the third point is already being addressed but currently lacks much of the data needed to validate and test the models but that will start to emerge as the first two issues are addressed.

In conclusion, although good progress has been made from a position of almost total ignorance about the higher-level effects of climate change just a couple of decades ago, there is still much work to be done in the coming years. The current gaps are being filled by a wide range of established and novel techniques, and traditional scientific disciplines are increasingly being grafted onto far more disparate disciplines, even including those from the social sciences, as we attempt to understand how freshwater ecosystems are valued in the socioeconomic sphere; this process will drive much of the debate and research effort in the near future.

REFERENCES

Adams, G.L., Pichler, D.E., Cox, E.J., O'Gorman, E.J., Seeney, A., Woodward, G., Reuman, D.C., 2013. Diatoms can be an important exception to temperature–size rules at species and community levels of organization. Glob. Change Biol. 19, 3540–3552.

Balvanera, P., Pfisterer, A.B., Buchmann, N., He, J.-S., Nakashizuka, T., Raffaelli, D., Schmid, B., 2006. Quantifying the evidence for biodiversity effects on ecosystem functioning and services. Ecol. Lett. 9, 1146–1156.

Bell, T., Newman, J.A., Silverman, B.W., Turner, S.L., Lilley, A.K., 2005. The contribution of species richness and composition to bacterial services. Nature 436, 1157–1160.

Benton, T.G., Solan, M., Travis, J.M.J., Sait, S.M., 2007. Microcosm experiments can inform global ecological problems. Trends Ecol. Evol. 22, 516–521.

Berger, S.A., Diehl, S., Stibor, H., Trommer, G., Ruhenstroth, M., 2010. Water temperature and stratification depth independently shift cardinal events during plankton spring succession. Glob. Change Biol. 16, 1954–1965.

Bersier, L.-F., 2007. A history of the study of ecological networks. In: Képès, F. (Ed.), Biological Networks, pp. 365–421.

Bolnick, D.I., Amarasekare, P., Araújo, M.S., Bürger, R., Levine, J.M., Novak, M., Rudolf, V.H.W., Schreiber, S.J., Urban, M.C., Vasseur, D.A., 2011. Why intraspecific trait variation matters in community ecology. Trends Ecol. Evol. 26, 183–192.

Bonada, N., Doledec, S., Statzner, B., 2007. Taxonomic and biological trait differences of stream macroinvertebrate communities between mediterranean and temperate regions: implications for future climatic scenarios. Glob. Change Biol. 13, 1658–1671.

Brown, J., Gillooly, J., Allen, A., Savage, V., West, G., 2004. Toward a metbaolic theory of ecology. Ecology 85, 1771–1789.

Brown, L.E., Hannah, D.M., Milner, A.M., 2007. Vulnerability of alpine stream biodiversity to shrinking glaciers and snowpacks. Glob. Change Biol. 13, 958–966.

Bullock, J.M., Aronson, J., Newton, A.C., Pywell, R.F., Rey-Benayas, J.M., 2011. Restoration of ecosystem services and biodiversity: conflicts and opportunities. Trends Ecol. Evol. 26, 541–549.

Cardinale, B.J., Duffy, J.E., Gonzalez, A., Hooper, D.U., Perrings, C., Venail, P., Narwani, A., Mace, G.M., Tilman, D., Wardle, D.a, Kinzig, A.P., Daily, G.C., Loreau, M., Grace, J.B., Larigauderie, A., Srivastava, D.S., Naeem, S., 2012. Biodiversity loss and its impact on humanity. Nature 486, 59–67.

Cardinale, B.J., Srivastava, D.S., Duffy, J.E., Wright, J.P., Downing, A.L., Sankaran, M., Jouseau, C., 2006. Effects of biodiversity on the functioning of trophic groups and ecosystems. Nature 443, 989–992.

Chave, J., 2013. The problem of pattern and scale in ecology: what have we learned in 20 years? Ecol. Lett. 16, 4–16.

Christensen, M.R., Graham, M.D., Vinebrooke, R.D., Findlay, D.L., Paterson, M.J., Turner, M.A., 2006. Multiple anthropogenic stressors cause ecological surprises in boreal lakes. Glob. Change Biol. 12, 2316–2322.

Costanza, R., d'Arge, R., de Groot, R., Farber, S., Grasso, M., Hannon, B., Limburg, K., Naeem, S., O'Neill, R.V., Paruelo, J., 1998. The Value of the World's Ecosystem Services and Natural Capital.

Daily, G.C., Matson, P.A., 2008. Ecosystem services: from theory to implementation. Proc. Natl. Acad. Sci. USA 105, 9455–9456.

Daufresne, M., Lengfellner, K., Sommer, U., 2009. Global warming benefits the small in aquatic ecosystems. Proc. Natl. Acad. Sci. USA 106, 12788–12793.

Demars, B.O.L., Russell Manson, J., Ólafsson, J.S., Gíslason, G.M., Gudmundsdóttir, R., Woodward, G., Reiss, J., Pichler, D.E., Rasmussen, J.J., Friberg, N., 2011a. Temperature and the metabolic balance of streams. Freshwater Biol. 56, 1106–1121.

Demars, B.O.L., Russell Manson, J., Olafsson, J.S., Gislason, G.M., Gudmundsdottír, R., Woodward, G.U.Y., Reiss, J., Pichler, D.E., Rasmussen, J.J., Friberg, N., 2011b. Temperature and the metabolic balance of streams. Freshwater Biol. 56, 1106–1121.

Dewson, Z.S., James, A.B.W., Death, R.G., 2007. Stream ecosystem functioning under reduced flow conditions. Ecol. Appl. 17, 1797–1808.

Dossena, M., Yvon-Durocher, G., Grey, J., Montoya, J.M., Perkins, D.M., Trimmer, M., Woodward, G., 2012. Warming alters community size structure and ecosystem functioning. Proc. Biol. Sci. Roy. Soc. 279, 3011–3019.

Durance, I., Ormerod, S.J., 2007. Climate change effects on upland stream macroinvertebrates over a 25-year period. Glob. Change Biol. 13, 942–957.

Durance, I., Ormerod, S.J., 2009. Trends in water quality and discharge confound long-term warming effects on river macroinvertebrates. Freshwater Biol. 54, 388–405.

Elton, C., 1927. Animal Ecology. Sidgwick and Jackson, London.

Feuchtmayr, H., Moss, B., Harvey, I., Moran, R., Hatton, K., Connor, L., Atkinson, D., 2010. Differential effects of warming and nutrient loading on the timing and size of the spring zooplankton peak: an experimental approach with hypertrophic freshwater mesocosms. J. Plankton Res. 32, 1715–1725.

Finlay, B.J., 2002. Global dispersal of free-living microbial eukaryote species. Science 296, 1061–1063.
Forster, J., Hirst, A.G., 2012. The temperature-size rule emerges from ontogenetic differences between growth and development rates. Funct. Ecol. 26, 483–492.
Forster, J., Hirst, A.G., Atkinson, D., 2012. Warming-induced reductions in body size are greater in aquatic than terrestrial species. Proc. Natl. Acad. Sci. USA 109, 19310–19314.
Friberg, N., Bonada, N., Bradley, D.C., Dunbar, M.J., Edwards, F.K., Grey, J., Hayes, R.B., Hildrew, A.G., Lamouroux, N., Trimmer, M., 2011. Biomonitoring of human impacts in freshwater ecosystems: the good, the bad and the ugly. Adv. Ecol. Res. 44, 1–68.
Friberg, N., Dybkjaer, J.B., Olafsson, J.S., Gislason, G.M., LARSEN, S.E., Lauridsen, T.L., 2009. Relationships between structure and function in streams contrasting in temperature. Freshwater Biol. 54, 2051–2068.
Gonzalez, A., Ronce, O., Ferriere, R., Hochberg, M.E., 2013. Evolutionary rescue: an emerging focus at the intersection between ecology and evolution. Phil.Trans. R. Soc. Biol. Sci. 368, 20120404.
Gray, C., Baumgartner, S., Jacob, U., Jenkins, G.B., O'Gorman, E.J., Lu, X., Ma, A., Pocock, M.J.O., Schuwirth, N., Thompson, M., Woodward, G., Baird, D.J., 2014. Ecological networks: the missing links in biomonitoring science. J. Appl. Ecol. 51, 1444–1449.
Greig, H.S., Kratina, P., Thompson, P.L., Palen, W.J., Richardson, J.S., Shurin, J.B., 2012. Warming, eutrophication, and predator loss amplify subsidies between aquatic and terrestrial ecosystems. Glob. Change Biol. 18, 504–514.
Hagen, M., Kissling, W.D., Rasmussen, C., Carstensen, D.W., Dupont, Y.L., Kaiser-Bunbury, C.N., O'Gorman, E.J., Olesen, J.M., De Aguiar, M.A.M., Brown, L.E., 2012. Biodiversity, species interactions and ecological networks in a fragmented world. Adv. Ecol. Res. 46, 89–120.
Hector, A., Bagchi, R., 2007. Biodiversity and ecosystem multifunctionality. Nature 448, 188–190.
Hildrew, A.G., Woodward, G., Winterbottom, J.H., Orton, S., 2004. Strong density dependence in a predatory insect: large-scale experiments in a stream. J. Anim. Ecol. 73, 448–458.
Hirst, A.G., Glazier, D.S., Atkinson, D., 2014. Body shape shifting during growth permits tests that distinguish between competing geometric theories of metabolic scaling. Ecol. Lett. 17, 1274–1281.
IPCC, 2007. Contribution of working groups I, II and III to the fourth assessment report of the Intergovernmental Panel on Climate Change. In: Pachauri, R.K., Reisinger, A. (Eds.), Intergovernmental Panel on …. IPCC, Geneva.
IPCC, 2013. Climate Change 2013: The Physical Science Basis. Working Group I Contribution to the Fifth Assessment Report of the Intergovernmental Panel on Climate Change. Summary for Policymakers (IPCC, 2013).
Jackson, M., Grey, J., 2013. Accelerating rates of freshwater invasions in the catchment of the River Thames. Biol. Invasions 15, 945–951.
Jeppesen, E., Moss, B., Bennion, H., Carvalho, L., De Meester, L., Feuchtmayr, H., Friberg, N., Gessner, M.O., Hefting, M., Lauridsen, T.L., 2010. Interaction of climate change and eutrophication. Clim. Change Impacts Freshwater Ecosyst. 119–151.
Jonsson, M., Malmqvist, B., 2003. Mechanisms behind positive diversity effects on ecosystem functioning: testing the facilitation and interference hypotheses. Oecologia 134, 554–559.
Lake, P.S., 2003. Ecological effects of perturbation by drought in flowing waters. Freshwater Biol. 48, 1161–1172.

Layer, K., Hildrew, A.G., Jenkins, G.B., Riede, J.O., Rossiter, S.J., Townsend, C.R., Woodward, G., 2011. Long-term dynamics of a well-characterised food web: four decades of acidification and recovery in the Broadstone Stream model system. Adv. Ecol. Res. 44, 69–117.

Ledger, M.E., Brown, L.E., Edwards, F.K., Milner, A.M., Woodward, G., 2012. Drought alters the structure and functioning of complex food webs. Nat. Clim. Change 3, 223–227.

Ledger, M.E., Brown, L.E., Edwards, F.K., Milner, A.M., Woodward, G., 2013. Drought alters the structure and functioning of complex food webs. Nat. Clim. Change 3, 223–227.

Leitch, A.R., Leitch, I.J., Trimmer, M., Guignard, M.S., Woodward, G., 2014. Impact of genomic diversity in river ecosystems. Trends Plant Sci. 19, 361–366.

Liboriussen, L., Lauridsen, T.L., Søndergaard, M., Landkildehus, F., Søndergaard, M., Larsen, S.E., Jeppesen, E., 2011. Effects of warming and nutrients on sediment community respiration in shallow lakes: an outdoor mesocosm experiment. Freshwater Biol. 56, 437–447.

Loreau, M., 2010. From Populations to Ecosystems: Theoretical Foundations for a New Ecological Synthesis (MPB-46). Princeton University Press.

McKee, D., Atkinson, D., Collings, S.E., Eaton, J.W., Gill, A.B., Harvey, I., Hatton, K., Heyes, T., Wilson, D., Moss, B., 2003. Response of freshwater microcosm communities to nutrients, fish, and elevated temperature during winter and summer. Limnol. Oceanogr. 48, 707–722.

McKie, B.G., Woodward, G., Hladyz, S., Nistorescu, M., Preda, E., Popescu, C., Giller, P.S., Malmqvist, B., 2008. Ecosystem functioning in stream assemblages from different regions: contrasting responses to variation in detritivore richness, evenness and density. J. Anim. Ecol. 77, 495–504.

Melián, C.J., Vilas, C., Baldó, F., González-Ortegón, E., Drake, P., Williams, R.J., 2011. Eco-evolutionary dynamics of individual-based food webs. Adv. Ecol. Res. 45, 225–268.

Moss, B., 2010. Climate change, nutrient pollution and the bargain of Dr Faustus. Freshwater Biol. 55, 175–187.

Moss, B., Kosten, S., Meerhof, M., Battarbee, R., Jeppesen, E., Mazzeo, N., Havens, K., Lacerot, G., Liu, Z., De Meester, L., 2011. Allied attack: climate change and eutrophication. Inland Waters 1, 101–105.

Moss, B., McKee, D., Atkinson, D., Collings, S.E., Eaton, J.W., Gill, A.B., Harvey, I., Hatton, K., Heyes, T., Wilson, D., 2003. How important is climate? Effects of warming, nutrient addition and fish on phytoplankton in shallow lake microcosms. J. Appl. Ecol. 40, 782–792.

Moya-Laraño, J., Bilbao-Castro, J.R., Barrionuevo, G., Ruiz-Lupión, D., Casado, L.G., Montserrat, M., Melián, C.J., Magalhães, S., 2014. Eco-evolutionary spatial dynamics: rapid evolution and isolation explain food web persistence. Adv. Ecol. Res. 50, 75–144.

O'Gorman, E.J., Benstead, J.P., Cross, W.F., Friberg, N., Hood, J.M., Johnson, P.W., Sigurdsson, B.D., Woodward, G., 2014. Climate change and geothermal ecosystems: natural laboratories, sentinel systems, and future refugia. Glob. Change Biol. 20, 3291–3299.

O'Gorman, E.J., Pichler, D.E., Adams, G., Benstead, J.P., Cohen, H., Craig, N., Cross, W.F., Demars, B.O.L., Friberg, N., Gislason, G.M., 2012. Impacts of Warming on the Structure and Functioning of Aquatic Communities: Individual-to Ecosystem-level Responses.

Odum, H.T., Odum, E.P., 2000. The energetic basis for valuation of ecosystem services. Ecosystems 3, 21–23.

Parmesan, C., Yohe, G., 2003. A globally coherent fingerprint of climate change impacts across natural systems. Nature 421, 37–42.

Perkins, D.M., Bailey, R.A., Dossena, M., Gamfeldt, L., Reiss, J., Trimmer, M., Woodward, G., 2014. Higher biodiversity is required to sustain multiple ecosystem processes across temperature regimes. Glob. Change Biol. 21, 396–406.

Perkins, D.M., McKie, B.G., Malmqvist, B., Gilmour, S., Reiss, J., Woodward, G., 2010. Environmental warming and biodiversity—ecosystem functioning in freshwater microcosms: partitioning the effects of species identity, richness and metabolism. Adv. Ecol. Res. 43, 177—209.

Perkins, D.M., Yvon-Durocher, G., Demars, B.O.L., Reiss, J., Pichler, D.E., Friberg, N., Trimmer, M., Woodward, G., 2012. Consistent temperature dependence of respiration across ecosystems contrasting in thermal history. Glob. Change Biol. 18, 1300—1311.

Petchey, O.L., McPhearson, P.T., Casey, T.M., Morin, P.J., 1999. Environmental warming alters food-web structure and ecosystem function. Nature 402, 69—72.

Raffaelli, D., 2004. How extinction patterns affect ecosystems. Science 306, 1141—1142.

Raffaelli, D., White, P.C.L., 2013. Ecosystems and their services in a changing world: an ecological perspective. Adv. Ecol. Res. 48, 1—70.

Rall, B.C., Vucic-Pestic, O., Ehnes, R.B., Emmerson, M., Brose, U., 2010. Temperature, predator—prey interaction strength and population stability. Glob. Change Biol. 16, 2145—2157.

Reiss, J., Bailey, R.A., Cássio, F., Woodward, G., Pascoal, C., 2010. Assessing the contribution of micro-organisms and macrofauna to biodiversity-ecosystem functioning relationships in freshwater microcosms. Adv. Ecol. Res. 43, 151—176.

Reiss, J., Bridle, J.R., Montoya, J.M., Woodward, G., 2009. Emerging horizons in biodiversity and ecosystem functioning research. Trends Ecol. Evol. 24, 505—514.

Reuman, D.C., Holt, R.D., Yvon-Durocher, G., 2014. A metabolic perspective on competition and body size reductions with warming. J. Anim. Ecol. 83, 59—69.

Rooney, N., McCann, K., Gellner, G., Moore, J.C., 2006. Structural asymmetry and the stability of diverse food webs. Nature 442, 265—269.

Rouse, W.R., Douglas, M.S.V., Hecky, R.E., Hershey, A.E., Kling, G.W., Lesack, L., Marsh, P., McDonald, M., Nicholson, B.J., Roulet, N.T., 1997. Effects of climate change on the freshwaters of arctic and subarctic North America. Hydrol. Process. 11, 873—902.

Scheffer, M., Carpenter, S., Foley, J.A., Folke, C., Walker, B., 2001. Catastrophic shifts in ecosystems. Nature 413, 591—596.

Schmid, B., Balvanera, P., Cardinale, B.J., Godbold, J., Pfisterer, A.B., Raffaelli, D., Solan, M., Srivastava, D.S., 2009. Consequences of species loss for ecosystem functioning: meta-analyses of data from biodiversity experiments. In: Naeem, S., Bunker, D.E., Hector, A., Loreau, M., Perrings, C. (Eds.), Biodiversity, Ecosystem Functioning and Human Wellbeing: An Ecological and Economic Perspective, pp. 14—29.

Sheridan, J.A., Bickford, D., 2011. Shrinking body size as an ecological response to climate change. Nat. Clim. Change 1, 401—406.

Stewart, R.I.A., Dossena, M., Bohan, D.A., Jeppesen, E., Kordas, R.L., Ledger, M.E., Meerhoff, M., Moss, B., Mulder, C., Shurin, J.B., 2013. Mesocosm experiments as a tool for ecological climate-change research. Adv. Ecol. Res. 48, 71—181.

Thackeray, S.J., Sparks, T.H., Frederiksen, M., Burthe, S., Bacon, P.J., Bell, J.R., Botham, M.S., Brereton, T.M., Bright, P.W., Carvalho, L., 2010. Trophic level asynchrony in rates of phenological change for marine, freshwater and terrestrial environments. Glob. Change Biol. 16, 3304—3313.

Thompson, R.M., Beardall, J., Beringer, J., Grace, M., Sardina, P., 2013. Means and extremes: building variability into community-level climate change experiments. Ecol. Lett. 16, 799—806.

Thomson, J.R., Bond, N.R., Cunningham, S.C., Metzeling, L., Reich, P., Thompson, R.M., Mac Nally, R., 2012. The influences of climatic variation and vegetation on stream biota: lessons from the Big Dry in southeastern Australia. Glob. Change Biol. 18, 1582—1596.

Travis, J., Reznick, D., Bassar, R.D., López-Sepulcre, A., Ferriere, R., Coulson, T., 2014. Do eco-evo feedbacks help us understand nature? Answers from studies of the Trinidadian guppy. Adv. Ecol. Res 50, 1–40.

Veraart, A.J., De Klein, J.J.M., Scheffer, M., 2011. Warming can boost denitrification disproportionately due to altered oxygen dynamics. PLoS One 6, e18508.

Vörösmarty, C.J., McIntyre, P.B., Gessner, M.O., Dudgeon, D., Prusevich, A., Green, P., Glidden, S., Bunn, S.E., Sullivan, C.A., Liermann, C.R., 2010. Global threats to human water security and river biodiversity. Nature 467, 555–561.

Walther, G.-R., 2010. Community and ecosystem responses to recent climate change. Phil.Trans. R. Soc. Biol. Sci. 365, 2019–2024.

White, E.P., Ernest, S.K.M., Kerkhoff, A.J., Enquist, B.J., 2007. Relationships between body size and abundance in ecology. Trends Ecol. Evol. 22, 323–330.

Woodward, G., Benstead, J.P., Beveridge, O.S., Blanchard, J., Brey, T., Brown, L.E., Cross, W.F., Friberg, N., Ings, T.C., Jacob, U., 2010a. Ecological networks in a changing climate. Adv. Ecol. Res. 42, 71–138.

Woodward, G., Dybkjaer, J.B., Ólafsson, J.S., Gíslason, G.M., Hannesdóttir, E.R., Friberg, N., 2010b. Sentinel systems on the razor's edge: effects of warming on Arctic geothermal stream ecosystems. Glob. Change Biol. 16, 1979–1991.

Woodward, G., Ebenman, B., Emmerson, M., Montoya, J.M., Olesen, J.M., Valido, A., Warren, P.H., 2005. Body size in ecological networks. Trends Ecol. Evol. 20, 402–409.

Woodward, G., Hildrew, A., 2002. Food web structure in riverine landscapes. Freshwater Biol. 777–798.

Woodward, G., Jones, J., Hildrew, A., 2002. Community persistence in Broadstone Stream (U.K.) over three decades. Freshwater Biol. 1419–1435.

Woodward, G., Papantoniou, G., Lauridsen, R.B., 2008. Trophic trickles and cascades in a complex food web: impacts of a keystone predator on stream community structure and ecosystem processes. Oikos 117, 683–692.

Woodward, G., Perkins, D.M., Brown, L.E., 2010c. Climate change and freshwater ecosystems: impacts across multiple levels of organization. Philos. Trans. R. Soc. Lond. Ser. B Biol. Sci. 365, 2093–2106.

Yvon-Durocher, G., Allen, A.P., Montoya, J.M., Trimmer, M., Woodward, G., 2010a. The temperature dependence of the carbon cycle in aquatic ecosystems. Adv. Ecol. Res. 43, 267–313.

Yvon-Durocher, G., Caffrey, J.M., Cescatti, A., Dossena, M., del Giorgio, P., Gasol, J.M., Montoya, J.M., Pumpanen, J., Staehr, P.A., Trimmer, M., 2012. Reconciling the temperature dependence of respiration across timescales and ecosystem types. Nature 487, 472–476.

Yvon-Durocher, G., Jones, J.I., Trimmer, M., Woodward, G., Montoya, J.M., 2010b. Warming alters the metabolic balance of ecosystems. Philos. Trans. R. Soc. Lond. Ser. B Biol. Sci. 365, 2117–2126.

Yvon-Durocher, G., Montoya, J.M., Trimmer, M., Woodward, G., 2011a. Warming alters the size spectrum and shifts the distribution of biomass in freshwater ecosystems. Glob. Change Biol. 17, 1681–1694.

Yvon-Durocher, G., Montoya, J.M., Woodward, G., Jones, I.J., Trimmer, M., 2011b. Warming increases the proportion of primary production emitted as methane from freshwater mesocosms. Glob. Change Biol. 17, 1225–1234.

Chapter 7

Global Aquatic Ecosystem Services Provided and Impacted by Fisheries: A Macroecological Perspective

Jonathan A.D. Fisher[1], Kenneth T. Frank[2] and Andrea Belgrano[3,4]

[1]*Centre for Fisheries Ecosystems Research, Fisheries and Marine Institute of Memorial University of Newfoundland, St. John's, NL, Canada;* [2]*Department of Fisheries and Oceans, Bedford Institute of Oceanography, Dartmouth, NS, Canada;* [3]*Department of Aquatic Resources, Institute of Marine Research, Swedish University of Agricultural Sciences, Lysekil, Sweden;* [4]*Swedish Institute for the Marine Environment (SIME), Göteborg, Sweden*

INTRODUCTION

Two and a half decades ago, macroecology was initially characterized as the study of the division of food and space among species on continents (Brown and Maurer, 1989). Since then, macroecological research has been expanded to include the study of relationships between organisms and their environments by characterizing and explaining patterns of abundance, distribution, and diversity (Brown, 1995; Gaston and Blackburn, 2000). Importantly, macroecology was meant to counter what was seen as increasingly reductionist and specialized approaches in ecology and to examine the roles and interactions among ecological and evolutionary explanations for large-scale patterns using integrated analyses (Brown and Maurer, 1989; Brown, 1995; Lawton, 1996; Blackburn and Gaston, 1999). Macroecology is necessarily built on advances in multiple fields of ecological and evolutionary biology (see Smith et al., 2014) and macroecological investigations occur at spatial scales increasingly relevant to biodiversity and ecosystem functioning research (Naeem, 2006), global change dynamics (Kerr et al., 2007), and fisheries (Witman and Roy, 2009; Fowler et al., 2013).

In fisheries research, despite the domination of reductionist approaches focused on the dynamics of single populations, a range of research approaches has also persisted. For example, some fisheries and aquatic scientists have long advocated for holistic, empirical, and large-scale comparative approaches that parallel the macroecological approach in terrestrial systems:

> *No doubt the reductionist approach appears more realistic. However, if it is the only way to proceed, it is questionable whether we will ever have enough information to model a lake or marine ecosystem well enough to answer such 'simple' questions as to what is the economically or biological optimal harvesting strategy of multispecies fishery? The only hope appears to be in the ability to find some abiotic and ecological master variables that govern the behavior of the whole system.*
>
> Paloheimo and Regier (1982).

Given such calls to identify the key drivers of patterns in complex aquatic systems, and similar calls echoed from analyses of multispecies, tropical fisheries (Jones, 1982; Marten and Polovina, 1982), macroecological research has recently been greatly extended into marine and freshwater systems in order to document, understand, and predict large-scale patterns of aquatic biodiversity, fisheries yields, and ecosystem functioning (Belgrano, 2004; Witman and Roy, 2009; Fowler et al., 2013). In combination, linking the key macroecological variables (Figure 1) in order to understand not only patterns but also processes, marine (27%) and limnic (8%) studies now contribute more than a third of all published macroecological research papers (Beck et al., 2012). This burgeoning growth can be attributed to a virtual explosion of large databases incorporating aquatic species' abundances, geographical distributions, key life history traits, together with environmental conditions, all of which have facilitated examinations of macroecological patterns of aquatic biodiversity (e.g., Worm et al., 2005; Abel et al., 2008; Sherman et al., 2009; Tisseuil et al., 2013). Remote sensing data have played an important role in this rapid expansion of research and have been used to further test and develop ecological theory including the relationship between size and abundance

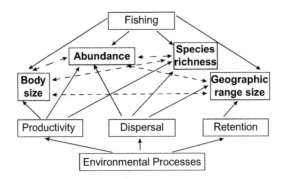

FIGURE 1 Schematic representation of key macroecological variables (in bold) linked to the climate and aquatic processes that drive their natural variability and also the effects of fishing. (*After Leichter and Witman (2009)*.) Solid lines represent driver–response relationships, while dashed lines illustrate the many ways that macroecological variables may be linked and portrayed.

(e.g., Jennings et al., 2008a), between biodiversity and functioning and resilience of fisheries ecosystems (e.g., Worm et al., 2006, 2009; Coll et al., 2008), and between the distribution and functional roles of trait variation within and among aquatic ecosystems (Hildrew et al., 2007; Fisher et al., 2010a,b; Barton et al., 2010, 2013). In the context of linking large-scale patterns with a greater understanding of ecosystem functioning, it is notable that building on the successes of the decade-long Census of Marine Life (a program designed to quantify the diversity of life in the seas), the next phase seeks to address questions of the functional roles of species (Pennisi, 2011).

Within these aquatic ecosystems that cover more than 70% of the globe and provide the overwhelming majority of Earth's habitable space, one key ecosystem function is the maintenance of fisheries yields, which has direct implications for food security and economies from local to global scales (Hilborn et al., 2003; Pauly et al., 2002, 2005). Globally, capture fisheries landings have leveled off around 90 million metric tons per year with approximately 89% derived from marine systems; aquaculture expansion provides an additional 55 million tons annually, primarily from freshwater ecosystems (FAO, 2012) (Figure 2). Capture fishery demands on primary productivity now exceed 6×10^9 tons globally (Watson et al., 2013), such that primary productivity increasingly constrains fisheries yields at higher trophic levels (Chassot et al., 2010). In a goods and services context, fish and fisheries productions provide 17% of animal protein to human populations and provide 3 billion people with almost 20% of their animal protein intake (FAO, 2012). In economic terms, capture fisheries alone generate gross revenues in the range of $80–85 billion annually, while employment (nearly 8% of the world's

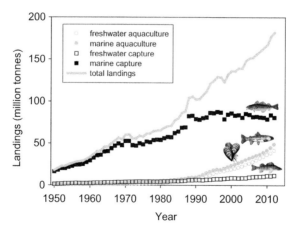

FIGURE 2 Global annual capture fisheries and aquaculture landings values from marine and freshwater systems. (*Data were derived from the Food and Agriculture Organization of the United Nations fisheries statistical collections via www.fao.org/fishery/statistics/global-production/en.*) These time series illustrate the relative contributions of four types of removals from aquatic systems and the coincident leveling off of marine capture fisheries and growth in marine and freshwater aquaculture c. 1990, with aquaculture driving the growth of total landings since 1990.

population), processing, and other services increase these monetary contributions by almost threefold (Sumaila et al., 2011).

Given the scale and magnitude of global fisheries, their impact has had both direct and indirect effects on aquatic ecosystem structure and functioning (Jackson et al., 2001) and has produced a profound global footprint (e.g., Hilborn et al., 2003; Dudgeon et al., 2006; Olden et al., 2007; Coll et al., 2008; Worm et al., 2006; Mora et al., 2011). For example, Halpern et al. (2008) quantified the scope and magnitude of cumulative human influences on marine systems and reported that impacts on the oceans were greatest in coastal and continental shelf systems. Further, while fishing effects exhibited a smaller spatial extent than climate change impacts, the combination of direct and indirect effects of fishing was equal to climate threats such as ocean acidification, directional temperature change, and ultraviolet radiation (Halpern et al., 2008). The indirect effects of fishing and their operative spatial scales have also been increasingly documented and linked to ecosystem functioning in both freshwater (Post et al., 2002; Dudgeon et al., 2006; Olden et al., 2007) and marine systems (Essington et al., 2006; Worm et al., 2006; Coll et al., 2008; Tittensor et al., 2009). Fisheries may even influence biogeochemical cycles (e.g., Katz et al., 2009; Wilson et al., 2009) and evolutionary patterns (Darimont et al., 2009; Fraser, 2013; Belgrano and Fowler, 2013), making increased understanding of connections among diversity, functioning, yields, and evolutionary impacts within these systems of paramount importance.

The expanded spatial and temporal scales at which these fisheries-induced changes in aquatic ecosystems are now documented have led to the application of macroecological approaches to understand the differential dynamics, resilience, and yields of fisheries ecosystems. In the context of understanding the complex links between biodiversity and ecosystem services at large scales, Raffaelli (2006) argued that ecologists should focus on ecosystem services and then work back to uncover processes linking those services to biodiversity. The largely applied nature of fisheries ecology coupled with systematic monitoring (in many developed nations) of ecosystem services such as landings, size structure, recruitment, and species spatial distributions suggests that the services-to-diversity direction should be the way the macroecological approach is applied to understand fisheries ecosystems. This is further necessitated by a focus on quantifying and improving the status of fisheries resources in the face of changing patterns of exploitation and biodiversity and differs somewhat from the perspective that preserving "diversity" should be a primary focus, with benefits derived thereafter (e.g., Palumbi et al., 2009).

Today, the macroecological perspective is increasingly being applied to research questions that directly link the functioning of aquatic ecosystems to fisheries yields and do so with an appreciation of both the macroecological dynamics of aquatic ecosystems (Fisher et al., 2010a,b) and the evolutionary constraints and consequences of overexploitation (Stergiou, 2002; Darimont et al., 2009; Fowler et al., 2013; Fraser, 2013). In this chapter, we first provide an overview of dominant macroecological patterns within aquatic ecosystems and their connections to fisheries productivity and dynamics, followed by an exploration of a central challenge in macroecology—identifying processes underlying patterns at inherently nonexperimental spatial scales. We then illustrate recent advances in trait-based approaches to understand fisheries ecosystem functioning. Finally, we outline the dominant eco-evolutionary effects of trait-specific selective fisheries on the diversity and functioning of aquatic ecosystems.

MACROECOLOGICAL VARIABLES AND THEIR INTERACTIONS WITHIN AQUATIC ECOSYSTEMS

The identification of aquatic macroecological patterns and their significance in the context of fisheries ecosystems has been realized for more than a century, though couched in somewhat different terms than those used today. For example, in his prescient section "The productivity of the sea in different latitudes," Johnstone (1908) described biogeographical patterns in fisheries landings and noted that tropical locations characterized by high species diversity were not, in fact, the major productive fisheries areas. Rather, the most productive regions were at higher latitudes characterized by lower fish diversity but larger-bodied fish species. Johnstone's observations of such spatial gradients represented one of the key variables in macroecological patterns (Figure 1) and their contributions to spatially variable fisheries productivity. Such observations capture one of the key overarching patterns, namely that the distribution of fisheries production is not uniform across the globe but rather quite heterogenous with 80—90% of landings biomass (Sherman et al., 2009) or 50% of total fish biomass (Jennings et al., 2008a) occurring within <25% of the global ocean, predominantly along upwelling regions and coastal shelves. Given such spatial variation and dependencies, explaining even "simple" fourfold differences in pelagic fishery catch rates between two sides of an ocean basin (e.g., Trenkel et al., 2014) would benefit from examinations of relationships within and among key macroecological variables including body sizes, range sizes, abundance, and their relationships to productivity and dispersal (Figure 1).

Fortunately, some of the most extensively quantified macroecological variables (i.e., abundance, geographic range, body size, species richness) are readily available due to the systematic monitoring of exploited fish species, some of which serve as basic inputs to fisheries stock assessments and to the evolving objective of developing ecosystem assessments (Levin et al., 2009; Link, 2010). Explanations for the differential rates of productivity also remain active areas of research involving an analysis of the linkages between ecological and evolutionary processes to patterns of aquatic diversity and dynamics of ecosystem functioning (Jennings et al., 2008a; Huston and Wolverton, 2009, 2011; Barton et al., 2013). In the context of biodiversity and ecosystem functioning research, since fisheries have the capacity to alter multiple macroecological variables simultaneously, the original objectives of macroecology investigating the division of food and space are relevant for adjustments made by populations, communities, and ecosystems (Figure 3). Four key macroecological variables are examined separately as well as their interaction, ordered simply in decreasing aggregation from properties of communities, to populations, and individuals.

Species Richness

Given its central role in many "biodiversity" studies, species richness is one diversity component that is relatively easily calculated and often analyzed. As such, it has been correlated with the functional characteristics (e.g., yield, productivity, collapsed taxa) of many large marine ecosystems (Worm et al., 2006). Also because of its prevalence, some of the most general aquatic macroecological patterns concern relationships involving species richness and area, latitude, and depth. For example, species area relationships (SARs) are one such pattern, where higher numbers of species tend to be found in systems characterized by larger areas. Cross-system comparisons among lakes

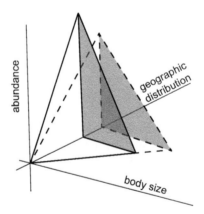

FIGURE 3 Schematic relationships among geographic distribution, population abundance, and body sizes among species within aquatic communities at two points in time (dashed and solid lines). It is assumed that a common, fixed amount of energy is available to be divided among community members. This limitation leads to a negative relationship between abundance and body size (depicted as two gray areas of equal size), while positive relationships are evident between body size and geographic distribution (larger species have larger home ranges) and distribution and abundance (more abundant species occupy a larger area). The dashed line represents a point in time in which the community exhibited the greatest range of body sizes and therefore lower abundances at a given geographic distribution extent. The solid lines show a different division of resources, where smaller bodied species have higher abundances, leading to a steeper abundance–distribution relationship. These schematic relationships stem from theoretical linkages between macroecological patterns (see Blackburn and Gaston, 2001) and empirical patterns within aquatic communities (Fisher and Frank, 2004).

(Barbour and Brown, 1974), coral reefs (Tittensor et al., 2007), and offshore marine fishing banks (Frank and Shackell, 2001) have revealed that aquatic SARs share similar patterns with those from terrestrial systems (Rosenzweig, 1995). Further, meta-analyses among hundreds of studies have revealed that SAR steepness is generally higher at low latitudes and within communities composed of large body-sized species (Drakare et al., 2006). In an ecosystem functioning context, whether the slopes of SAR curves vary systematically through time or space as a function of exploitation provides potential diagnostics of exploitation effects (Blanchard, 2001; Tittensor et al., 2007). As in terrestrial systems (Rosenzweig, 1995), aquatic SARs have relevance for the conservation of local populations and the distribution, locations, and configuration of aquatic protected areas (Niegel, 2003).

Species richness in many systems also varies markedly with latitude and longitude, with much emphasis on the so-called latitudinal diversity gradient. This description is somewhat misleading because the diversity gradient is not caused by latitude per se, but instead by such factors as temperature, energy availability, habitat area, species interactions, and rates of evolution. Based on meta-analyses among aquatic and terrestrial systems, Hillebrand (2004) reported that latitudinal diversity gradients were weaker and less steep in freshwater than in marine systems, whereas larger-bodied, or higher trophic-level species tended to exhibit steeper latitudinal gradients in both freshwater and marine systems. Heino (2011) identified the scale-dependent influences on freshwater LDGs, whereas Fisher and Frank (2013) reviewed the extent to which species groups ranging from bacteria to whales conform to, or run counter to, this pattern in marine systems.

Water temperature remains the most widely documented contemporary correlate of marine species richness (Tittensor et al., 2010). Across three large marine ecosystems, Fisher et al. (2008) demonstrated that interannual variation in water temperature induced by environmental forcing drove rapid changes in the relative steepness of an Atlantic marine fish latitudinal diversity gradient. This confirmed the role of water temperature as an influential, dynamic driver of a contemporary macroecological pattern. Within freshwater systems, global fish species richness is best explained by habitat area and environmental heterogeneity (Guegan et al., 1998; Tisseuil et al., 2013), while taxonomic patterns of other global freshwater species richness are primarily associated with climate/productivity factors (Tisseuil et al., 2013). A marked longitudinal pattern also exists for marine fishes, with highest diversity areas centered on the Indo-Pacific and Caribbean Atlantic regions, while centers of origin for cold temperate species are associated with the North Pacific and Antarctic (Briggs, 2007). In combination, these areas represent important evolutionary centers from which much species dispersal and functional differentiation originated (Mora et al., 2003; Briggs, 2007).

Analogous to the inverse relationship between species richness and altitude in terrestrial systems (Nogués-Bravo et al., 2008), changes in species richness with depth are patterns that have commanded the attention of deep-sea biologists and fisheries oceanographers for several decades. Macpherson and Duarte (1994) illustrated that northeast Atlantic fish diversity showed a peak at intermediate depths, while Smith and Brown (2002) reported similar patterns within Pacific pelagic fish assemblages. These patterns, however, are not independent of species traits such as body size, or latitude (Macpherson and Duarte, 1994). As with the species—area patterns, the link between latitudinal and depth gradients and ecosystem functioning is based on the assumption that species richness per se provides an informative metric of local biodiversity, with more species equating to more potential for functional compensation to buffer potential biodiversity losses (e.g., Worm et al., 2006). However, given the multiple ways in which taxonomic diversity may be related to functional diversity (see Naeem and Wright, 2003), species richness alone may be less informative as a diversity metric than when trait diversity information is also included.

Abundance

Globally, fish abundances and fisheries catches (most often reported as biomass: Figure 2) are ultimately constrained by available primary production, which itself shows striking spatial variation across aquatic habitats (Marten and Polovina, 1982; Nixon, 1988). A synthesis of these cross-system differences has been expressed in a recent relationship linking satellite-derived ocean color data and landings data, revealing that the global fisheries are being increasingly constrained by primary productivity (Chassot et al., 2010; Watson et al., 2013). Within systems, aquatic organism abundances are also strongly and negatively related to body sizes (Duarte et al., 1987; White et al., 2007; see also Figure 3). Against this background of abundance limitations, the frequency distribution of abundances among marine species ("species abundance distribution": SAD) displays a predominantly lognormal SAD, similar to that observed in many terrestrial systems (Gray et al., 2006). Comparing the shape of such distributions has long been used to test one or more of the many hypotheses proposed to underlie SADs (McGill and Nekola, 2010; Connolly et al., 2014).

In the context of fisheries ecosystem functioning, the form of the SAD is also related to the relative dominance of individual species. High relative dominance of a few species, characteristic of temperate and sub-Arctic systems, supported the establishment of many single-species-dominant fisheries there (Johnstone, 1908). It is notable that species dominance has been identified as a key diversity characteristic, one expected to respond more rapidly to anthropogenic forcing than, for example, species richness (Hillebrand et al., 2008). Dominance is also a characteristic that highlights the differential importance of a species when dominants monopolize resources and have larger, possibly ecosystem-level implications when those species decline (see Hillebrand et al., 2008). However, given the relatively recent quantitative treatment of the implications of different patterns of dominance in the context of fisheries resilience, this is an area that is ripe for new ideas and analyses.

The general pattern of declining abundance with increasing depth is another feature of freshwater (Jones, 1982) and marine ecosystems (Marten and Polovina, 1982; Rex et al., 2006), such that in the marine realm, the largest benthic areas are also among the least productive (Norse et al., 2012). These patterns appear to be related to declines in food availability at depth. Rex et al. (2006) illustrated the generality of this pattern. Abundance declined with depth within diverse taxonomic groups, with declines in local abundance in three out of four benthic groups (meiofauna, macrofauna, megafauna; bacteria were the exception) across 10 orders of magnitude in abundance at depths ranging to 6000 m. Similar trends existed when biomass rather than abundance was evaluated, but there was much greater overlap in biomass among benthic groups (Rex et al., 2006). These contrasting patterns illustrate the need for biodiversity and ecosystem functioning (B-EF) studies to consider the most appropriate metric (either abundance or biomass) as well as characterization of the habitat (three-dimensional in aquatic systems). In addition to these univariate patterns, given the associations between macroecological variables, local abundance has been linked to geographical distribution and body sizes (e.g., Figure 3). Those relationships are described in the following sections.

Geographical Distribution

When considering aquatic macroecological patterns involving geographic distributions, it is notable to first highlight some differences and similarities with terrestrial ecology. In comparative analyses, lakes and rivers have long been characterized as analogous to "islands" in a terrestrial matrix, while ocean basins have been considered more like "continents" (e.g., Magnusson, 1988). These perceived differences have implications for the relative roles of isolation (higher in freshwater systems), extinction dynamics (higher in freshwater systems), and connectivity (higher in marine systems) in structuring large-scale patterns (Magnusson, 1988). However, large-scale analyses of marine dispersal estimates have illustrated the importance of taxonomic differences, geographic location, depth, body size, and egg types in regulating marine dispersal rates (Bradbury et al., 2008), while population genetic analyses illustrate the restricted spatial scales at which even widespread, broadcast spawners show a fine scale population structure (Hauser and Carvalho, 2008; Bradbury et al., 2013). Taken together, these emerging constraints on marine geographic distributions suggest a shift in thinking about marine systems toward the importance of measuring and characterizing isolation (Hauser and Carvalho, 2008) and the effects of fisheries in eroding population spatial structure (Ciannelli et al., 2013). Within freshwater systems, the unique signatures of isolation and connectivity resulting from the last glacial maximum account for much of the differences in diversity and turnover among systems, where basins that had been connected during that time now feature higher species richness but lower endemisim and species turnover among water bodies (Dias et al., 2014). Hence, in the context of Magnusson's (1988) depictions, it is necessary to not only examine current connectivity among freshwater "islands" but also to understand the historical processes that influence contemporary patterns.

Similar to the species abundance distributions described above, the frequency distributions of species' geographic ranges are often dominated by species with relatively narrow geographic ranges, with comparatively few species geographically widespread (Brown, 1995; Gaston and Blackburn, 2000). Among global comparisons of freshwater and marine fishes, these distributions conform to this pattern and show striking similarities in their form (Strona et al., 2012). In both systems, body size was positively correlated with metrics of geographic range size, while traits including trophic level had more of a variable association with geographic range size (Strona et al., 2012).

Patterns commonly reported based on geographic ranges are positive relationships between local abundance and geographic range (based on presence/absence or calculated occupancy). Such bivariate abundance–range relationships are reported as "interspecific" patterns among species or as "intraspecific" patterns for one species, often through time (Gaston et al., 2000; Blackburn et al., 2006). In both cases, species (or time periods) of high local occupancy are also characterized by widespread geographic distributions and both patterns have implications for fisheries. During times of declining abundance and geographic distribution in the intraspecific case, the nonrandom search behavior of fishers may make stocks susceptible to sudden collapse as the few remaining concentrations sustain high catch rates until the last concentration is depleted. This is one reason why catch rate statistics provide inaccurate assessments of stock abundance (so-called hyperstability of catch rates), particularly for species that conform to the abundance–range relationship (reviewed by Fisher and Frank, 2004). In the interspecific case, observed shifts in species relative positions in this relationship—driven largely by fishing—have been sufficient to change the slopes of marine fish abundance–range relationships (Fisher and Frank, 2004; Frisk et al., 2011). One hypothesis to explain the steepening slopes is that reductions in body size associated with overfishing results in a greater number of individuals occupying a given area and this, in turn, causes a steeper slope—an example illustrating the linkages among abundance, geographic ranges, and body sizes is seen in Figure 3.

In addition to this proposed mechanistic relationship involving a greater packing of individuals per unit area with declining body size, Borregaard (2010) reviewed the more than a dozen mechanisms proposed to underlie abundance–distribution relationships and mapped the potential links between the different types of relationships (intraspecific and interspecific) and their multiple causes. Rindorf and Lewy (2012) examined the consequences of applying different indices of geographic distribution and provided recommendations for fisheries researchers interested in applying these methods. It is also notable that the proposed differences noted by Magnusson (1988) between freshwater and marine ecosystems due to contrasting connectivity may contribute to reported differences in abundance—range relationships and latitudinal diversity gradients described in this section. For example, in their meta-analysis, Blackburn et al. (2006) reported that freshwater abundance—range relationships tended to be relatively weak compared to patterns derived from marine systems, which they suggested provided some support for the hypothesis that limited dispersal leads to compositional differences. Similarly, as noted above, Hillebrand (2004) reported weaker and less steep latitudinal diversity gradients in freshwater systems, which could result from their relative isolation or their smaller areas, or smaller body sizes of inhabitants—which also illustrates the multiple potential mechanisms underlying many macroecological relationships.

Body Size

In terrestrial and aquatic ecosystems, body size has long been considered a key trait, given the relationships between size and constraints on strength, speed, diffusive processes, complexity, rates of living processes, and abundance (Peters, 1983; Bonner, 2006; Blanchard, 2011). Given that ontogenetic development in the majority of aquatic species progresses from eggs through larval, juvenile, and adult sizes, often spanning orders of magnitude, aquatic food web interactions are often constrained more by body size differences among individuals rather than species-level characteristics. Raffaelli (2007) cautioned that focusing exclusively on species-level descriptions of this key functional trait limits some studies that do not account for extensive intraspecific variation in sizes and termed this reliance on taxonomic identifiers alone as the "curse of the Latin binomial." This has led to the important and remarkably consistent characterization of size-structured food webs and size-dependent predator—prey interactions within and among aquatic systems (Hildrew et al., 2007). Such size-structuring is one of the features of aquatic ecosystems that show remarkably consistent patterns at large scales; such functional similarities link fisheries productivity, and aquatic ecosystem functioning at large spatial scales. Among global aquatic systems, body size has been shown to be strongly associated with extinction vulnerability in freshwater ($n = 207$ species) and marine ($n = 50$) fishes that were listed as "endangered" or "critically endangered" by the International Union for the Conservation of Nature (Olden et al., 2007). However, while the listed freshwater species tend to be skewed toward small-bodied species, the opposite is true within marine systems, where extinction risk increases with fish size (Olden et al., 2007). Against this background of potential species loss, the addition of nonnative fishes within freshwater systems has altered biogeographic patterns of size variation in some regions, given the addition of predominantly large-bodied, nonnative species (Blanchet et al., 2010).

Relationships between body size and abundance can take many forms but are often inversely related among members of ecological assemblages (White et al., 2007). For example, Figure 4 provides an overview of the importance and variability in the relationship between body sizes and abundance or "abundance size spectra," where steeper slopes in this relationship for planktonic organisms indicate a lack of large-bodied plankton (Figure 4(a)). The slopes of such relationships have been examined empirically among systems of varying productivity and revealed that differences in the abundance of large zooplankton alter these relationships markedly (Sprules and Munawar, 1986). These slopes have also been modeled at a global scale and indicate that within marine systems, North Atlantic and subarctic plankton communities tend to have an abundance of large-bodied individuals and species, while areas such as mid-ocean gyres tend to be dominated by small-bodied organisms (Barton et al., 2013; Figure 4(b)). These findings show a remarkable degree of variation across the world's oceans and are also notable in relation to independent findings from higher levels of the food chain. In a separate investigation of marine fish size structure across the world's large marine ecosystems, species-level mean fish sizes were also highest in the North Atlantic (Figure 4(c)). This close association deserves more study to determine whether a common driver and/or trophic linkage contributes to these similar body size patterns among marine taxa that span much of the range of observed body sizes. As a key functional trait, body sizes of aquatic organisms have received increasing treatment in linking macroecological characteristic to ecosystem functioning.

A CENTRAL CHALLENGE: IDENTIFYING PROCESSES UNDERLYING MACROECOLOGICAL PATTERNS

Ever since macroecology was considered a distinct branch of ecological study, it has progressed through several stages. It first attempted to more fully document existing patterns, which was followed by the development of novel approaches to the study of the derived patterns and linking them to mechanisms that may operate at somewhat smaller spatial scales. Finally, macroecological investigations have led to the discovery and documentation of new patterns (Brown, 1999; Blackburn and Gaston, 1999). The field of biodiversity ecosystem functioning research has undergone a similar evolution but at expanding spatial scales, with researchers also calling for larger-scale investigations in order to test the generality of results derived largely from small-scale experimental investigations (Naeem, 2006; Raffaelli, 2006; Kerr et al., 2007). Multiple avenues have been advocated and explored to better link macroecological patterns to underlying processes (Brown, 1999; Blackburn and Gaston, 1999; McGill, 2003; Blackburn, 2004; Fisher et al., 2010c; McGill and Nekola, 2010; Beck et al., 2012). Here we examine a subset of three empirical

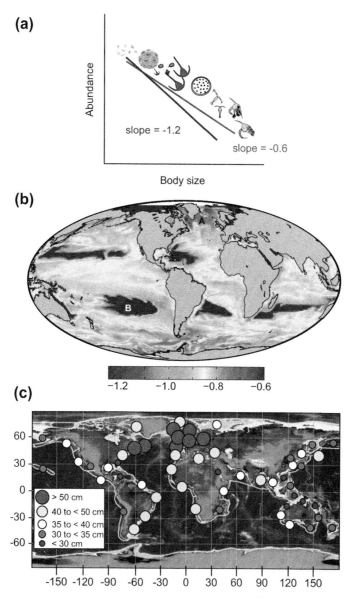

FIGURE 4 (a) A schematic illustration of two aquatic size spectra differing twofold in slope due to the increased abundance of larger organisms contributing to drive a shallower slope. (b) Global model illustration of phytoplankton size–spectra slopes, ranging twofold from high in some high latitude areas including a subpolar location with many large cells (e.g., point "A") and lower in a subtropical location with relatively few large cells (e.g., point "B"). (*After Barton et al. (2013); used with permission from John Wiley & Sons, Inc.*) (c) Global variation in geometric mean fish lengths from 57 large marine ecosystems (outlined by yellow lines), illustrating a pattern in which higher average fish lengths occur in areas of expected larger phytoplankton cell size concentrations. (*After Fisher et al. (2010a).*) These similar macroecological patterns between phytoplankton (b) and fishes (c) point to the need for investigations of potential linkages between global trait variation in plankton communities and their influence on trait variation at higher trophic levels.

research approaches that have been applied within macroecological aquatic fisheries investigations to identify natural and anthropogenic drivers of ecosystem functioning. These include (1) physical drivers of macroecological patterns; (2) structural relationships among key variables and predictions at multiple scales; and (3) dynamic macroecological patterns driven by natural and anthropogenic forces through time.

Physical and Biological Associations with Macroecological Patterns

In macroecological analyses, the need to link large-scale patterns to well-defined mechanisms that may operate at smaller scales but are underlain by physical or biological laws and principles with predictable consequences has long been advocated (Brown, 1999). In aquatic systems, such approaches have a rich history in the discipline of "fisheries oceanography," a term that applies generally to aquatic systems and seeks to understand relationships between the dynamics of fisheries resources and the dynamics of their environments (see Leggett and Frank, 2008). At macroecological scales, fisheries oceanography patterns have often been studied using spatial comparisons of key traits, productivity, etc. at high trophic levels either with the physical environment directly or with patterns of biological productivity at lower trophic levels. For example, as noted above, the primary productivity of aquatic systems generally limits fisheries yields at regional to global scales (Chassot et al., 2010), and the combination of primary productivity, body size distributions, and life history theory has produced regional-to-global scale estimates of fisheries productivity (Jennings et al., 2008a).

In a physical environmental context, for example, temperature remains the most general predictor of species richness among marine organisms (Tittensor et al., 2010), while habitat area and heterogeneity explain most of the spatial variation in freshwater systems (Tisseiu et al., 2013). Quantifying associations between biophysical features and macroecological patterns are increasing and may be the sole method for investigating relationships in systems characterized by cumulative data and/or time-averaged environmental conditions. However, additional approaches have recently emerged due to the widespread recognition both that contemporary macroecological analyses benefit from evolutionary considerations and constraints (Beck et al., 2012) and that, more general, multiple potential mechanisms can lead to similar patterns in nature (McGill, 2003; Nekola and Brown, 2007).

Structural Relationships Among Key Variables and Predictions at Multiple Scales

Given the limited scope for biophysical associations and correlations to completely explain macroecological relationships, additional methods have been advocated including examining key relationships between macroecological variables and quantifying whether predicted relationships are observed at multiple spatial scales. For example, macroecologists now acknowledge the need to integrate analyses across a broad range of scales (Kerr et al., 2007) and are now seeking to deal explicitly with scale-dependent variation in the principal determinants of the patterns they seek to understand, principally through the use of modeling (McGill, 2008). For example, Blackburn and Gaston (2001) outlined scenarios in which abundance, geographic distributions, and body size are functionally linked via energy availability and allocation among members of ecological communities (see also Figure 3). Importantly, their approach allows for a series of falsifiable predictions of the scaling of relationships among key macroecological variables. Such falsifiable predictions, along with novel predictions, can be used to guide future research questions and improve the power of macroecological analyses (McGill and Nekola, 2010). The influence of spatial scale on macroecological patterns has also been used to identify probable factors controlling relationships between macroecological variables. In one aquatic case study, the scale dependence of the positive relationship between energy availability and fish species richness was investigated among 7885 lakes. The study revealed that increasing spatial scale leads to stronger relationships, partly due to the ability of high-energy lakes to support regionally rare species (Gardezi and Gonzalez, 2008). Such types of analyses are key to resolving relationships between global change and macroecological patterns, yet have largely been lacking in global change studies that often attempt to scale up from very small study areas (Kerr et al., 2007).

Dynamic Macroecological Patterns Driven by Anthropogenic and Natural Forces

A third avenue used to reveal potential drivers underlying macroecological patterns is through analyses of "natural experiments" (changes in the system brought about by natural events) and "experiments in nature" (where changes in the system are brought about by human activities); both have long been advocated in ecological research (see Diamond, 1986; Blackburn, 2004). What these approaches have in common is the utilization of a perturbation to a system to assess the resulting effects that may spread within biological communities. Fisher et al. (2010c) reviewed examples of dynamic macroecology patterns at temporal scales from seasons to centuries resulting from drivers as habitat fragmentation, climate change, exploitation, biotic homogenization, changes in body sizes, changing trophic relationships, and altered phenology. While these drivers are not unique to aquatic systems, many have occurred as a result of exploitation, yet their application in understanding ecosystem functioning has not been widely recognized. For example, in their prognosis of the future of biodiversity and ecosystem functioning research,

Gamfeldt and Hillebrand (2008) noted that to date, removal experiments were not widely used to test the effects of aquatic biodiversity on ecosystem functioning and that removing consumer species is one way to test whether their biomass and functional roles are compensated by remaining species. However, while this view may be true in the context of controlled experimental manipulations, large-scale removals of consumer species remain hallmarks of commercial fisheries (Venturelli et al., 2009; Hixon et al., 2014; but see Figure 5), such that these types of approaches, if performed inadvertently, continue to contribute much to understanding the dynamics and functioning of exploited aquatic ecosystems in the context of "experiments in nature" as defined above (Jennings and Blanchard, 2004; Fisher et al., 2010b).

As an example of structural relationships, it has been demonstrated that in comparisons among marine systems large body size in fishes and extinction risk are strongly related (Olden et al., 2007). From such a pattern, it can be hypothesized that assemblages and regions characterized by an abundance of large-bodied fishes where intensive size-selective fisheries have been operating may be at most risk of functional changes through time. Data now exist to test this hypothesis. An example of the potential to link body size and a key marine ecosystem output is illustrated in Figure 5, where the percentage change in catch from high trophic level species in fisheries has been quantified and may

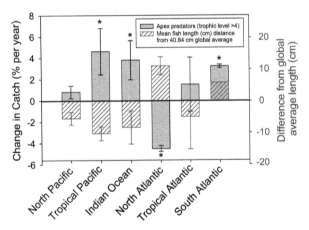

FIGURE 5 The bigger they were the harder they fell. Reciprocal relationships between metrics of decline in apex predatory marine fishes based on rates of change between 1950 and 2001 (*data from Essington et al. (2006); copyright (2006) National Academy of Sciences, USA*) and mean fish lengths (*data from Fisher et al. (2010a)*) among six ocean regions based on data from 30 large marine ecosystems (LMEs) in which a change in mean trophic level of fisheries was present; Essington et al. (2006) provide details. One pattern that is evident is that in LMEs characterized by lower than average fish lengths, changes in catch tended to be positive (* indicates significant differences in catch rates), while in the North Atlantic, where the largest average fish sizes were found, significant declines in high trophic level fishes occurred.

serve as an index of ecosystem functioning (data from Essington et al., 2006). It is clear from Figure 5 that the North Atlantic stands out as having the greatest declines in catch per year, whereas high trophic level catches increased in other regions. However, by overlaying an independent estimate of fish size within these ocean regions (data from Fisher et al., 2010a), it is evident that the North Atlantic is also characterized by a predominance of anomalously large, late maturing species (see also Figure 4), a trait that predisposes Atlantic assemblages to overexploitation by fisheries that directly or indirectly target larger individuals. Such declines in size structure are also evident within the entire Northwest Atlantic continental fish community (Fisher et al., 2010b). These apparent links between fisheries landings dynamics and body size characteristics among global large marine ecosystems provide a link between a key functional trait and an ecosystem change (Fisher et al., 2010a,b).

In addition to exploitation, dynamic physical environments have long been characterized as driving changes in biological oceanographic and limnetic patterns, with scale-dependent effects (Legendre and Demers, 1984). At macroecological spatial scales, such drivers as global climate change have been described as a means of "pseudo-experimental" analyses at macroecological spatial scales (Kerr et al., 2007). Examples include rapid changes in species richness patterns with changing ocean conditions (Fisher et al., 2008), shifts in depth and latitude of exploited ocean species (Pinsky et al., 2013), and the observed (Boyce et al., 2014) and projected (Blanchard et al., 2012) responses of fisheries ecosystems to changes in primary productivity.

Whether anthropogenic or environmental, a dynamic macroecological approach can involve re-evaluating macroecological patterns from a temporal perspective where data allow, identifying indicator metrics of large-scale changes, identifying links between macroecological patterns (Blackburn and Gaston, 2001), and projecting future states (Fisher et al., 2010c). Given the increase in freshwater and marine time series data and amalgamations at regional to global spatial scales, the scope for dynamic approaches has increased greatly.

A TRAITS-BASED FOCUS ON AQUATIC FUNCTIONAL DIVERSITY

Quantifying the diversity of measurable morphological, physiological, or phenological features of individuals—or "functional traits"—that influence growth, reproduction, and survival at the individual level (Violle et al., 2007) has increasingly been advocated as a means to link biodiversity and ecosystem functioning (Petchey and Gaston, 2006; Reiss et al., 2009; Webb et al., 2010; Mouillot et al., 2013; Verberk et al., 2013). In aquatic systems, such trait-based analyses are important in order to resolve drivers underlying the changes in aquatic ecosystem functioning we have described above, as commercial and recreational fisheries exploitation and changing aquatic conditions have had marked impacts on the traits and functioning of many aquatic species (Palkovacs et al., 2013). As recounted in the previous section, given the size-structured nature of aquatic ecosystems much emphasis has been placed on body size as an important individual-level trait (Hildrew et al., 2007). For example, adult body size and/or older aged (repeat) spawners have been shown to influence reproductive rates (Venturelli et al., 2009), egg quality and larval growth (Berkley et al., 2004), energy transfer efficiencies (Jennings et al., 2008a; Barnes et al., 2010), predation abilities (Scharf et al., 2000), nutrient cycling (Taylor et al., 2006), and other features (Woodward et al., 2005; see also Figure 3). These relationships have revealed the functional value in preserving the size (and age) structure of aquatic populations (Berkeley et al., 2004; Hixon et al., 2014). Given the utility of detailed trait inventories, in recent years, there have been repeated calls for databases of trait information at the individual level (e.g., McGill et al., 2006; Barton et al., 2013; Trebilco et al., 2013) and data collections in aquatic systems can greatly increase trait information with only marginal increases in processing time (Link et al., 2008).

As Figures 4(b) and (c) and 5 illustrate, it is clear that traits such as body size are not randomly distributed in space, but species with similar traits cluster within particular regions such that marine community-level metrics show clear spatial variation. The same finding has been reported from freshwater systems (Heino et al., 2013). In the context of fisheries development, these trait differences have contributed to the differential development of fisheries at different locations. These patterns have led to the recent revitalization of traits-based approaches as an avenue for predicting changes in ecosystem functioning and community composition, given increased understanding of how traits are "selected" within particular environments (Webb et al., 2010), and what the effects of a loss of trait diversity means for ecosystem functioning (Olden et al., 2004; McGill et al., 2006). However, it has been pointed out that individual traits are seldom the target of selection, and that suites of traits can interact, necessitating multivariate analyses (Olden et al., 2004; McGill et al., 2006; Heino et al., 2013; Verberk et al., 2013). This line of reasoning, developed by freshwater and terrestrial ecologists, shares some similarities with calls from marine researchers who have advocated a biological traits analysis at the level of species (Bremner et al., 2003; Mouillot et al., 2013). Together, these studies point toward the need to further developing a suite of approaches for quantifying trait variation and its connections to ecosystem structure and functioning in the face of anthropogenic, biotic, and abiotic filters on traits within communities (Mouillot et al., 2013). McGill et al. (2006) listed similarities in goals between the macroecological research program and a traits-based program, where both seek to find general rules governing the structure and functioning of ecosystems. They cautioned too, however, that emphasis on a single variable (as is often the case in macroecological investigations) may yield less powerful insights than those centered on a suite of performance traits.

At macroecological spatial scales, examinations of available reef fish data have shown contrasts between local maxima of species richness and functional diversity. For example, comparisons of site-specific species richness with metrics that include evenness and functional traits show changing species diversity—functional diversity relationships between the tropics and the temperate areas, such that temperate areas have higher functional diversity per unit species diversity than tropical systems (Stuart-Smith et al., 2013). These patterns indicate that species richness hotspots and functional diversity hotspots do not always overlap, such that examinations of species richness alone as a metric linking "diversity" and functioning may ignore important trait variations that may be more relevant to the structure and functioning of aquatic ecosystems (see also Fisher et al., 2010a,b).

Also in the context of linking trait variation to ecosystem functioning, Bruno and Cardinale (2008) outlined how the potential for effects of changes in predator richness on lower trophic levels is variable and largely dependent on the traits of the predators, such that future work should involve quantifying the functional roles of predators to link their diversity to ecosystem functioning. Similarly, in their synthesis on the ecological implications of numerical dominance, Hillebrand et al. (2008) stated that: "the consequences of dominance for ecosystem processes or species interactions may strongly differ, depending on whether the dominant species reflects average trait values or has traits dissimilar to the rest of the community." Following closely on these opinions regarding the importance of trait variation, Shackell et al. (2010) examined the dynamics of multiple trophic levels within a large marine ecosystem characterized by anomalous bursts in prey fish populations and an apparent trophic cascade that affected abundances of zooplankton and phytoplankton—but all within a system in which the top predator biomass remained relatively stable. After quantifying the potential influences of water column stratification, sea surface temperature, top predator size, and top predator biomass, top predator size had the greatest influence on the structure of the system, as body mass declined by ~60% over 38 years (Figure 6). This example illustrates a different type of cascade than that of predator richness declines (Bruno and Cardinale, 2008). Rather, a functional cascade was evident—and driven by trait changes within functional groups—where by the end of the time series, the body masses of piscivores became more like those of the next-smallest group members (large benthivores) at the start of the time series, while large benthivores became more like those of the next smallest (small benthivores), and so on (Figure 6). Given that the assigned roles of species within these functional groups were originally largely driven by their body sizes, it is likely that these earlier differences among functional groups have not persisted. The functional group body size dynamics depicted in Figure 6 support the contention that functional groups defined a priori may be less informative than their dynamic, constituent species (see also Naeem and Wright, 2003). A trait-based mechanism also appears to explain shifts in functional efficiency of predators (Shackell et al., 2010). For example, while piscivores prey primarily on planktivores, declines in piscivore mass and relative constancy in planktivore mass (Figure 6) alter the predator:prey mass ratio and may decrease predatory efficiency (Shackell et al., 2010).

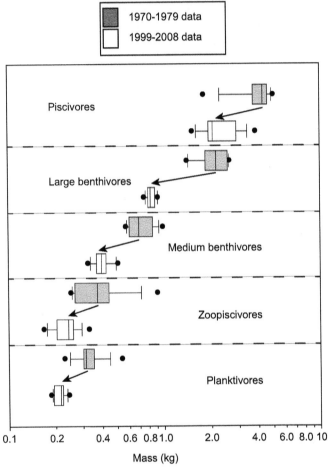

FIGURE 6 An example of trait homogenization (body mass) through time within marine fish functional groups (defined by species) as a result of size-selective fishing on the western Scotian Shelf. (*Data from Shackell et al. (2010).*) Gray box plots are based on annual data from 1970 to 1979 while open box plots illustrate sizes from 1999 to 2008. Median values are shown by lines within boxes, boxes denote the 25th and 75th percentiles and whiskers illustrate the 10th and 90th percentiles. The arrows show the extent of trait homogenization within functional groups through time, such that in the larger-bodied species, the range of body sizes in the later period approximates the body sizes of the next-largest-bodied functional group in the earlier period. Such shifts and homogenization of traits suggest that functional groupings based on species identities alone can mask much functional trait variation in exploited ecosystems.

Analogous observations have been made on the role of larger fish body sizes (and ages) as contributors to higher maximum reproductive rates, illustrating that biomass metrics alone masks the potential for significant variation in reproductive performance and that fish size matters (Venturelli et al., 2009; Hixon et al., 2014). These examples of trait variation again bolster Raffaelli's (2007) caution against using species-based functional descriptions where key traits vary. It also indicates the apparent utility of "biological traits analysis" where a series of traits among organisms is used to characterize functional diversity, rather than quantifying diversity by taxonomic or trophic groups (see Bremner et al., 2003).

Shifts in functioning and key trait changes with exploitation and directional trait changes are not exclusive to marine systems. Taylor et al. (2007) experimentally demonstrated the functional role of a freshwater detritovore fish in a diverse system, by removing the migratory flannelmouth characin (*Prochilodus mariae*) from its native habitat and quantifying the effects on carbon flow in a South American river. In areas without these fishes, carbon flux decreased, which increased benthic particulate levels, increased respiration, and increased gross primary productivity. Furthermore, given a decline by more than 50% in body mass of *P. mariae* since the 1970s, induced by size-selective fisheries, their functional roles have declined in this freshwater system even in areas where they remain (Taylor et al., 2007). More extensive examinations of nutrient cycling in species-rich freshwater systems in both South America and Africa similarly reported that nutrient recycling was dominated by few species (McIntyre et al., 2007), while in the North Sea, up to a third of primary production may now cycle through a small, noncommercial species (Jennings et al., 2008b). Notably, simulated overfishing of targeted species yielded stronger declines in nutrient cycling than those ranked by rarity, body size, or trophic position (McIntyre et al., 2007). This suggests a need to examine the key traits of species and the species targeted by fisheries, and where they intersect.

ECOLOGICAL AND EVOLUTIONARY EFFECTS OF SELECTIVE FISHERIES ON AQUATIC ECOSYSTEM FUNCTIONING

Into this complex and delicately adjusted struggle for existence man now throws the weight of his influence, and as the captor of hosts of marine organisms becomes a disturbing factor of importance.

Johnstone (1908, p. 102).

The trait-based focus of aquatic functional diversity research outlined above has clear connections with, and implications for, understanding the functioning of fisheries ecosystems and linking the ecological and evolutionary impacts of fisheries. Given the inherent size selectivity of most fisheries (where larger/older individuals are sought and/or more likely to be captured by large hooks or net-based fisheries often regulated with minimum mesh sizes), fishery-induced declines have been quantified in size structure and associated life history traits including size and age at maturity within populations (Darimont et al., 2009; Palkovacs et al., 2012), as well as changes in community composition (Fisher et al., 2010b). Changes in phenotypes resulting from plasticity and genetic changes are often considered together as eco-evolutionary factors (Palkovacs et al., 2012). However, quantifying their interactions and relative roles is critical for identifying potential time scales for functional trait change and recovery (Fraser, 2013). As shown in Figure 7(a), natural mortality and harvest-induced mortality affect different sized individuals oppositely, with high natural mortality at small, young sizes, while harvest pressure affects larger, older individuals (Stenseth and Dunlop, 2009). This incongruity in selective pressures may drive populations away from their naturally evolved life history configuration. For example, Figure 7(b) illustrates that when harvest-induced mortality on larger, older individuals is added to a population, there is often a decline in key life history traits through time. A key question, therefore, related to maintaining the functioning of size-structured ecosystems is whether the changes are of ecological origin alone (and therefore rapidly reversible after mortality is removed: gray line in Figure 7(b)) or due to evolutionary changes in the structure of populations, where strong evolutionary selection for smaller/younger age at maturity implies much greater recovery times (black line in

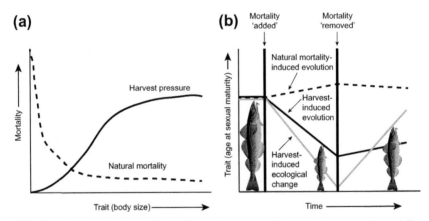

FIGURE 7 Schematic relationships linking evolutionary and ecological responses to mortality from human harvesting and natural mortality. (*Adapted by permission from* Stenseth and Dunlop, 2009.) (a) Harvesting pressure and natural mortality in fisheries often operate in opposite directions; this can lead to a mismatch in mortality rates where harvesting can induce trait changes as depicted in (b). (b) Ecological trait changes induced by harvesting (e.g., age at sexual maturity) occur more quickly than evolutionary trait changes, although evolutionary trait changes are expected to be slower to recover when harvesting mortality is removed.

Figure 7(b)). Strong selection against large individuals in experimental populations over multiple generations has shown genetically based changes in fecundity, egg volume, larval traits, behavioral traits (Walsh et al., 2006), and changes at genetic loci associated with body size (van Wijk et al., 2013). While such examinations of interacting ecological and evolutionary effects of fishing may hold much promise (Fraser, 2013), there is room to improve on existing approaches to better resolve the effects of selection on large, open, exploited populations (Diaz Pauli and Heino, 2014).

Fowler et al. (2013) provide one way in which macroecological distributions of predation rates on species targeted by fisheries can be compared with fisheries exploitation rates and expected patterns of biodiversity. Those approaches illustrate the potential gulf between high exploitation (predation) rates by humans versus lower predation rates by other species (or on size classes) within ecosystems and thereby suggest fishery management limits and/or targets for exploitation that are more closely aligned with those of predators thought to exploit stocks over evolutionary time scales. Such methods provide one means of addressing the "simple" question as to what is the biologically optimal harvesting strategy of multispecies fisheries suggested by Paloheimo and Regier (1982, as quoted above). Law (2000) discussed these concerns earlier in the context of commercially exploited stocks and highlighted the need to consider the effects of fishing selectivity and phenotypic change in relation to genetic variation. Macroecological functional traits such as body size need to be included in models investigating fishing size selectivity with particular reference to changes in the size composition of the predatory fish stocks (Blanchard et al., 2009; Law et al., 2009).

Macroecological patterns form the base for a systemic management approach that has its applications to questions regarding sustainable genetic effects (Etnier and Fowler, 2005), biomass harvests, and allocation of harvests over space and season—issues that span any subdivision of nature. Examples of this approach have resulted in estimates of what is sustainable—exemplified by what can be called the Ecologically Allowable Take for the Bering Sea and Georges Bank (Fowler and Hobbs, 2002) and also for the management of multispecies fisheries (Fowler, 1999).

We are currently faced with the urgency to move toward Integrated Ecosystem Assessments (IEAs) for marine ecosystems worldwide (Levin et al., 2009). A macroecological perspective needs to be a central part of any attempt to produce IEAs that need to include full consideration of the ecological and evolutionary effects of selective fisheries. Only with the inclusion of these components can IEAs be informative for Management Strategy Evaluations that try to develop sustainable management solutions that link ecological outcomes, ecosystem services, and socio-economic considerations. These ingredients are necessary to achieve a healthy ecosystem-based management of the world oceans.

ACKNOWLEDGMENTS

J.A.D.F. was supported by a Discovery Grant from the Natural Sciences and Engineering Research Council of Canada. The Center for Fisheries Ecosystems Research is supported by the Government of Newfoundland and Labrador and by the Research & Development Corporation of Newfoundland and Labrador.

REFERENCES

Abel, R., Thieme, M.L., Revenga, C., et al., 2008. Freshwater ecoregions of the World: a new map of biogeographic units for freshwater biodiversity conservation. Bioscience 58, 403–414.

Barbour, C.D., Brown, J.H., 1974. Fish species diversity in lakes. Am. Nat. 108, 473–489.

Barnes, C., Maxwell, D., Reuman, D.C., Jennings, S., 2010. Global patterns in predator-prey size relationships reveal size dependency of trophic transfer efficiency. Ecology 91, 222–232.

Barton, A.D., Dutkiewicz, S., Flierl, G., Bragg, J., Follows, M.J., 2010. Patterns of diversity in marine phytoplankton. Science 327, 1509–1511.

Barton, A.D., Pershing, A.J., Litchman, E., Record, N.R., Edwards, K.F., Finkel, Z.V., Kiorboe, T., Ward, B.A., 2013. The biogeography of marine plankton traits. Ecol. Lett. 16, 522–534.

Beck, J., Ballesteros-Mejia, L., Buchmann, C.M., Dengler, J., Fritz, S.A., Bruber, B., Hof, C., Jansen, F., Knapp, S., Kreft, H.l., Schneider, A.-K., Winter, M., Dormann, C.F., 2012. What's on the horizon for macroecology? Ecography 35, 673–683.

Belgrano, A., 2004. Foreward to: emergent properties of complex marine systems: a macroecological perspective. Mar. Ecol. Prog. Ser. 273, 227.

Belgrano, A., Fowler, C.W., 2013. How fisheries affect evolution. Science 342, 1176–1177.

Berkeley, S.A., Chapman, C., Sogard, S.M., 2004. Maternal age as a determinant of larval growth and survival in a marine fish, Sebastes melanops. Ecology 1258–1264.

Blackburn, T.M., 2004. Method in macroecology. Basic Appl. Ecol. 5, 401–412.

Blackburn, T.M., Gaston, K.J., 1999. A critique for macroecology. Oikos 84, 353–368.

Blackburn, T.M., Gaston, K.J., 2001. Linking patterns in macroecology. J. Anim. Ecol. 70, 338–352.

Blackburn, T.M., Cassey, P., Gaston, K.J., 2006. Variations on a theme: sources of heterogeneity in the form of the interspecific relationship between abundance and distribution. J. Anim. Ecol. 75, 1426–1439.

Blanchard, F., 2001. The effect of fishing on demersal fish community dynamics: an hypothesis. ICES J. Mar. Sci. 58, 711–718.

Blanchard, J.L., 2011. Body size and ecosystem dynamics: an introduction. Oikos 120, 481–482.

Blanchard, J.L., Jennings, S., Law, R., Castle, M.D., McCloghrie, P., Rochet, M.-J., Benoit, E., 2009. How does abundance scale with body size in coupled size-structured food webs? J. Anim. Ecol. 78, 270–280.

Blanchard, J.L., Jennings, S., Holmes, R., Harle, J., Merino, G., Allen, J.I., Holt, J., Dulvy, N.K., Barange, M., 2012. Potential consequences of climate change for primary production and fish production in large marine ecosystems. Phil. Trans. R. Soc. B 367, 2979–2989.

Blanchet, S., Grenouillet, G., Beauchard, O., Tedesco, P.A., Leprieur, F., Durr, H.H., Busson, F., Oberdorff, T., Brosse, S., 2010. Non-native fish species disrupt the worldwide pattern of freshwater fish body size: implications for Bergmann's rule. Ecol. Lett. 13, 421–431.

Bonner, J.T., 2006. Why Size Matters: from Bacteria to Blue Whales. Princeton Press, Princeton, USA.

Borregaard, M.K., 2010. Causality of the relationship between geographic distribution and species abundance. Q. Rev. Biol. 85, 3–25.

Boyce, D.G., Dowd, M., Lewis, M.R., Worm, B., 2014. Estimating global chlorophyll changes over the past century. Prog. Oceanogr. 122, 163–173.

Bradbury, I.R., Laurel, B., Snelgrove, P.V.R., Bentzen, P., Campana, S.E., 2008. Global patterns in marine dispersal estimates: the influence of geography, taxonomic category and life history. Proc. R. Soc. B 275, 1803–1809.

Bradbury, I.R., Hubert, S., Higgins, B., Bowman, S., Borza, T., Paterson, I.G., Snelgrove, P.V.R., Morris, C.J., Gregory, R.S., Hardie, D., Hutchings, J.A., Ruzzante, D.E., Taggart, C.T., Bentzen, P., 2013. Genomic islands of divergence and their consequences for the resolution of spatial structure in an exploited marine fish. Evol. Appl. 6, 450–461.

Bremner, J., Rogers, S.I., Frid, C.L.J., 2003. Assessing functional diversity in marine benthic ecosystems: a comparison of approaches. Mar. Ecol. Prog. Ser. 254, 11–25.

Briggs, J.C., 2007. Marine longitudinal biodiversity: causes and conservation. Diversity Distrib. 13, 544–555.

Brown, J.H., 1995. Macroecology. University of Chicago Press, Chicago, USA.

Brown, J.H., 1999. Macroecology: progress and prospect. Oikos 87, 3–14.

Brown, J.H., Maurer, B.A., 1989. Macroecology: the division of food and space among species on continents. Science 243, 1145–1150.

Bruno, J.F., Cardinale, B.J., 2008. Cascading effects of predator richness. Front. Ecol. Environ. 6, 539–546.

Chasot, E., Bonhommeau, S., Dulvy, N.K., Mélin, F., Watson, R., Gascuel, D., Le Pape, O., 2010. Global marine primary production constrains fisheries catches. Ecol. Lett. 13, 495–505.

Ciannelli, L., Fisher, J.A.D., Skern-Mauritzen, M., Hunsicker, M.E., Hildalgo, M., Frank, K.T., Bailey, K.M., 2013. Theory, consequences and evidence of eroding population spatial structure in harvested marine fishes: a review. Mar. Ecol. Prog. Ser. 480, 227–243.

Coll, M., Libralto, S., Tudela, S., Palomera, I., Pranovi, F., 2008. Ecosystem overfishing in the ocean. PLoS One 3, e3881. http://dx.doi.org/10.1371/journal.pone.0003881.

Connolly, S.R., et al., 2014. Commonness and rarity in the marine biosphere. Proc. Natl. Acad. Sci. USA 111, 8524–8529.

Darimont, C.T., Carlson, S.M., Kinnison, M.T., Paquet, P.C., Reimchen, T.E., Wilmers, C.C., 2009. Human predators outpace other agents of trait change in the wild. Proc. Natl. Acad. Sci. USA 106, 952–954, 21 of 29 species were fishes in their analyses!

Diamond, J., 1986. Overview: laboratory experiments, field experiments, and natural experiments. In: Diamond, J., Case, T.J. (Eds.), Community Ecology. Harper Row, New York, pp. 3–22.

Dias, M.S., et al., 2014. Global imprint of historical connectivity on freshwater fish biodiversity. Ecol. Lett. 17, 1130–1140.

Diaz Pauli, B., Heino, M., 2014. What can selection experiments teach us about fisheries-induced evolution? Biol. J. Linn. Soc. 111, 485–503.

Drakare, S., Lennon, J.J., Hillebrand, H., 2006. The imprint of the geographical, evolutionary and ecological context on species-area relationships. Ecol. Lett. 9, 215–227.

Duarte, C.M., Agusti, S., Peters, H., 1987. An upper limit to the abundance of aquatic organisms. Oecologia 74, 272–276.

Dudgeon, D., Arthington, A.H., Gessner, M.O., Kawabata, Z.-I., Knowler, D.J., Leveque, C., Naiman, R.J., Prieur-Richard, A.-H., Soto, D., Stiassny, M.L.J., Sullivan, C.A., 2006. Freshwater biodiversity: importance, threats, status and conservation challenges. Biol. Rev. 81, 163–182.

Essington, T.E., Beaudreau, A.H., Wiedenmann, J., 2006. Fishing through marine food webs. Proc. Natl. Acad. Sci. USA 103, 3171–3175.

Etnier, M.A., Fowler, C.W., 2005. Comparison of size selectivity between marine mammals and commercial fisheries with recommendations for restructuring management policies. In: U.S. Dep. Commer., NOAA Technical Memorandum, NMFS-afsc-159, p. 274.

FAO (Food and Agriculture Organization), 2012. The State of World Fisheries and Aquaculture 2012. Food and Agriculture Organization of the United Nations, Rome.

Fisher, J.A.D., Frank, K.T., 2004. Abundance-distribution relationships and the conservation of exploited marine fishes. Mar. Ecol. Prog. Ser. 279, 201–213.

Fisher, J.A.D., Frank, K.T., 2013. Pelagic communities. In: Bertness, M.D., Bruno, J.F., Silliman, B.R., Stachowicz, J.J. (Eds.), Marine Community Ecology and Conservation. Sinauer, Sunderland, pp. 337–363.

Fisher, J.A.D., Frank, K.T., Petrie, B., Leggett, W.C., Shackell, N.L., 2008. Temporal dynamics within a contemporary latitudinal diversity gradient. Ecol. Lett. 11, 883–897.

Fisher, J.A.D., Frank, K.T., Leggett, W.C., 2010a. Global variation in marine fish body size and its role in biodiversity-ecosystem functioning. Mar. Ecol. Prog. Ser. 405, 1–13.

Fisher, J.A.D., Frank, K.T., Leggett, W.C., 2010b. Breaking Bergmann's rule: truncation of Northwest Atlantic marine fish body sizes. Ecology 91, 2499–2505.

Fisher, J.A.D., Frank, K.T., Leggett, W.C., 2010c. Dynamic macroecology on ecological timescales. Glob. Ecol. Biogeogr. 19, 1–15.

Fowler, C.W., 1999. Management of multi-species fisheries: from overfishing to sustainability. ICES J. Mar. Sci. 56 (6), 927–932.

Fowler, C.W., Hobbs, L., 2002. Limits to natural Variation: Implications for systemic management. Anim. Biodiversity Conserv. 25 (2), 7–45.

Fowler, C.W., Belgrano, A., Casini, M., 2013. Holistic fisheries management: combining macroecology, ecology, and evolutionary biology. Mar. Fish. Rev. 75, 1–36.

Frank, K.T., Shackell, N.L., 2001. Area-dependent patterns of finfish diversity in a large marine ecosystem. Can. J. Fish. Aquat. Sci. 58, 1703–1707.

Fraser, D.J., 2013. The emerging synthesis of evolution with ecology in fisheries science. Can. J. Fish. Aquat. Sci. 70, 1417–1428.

Frisk, M.G., Duplisea, D.E., Trenkel, V.M., 2011. Exploring the abundance-occupancy relationships for the Georges Bank finfish and shellfish community from 1963 to 2006. Ecol. Appl. 21, 227–240.

Gamfeldt, L., Hillebrand, H., 2008. Biodiversity effects on aquatic ecosystem functioning—maturation of a new paradigm. Int. Rev. Hydrobiol. 93, 550–564.

Gardezi, T., Gonzalez, A., 2008. Scale dependence of species-energy relationships: evidence from fishes in thousands of lakes. Am. Nat. 171, 800–815.

Gaston, K.J., Blackburn, T.M., 2000. Pattern and Process in Macroecology. Blackwell Publishing, Oxford, UK.

Gaston, K.J., Blackburn, T.M., Greenwood, J.J.D., Gregory, R.D., Quinn, R.M., Lawton, J.H., 2000. Abundance-occupancy relationships. J. Appl. Ecol. 37 (Suppl. 1), 39–59.

Gray, J.S., Bjørgesæter, A., Ugland, K.I., Frank, K., 2006. Are there differences in structure between marine and terrestrial assemblages? J. Exp. Mar. Biol. Ecol. 330, 19–26.

Guégan, J.-F., Lek, S., Oberdorff, T., 1998. Energy availability and habitat heterogeneity predict global riverine fish diversity. Nature 391, 382–384.

Halpern, B.S., Walbridge, S., Selkoe, K.A., Kappel, C.V., Micheli, F., D'Agrosa, C., Bruno, J.F., Casey, K.S., Ebert, C., Fox, H.E., Fujita, R., Heinemann, D., Lenihan, H.S., Madin, E.M.P., Perry, M.T., Selig, E.R., Spalding, M., Steneck, R., Watson, R., 2008. A global map of human impact on marine ecosystems. Science 319, 948–952.

Hauser, L., Carvalho, G.R., 2008. Paradigm shifts in marine fisheries genetics: ugly hypotheses slain by beautiful facts. Fish Fish. 9, 333–362.

Heino, J., 2011. A macroecological perspective of diversity patterns in the freshwater realm. Freshwater Biol. 56, 1703–1722.

Heino, J., Schmera, D., Eros, T., 2013. A macroecological perspective of trait patterns in stream communities. Freshwater Biol. http://dx.doi.org/10.1111/fwb.12164.

Hilborn, R., Branch, T.A., Ernst, B., Magnusson, A., Minte-Vera, C.V., Scheuerell, M.D., Valero, J.L., 2003. State of the world's fisheries. Ann. Rev. Environ. Resour. 28, 359–399.

Hildrew, A., Raffaelli, D., Edmonds-Brown, R. (Eds.), 2007. Body Size: The Structure and Function of Aquatic Ecosystems. Cambridge University Press, Cambridge, UK.

Hillebrand, H., 2004. Strength, slope and variability of marine latitudinal gradients. Mar. Ecol. Prog. Ser. 273, 251–267.

Hillebrand, H., Bennett, D.M., Cadotte, M.C., 2008. Consequences of dominance: a review of evenness effects on local and regional ecosystem processes. Ecology 89, 1510–1520.

Hixon, M.A., Johnson, D.W., Sogard, S.M., 2014. BOFFFFs: on the importance of conserving old-growth age structure in fisheries populations. ICES J. Mar. Sci. http://dx.doi.org/10.1093/icesjms/fst200.

Huston, M.A., Wolverton, S., 2009. The global distribution of net primary production: resolving the paradox. Ecol. Monogr. 79, 343–377.

Huston, M.A., Wolverton, S., 2011. Regulation of animal size by eNPP, Bergmann's rule, and related phenomena. Ecol. Monogr. 81, 349–405.

Jackson, J.B.C., Kirby, M.X., Berger, W.H., Bjorndal, K.A., Botsford, L.W., Bourque, B.J., 2001. Historical overfishing and the recent collapse of coastal ecosystems. Science 293, 629–638.

Jennings, S., Blanchard, J.L., 2004. Fish abundance with no fishing: predictions based on macroecological theory. J. Anim. Ecol. 73, 632–642.

Jennings, S., Mélin, F., Blanchard, J.L., Forster, R.M., Dulvy, N.K., Wilson, R.W., 2008a. Global-scale predictions of community and ecosystem properties from simple ecological theory. Proc. R. Soc. B 275, 1375–1383.

Jennings, S., van Hal, R., Hiddink, J.G., Maxwell, T.A.D., 2008b. Fishing effects on energy use by North Sea fishes. J. Sea Res. 60, 74–88.

Johnstone, J., 1908. Conditions of Life in the Sea. Cambridge University Press, Cambridge, UK.

Jones, R., 1982. Ecosystems, food chains and fish yields. In: Pauly, D., Murphy, G.I. (Eds.), Theory and Management of Tropical Fisheries. ICLARM CISRO Press, Manila, pp. 195–237.

Katz, T., Yahel, G., Yahel, R., Tunnicliffe, V., Herut, B., Snelgrove, P., Crusius, J., Lazar, B., 2009. Groundfish overfishing, diatom decline, and the marine silica cycle: lessons from Saanich Inlet, Canada, and the Baltic Sea cod crash. Glob. Biogeochem. Cycles 23, GB4032. http://dx.doi.org/10.1029/2008GB003416.

Kerr, J.T., Kharouba, H.M., Currie, D.J., 2007. The macroecological contribution to global change solutions. Science 316, 1581–1584.

Law, R., 2000. Fishing, selection, and phenotypic evolution. ICES J. Mar. Sci. 57, 659–668.

Law, R., Plank, M.J., James, A., Blanchard, J.L., 2009. Size-spectra dynamics from stochastic predation and growth of individuals. Ecology 90 (3), 802–811.

Lawton, J.H., 1996. Patterns in ecology. Oikos 75, 145–146.

Legendre, L., Demers, S., 1984. Towards dynamic biological oceanography and limnology. Can. J. Fish. Aquat. Sci. 41, 2–19.

Leggett, W.C., Frank, K.T., 2008. Paradigms in fisheries oceanography. Oceanogr. Mar. Biol. Ann. Rev. 46, 331–363.

Leichter, J.J., Witman, J.D., 2009. Basin-scale oceanographic influences on marine macroecological patterns. In: Witman, J.D., Roy, K. (Eds.), Marine Macroecology. University of Chicago Press, Chicago, pp. 205–226.

Levin, P.S., Fogarty, M.J., Murawski, S.A., Fluharty, D., 2009. Integrated ecosystem assessments: developing the scientific basis for ecosystem-based management of the ocean. PLoS Biol. 7 (1), 23–28.

Link, J.S., 2010. Ecosystem-based Fisheries Management: Confronting Tradeoffs. Cambridge University Press, Cambridge.

Link, J.S., Burnett, J., Kostovick, P., Galbaith, J., 2008. Value-added sampling for fishery independent surveys: don't stop after you're done counting and measuring. Fish. Res. 93, 229–233.

Macpherson, E., Duarte, C.M., 1994. Patterns in species richness, size, and latitudinal range of East Atlantic fishes. Ecography 17, 242–248.

Magnuson, J.J., 1988. Two worlds for fish recruitment: lakes and oceans. Am. Fish. Soc. Symp. 5, 1–6.

Marten, G.G., Polovina, J.J., 1982. A comparative study of fish yields from various tropical ecosystems. In: Pauly, D., Murphy, G.I. (Eds.), Theory and Management of Tropical Fisheries. ICLARM CISRO Press, Manila, pp. 255–285.

McGill, B., 2003. Strong and weak tests of macroecological theory. Oikos 102, 679–685.

McGill, B.J., 2008. Exploring predictions of abundance from body mass using hierarchical comparative approaches. Am. Nat. 172, 88–101.

McGill, B.J., Nekola, J.C., 2010. Mechanism in macroecology: AWOL or purloined letter? towards a pragmatic view of mechanism. Oikos 119, 591–603.

McGill, B.J., Enquist, B.J., Weiher, E., Westoby, M., 2006. Rebuilding community ecology from functional traits. Trends Ecol. Evol. 21, 178–185.

McIntyre, P.B., Jones, L.E., Flecker, A.S., Vanni, M.J., 2007. Fish extinctions alter nutrient recycling in tropical freshwaters. Proc. Natl. Acad. Sci. USA 104, 4461–4466.

Mora, C., Chittaro, P.M., Sale, P.F., Kritzer, J.P., Ludsin, S.A., 2003. Patterns and processes in reef fish diversity. Nature 421, 933–936.

Mora, C., et al., 2011. Global human footprint on the linkage between biodiversity and ecosystem functioning in reef fishes. PLoS Biol. 9, e1000606.

Mouillot, D., Graham, N.A.J., Villeger, S., Mason, N.W.H., Bellwood, D.R., 2013. A functional approach reveals community responses to disturbances. Trends Ecol. Evol. 28, 167–177.

Naeem, S., 2006. Expanding scales in biodiversity-based research: challenges and solutions for marine systems. Mar. Ecol. Prog. Ser. 311, 273–283.

Naeem, S., Wright, J.P., 2003. Disentangling biodiversity effects on ecosystem functioning: deriving solutions to a seemingly insurmountable problem. Ecol. Lett. 6, 567–579.

Nekola, J.C., Brown, J.H., 2007. The wealth of species: ecological communities, complex systems and the legacy of Frank Preston. Ecol. Lett. 10, 188–196.

Niegel, J.E., 2003. Species-area relationships and marine conservation. Ecol. Appl. 13, S139–S145.

Nixon, S.W., 1988. Physical energy inputs and the comparative ecology of lake and marine ecosystems. Limnol. Oceanogr. 33, 1005–1025.

Nogués-Bravo, D., Araújo, M.B., Romdal, T., Rahbek, C., 2008. Scale effects and human impact on the elevational species richness gradients. Nature 453, 216–219.

Norse, E.A., et al., 2012. Sustainability of deep-sea fisheries. Mar. Policy 36, 307–320.

Olden, J.D., Poff, N.L., Douglas, M.R., Douglas, M.E., Fausch, K.D., 2004. Ecological and evolutionary consequences of biotic homogenization. Trends Ecol. Evol. 19, 18–24.

Olden, J.D., Hogan, Z.S., Vander Zanden, M.J., 2007. Small fish, big fish, red fish, blue fish: size-based extinction risk of the world's freshwater and marine fishes. Glob. Ecol. Biogeogr. 16, 694–701.

Palkovacs, E.P., Kinnison, M.T., Correa, C., Dalton, C.M., Hendry, A.P., 2012. Fates beyond traits: ecological consequences of human-induced trait change. Evol. Appl. 5, 183–191.

Palkovacs, E.P., Kinnison, M.T., Correa, C., Dalton, C.M., Hendry, A.P., 2013. Fates beyond traits: ecological consequences of human-induced trait change. Evol. Appl. http://dx.doi.org/10.1111/j.1752-4571.2011.00212.x.

Paloheimo, J.E., Regier, H.A., 1982. Ecological approaches to stressed multispecies fisheries resources. In: Mercer, C. (Ed.), Multispecies Approaches to Fisheries Management Advice, Can. Spec. Publ. Fish. Aquat. Sci, vol. 59, pp. 127–132. Ottawa.

Palumbi, S.R., Sandifer, P.A., Allan, J.D., Beck, M.W., Fautin, D.G., Fogarty, M.J., Halpern, B.S., Incze, L.S., Leong, J.-A., Norse, E., Stachowicz, J.J., Wall, D.H., 2009. Managing for ocean biodiversity to sustain marine ecosystem services. Front. Ecol. Environ. 7, 204–211.

Pauly, D., Christiensen, V., Guenette, S., Pitcher, T.J., Sumalia, U.R., Walters, C.J., Zeller, D., 2002. Towards sustainability in world fisheries. Nature 418, 689–695.

Pauly, D., Watson, R., Adler, J., 2005. Global trends in world fisheries: impacts on marine ecosystems and food security. Phil. Trans. R. Soc. B 360, 5–12.

Pennisi, E., 2011. Marine census scrambles to fund a second phase with expanded focus. Science 333, 686.

Petchey, O.L., Gaston, K.J., 2006. Functional diversity: back to basics and looking forward. Ecol. Lett. 9, 741–758.

Peters, R.H., 1983. The Ecological Implications of Body Size. Cambridge University Press, Cambridge, UK.

Pinsky, M.L., Worm, B., Fogarty, M.J., Sarmiento, J.L., Levin, S.A., 2013. Marine taxa track local climate velocities. Science 341, 1239–1242.

Post, J.R., Sullivan, M., Cox, S., Lester, N.P., Walters, C.J., Parkinson, E.A., Paul, A.J., Jackson, L., Shuter, B.J., 2002. Canada's recreational fisheries: the invisible collapse? Fisheries 27, 6–17.

Raffaelli, D., 2007. Food webs, body size and the curse of the Latin binomial. In: Rooney, N., McCann, K.S., Noakes, D.L. (Eds.), From Energetics to Ecosystems: The Dynamics and Structure of Ecological Systems. Springer, Dordrecht, pp. 53–64.

Raffaelli, D.G., 2006. Biodiversity and ecosystem functioning: issues of scale and tropic complexity. Mar. Ecol. Prog. Ser. 311, 285–294.

Reiss, J., Bridle, J.R., Montoya, J.M., Woodward, G., 2009. Emerging horizons in biodiversity and ecosystem functioning research. Trends Ecol. Evol. 24, 505–514.

Rex, M.A., Etter, R.J., Morris, J.S., Crouse, J., McClain, C.R., Johnson, N.A., Stuart, C.T., Deming, J.W., Thies, R., Avery, R., 2006. Global bathymetric patterns of standing stock and biomass in the deep-sea benthos. Mar. Ecol. Prog. Ser. 317, 1–8.

Rindorf, A., Lewy, P., 2012. Estimating the relationship between abundance and distribution. Can. J. Fish. Aquat. Sci. 69, 382–397.

Rosenzweig, M.A., 1995. Species Diversity in Space and Time. Cambridge University Press, Cambridge.

Scharf, F.S., Juanes, F., Rountree, R.A., 2000. Predator size – prey size relationships of marine fish predators: interspecific variation and effects of ontogeny and body size on trophic-niche breadth. Mar. Ecol. Prog. Ser. 208, 229–248.

Shackell, N.L., Frank, K.T., Fisher, J.A.D., Petrie, B., Leggett, W.C., 2010. Decline in top predator body size and changing climate alter trophic structure in an oceanic ecosystem. Proc. R. Soc. B 277, 1353–1360.

Sherman, K., Belkin, I.M., Friedland, K.D., O'Reilly, J., Hyde, K., 2009. Accelerated warming and emergent trends in fisheries biomass yields of the World's large marine ecosystems. Ambio 38, 215–224.

Smith, F.A., Gittleman, J.L., Brown, J.H., 2014. Foundations of Macroecology. University of Chicago Press, Chicago.

Smith, K.F., Brown, J.H., 2002. Patterns of diversity, depth range and body size among pelagic fishes along a gradient of depth. Glob. Ecol. Biogeogr. 11, 313–322.

Sprules, W.G., Munawar, M., 1986. Plankton size spectra in relation to ecosystem productivity, size, and perturbation. Can. J. Fish. Aquat. Sci. 43, 1789–1794.

Stenseth, N.C., Dunlop, E.S., 2009. Unnatural selection. Nature 457, 803–804.

Stergiou, K.I., 2002. Overfishing, tropicalization of fish stocks, uncertain and ecosystem management: resharpening Ockham's razor. Fish. Res. 55, 1–9.

Strona, G., Galli, P., Montano, S., Seveso, D., Fattorini, S., 2012. Global-scale relationships between colonization ability and range size in marine and freshwater fish. PLoS One 7 (11), e49465.

Stuart-Smith, R.D., et al., 2013. Integrating abundance and functional traits reveals new global hotspots of fish diversity. Nature 501, 539–542.

Sumaila, U.R., Cheung, W.W.L., Lam, V.W.Y., Pauly, D., Herrick, S., 2011. Climate change impacts on the biophysics and economics of the world fisheries. Nat. Clim. Change. http://dx.doi.org/10.1038/NCLIMATE1301.

Taylor, B.W., Flecker, A.S., Hall Jr., R.O., 2006. Loss of harvested fish species disrupts carbon flow in a diverse tropical river. Science 313, 833–836.

Taylor, B.W., Flecker, A.S., Hall Jr., R.O., 2007. Loss of a harvested fish species disrupts carbon flow in a diverse tropical river. Science 313, 833–836.

Tisseuil, C., Cornu, J.-F., Beauchard, O., Brosse, S., Darwall, W., Holland, R., Hugueny, B., Tedesco, P.A., Oberdorff, T., 2013. Global diversity patterns and cross-taxa convergence in freshwater systems. J. Anim. Ecol. 82, 365–376.

Tittensor, D.P., Micheli, F., Nyström, M., Worm, B., 2007. Human impacts on the species-area relationship in reef fish assemblages. Ecol. Lett. 10, 760–772.

Tittensor, D.P., Worm, B., Myers, R.A., 2009. Macroecological changes in exploited marine systems. In: Witman, J.D., Roy, K. (Eds.), Marine Macroecology. University of Chicago Press, Chicago, pp. 310–337.

Tittensor, D.P., Mora, C., Jetz, W., et al., 2010. Global patterns and predictors of marine biodiversity across taxa. Nature 466, 1098–1101.

Trebilco, R., Baum, J.K., Salomon, A.K., Dulvy, N.K., 2013. Ecosystem ecology: size-based constraints on the pyramids of life. Trends Ecol. Evol. 28, 423–431.

Trenkel, V.M., et al., 2014. Comparative ecology of widely distributed pelagic fish species in the North Atlantic: implications for modelling climate and fisheries impacts. Prog. Oceanogr. http://dx.doi.org/10.1016/j.pocean.2014.04.030.

Venturelli, P.A., Shuter, B.J., Murphy, C.A., 2009. Evidence for harvest-induced maternal influences on the reproductive rates of fish populations. Proc. R. Soc. Lond. B 276, 919–924.

Verberk, W.C.E.P., van Noordwijk, C.G.E., Hildrew, A.G., 2013. Delivering on a promise: integrating species traits to transform descriptive community ecology into a predictive science. Freshwater Sci. 32, 531–547.

Violle, C., Navas, M.-L., Vile, D., Kazakou, E., Fortunel, C., Hummel, I., Garnier, E., 2007. Let the concept of trait be functional! Oikos 116, 882–892.

van Wijk, S.J., et al., 2013. Experimental harvesting of fish populations drives genetically based shifts in body size and maturation. Front. Ecol. Environ. 11, 181–187.

Walsh, M.R., Munch, S.B., Chiba, S., Conover, D.O., 2006. Maladaptive changes in multiple traits caused by fishing: impediments to population recovery. Ecol. Lett. 9, 142–148.

Watson, R., Zeller, D., Pauly, D., 2013. Primary productivity demands of global fishing fleets. Fish Fish. 15, 231–241.

Webb, C.T., Hoeting, J.A., Ames, G.M., Pyne, M.I., Poff, N.L., 2010. A structured and dynamic framework to advance traits-based theory and prediction in ecology. Ecol. Lett. 13, 267–283.

White, E.P., Morgan Ernest, S.K., Kerkhoff, A.J., Enquist, B.J., 2007. Relationships between body size and abundance in ecology. Trends Ecol. Evol. 22, 323–330.

Wilson, R.W., Millero, F.J., Taylor, J.R., Walsh, P.J., Christensen, V., Jennings, S., Grosell, M., 2009. Contribution of fish to the marine inorganic carbon cycle. Science 323, 359–362.

Witman, J.D., Roy, K. (Eds.), 2009. Marine Macroecology. University of Chicago Press, Chicago.

Woodward, G., Ebenman, B., Emmerson, M., Montoya, J.M., Olesen, J.M., Valido, A., Warren, P.H., 2005. Body size in ecological networks. Trends Ecol. Evol. 20, 402–409.

Worm, B., Sandow, M., Oschlies, A., Lotze, H.K., Myers, R.A., 2005. Global patterns of predator diversity in the open oceans. Science 309, 1365–1369.

Worm, B., et al., 2006. Impacts of biodiversity loss on ocean ecosystem services. Science 314, 787–790.

Worm, B., et al., 2009. Rebuilding global fisheries. Science 325, 578–585.

Chapter 8

Valuing Biodiversity and Ecosystem Services in a Complex Marine Ecosystem

Ute Jacob[1], Tomas Jonsson[2], Sofia Berg[3], Thomas Brey[4], Anna Eklöf[5], Katja Mintenbeck[4], Christian Möllmann[1], Lyne Morissette[6], Andrea Rau[7], Owen Petchey[8]

[1]*Institute for Hydrobiology and Fisheries Science, University of Hamburg, Hamburg, Germany;* [2]*Population Ecology Unit, Institute for Ecology, Uppsala, Sweden;* [3]*EnviroPlanning AB, Göteborg, Sweden;* [4]*Alfred Wegener Institute for Polar and Marine Research, Bremerhaven, Germany;* [5]*Department of Physics, Chemistry and Biology (IFM), Linköping University, Linköping, Sweden;* [6]*M-Expertise Marine, Sainte-Luce, Canada;* [7]*Johann Heinrich von Thünen Institute for Baltic Sea Fisheries, Rostock, Germany;* [8]*Institute of Evolutionary Biology and Environmental Studies, University of Zurich, Zurich, Switzerland*

INTRODUCTION

Human activity is affecting ecosystems worldwide and there is concern that the threatening biodiversity crisis will affect provisioning of important ecosystems services and therefore human well-being. We already know how some single species are known as providing multiple services arising from different traits—for example, in coastal habitats like salt marshes or mangroves where the structural species provide shoreline protection, and may serve as a nursery for other species. Each of these services arises from different traits of the species. A deeper understanding of how services relate to traits and/or the functional diversity will allow for a much better understanding of the roles species have in ecosystem service provision.

However, for most services in most ecosystems, the details of the roles of single species are poorly resolved. This uncertainty is even greater when considering that species interact in complex networks. The species directly providing an ecosystem service may be known but not how they depend on and interact with other species. Recent research has shown that there is not necessarily a simple linear relationship between biodiversity and ecosystem function (Tilman, 1999; Loreau et al., 2001, 2002; Petchey et al., 2004a), implying that every species is not equally important. Instead some species may be characterized as drivers (by being able to compensate for the loss of other species) and others as passengers (Walker, 1992), suggesting that it is increasing functional diversity (Tilman, 2001) rather than species richness per se that facilitates ecosystem functioning. This has led to a focus on functional diversity and the characterization of functionally significant components of biodiversity (Díaz and Cabido, 2001; Petchey et al., 2004b) instead of species and species richness. The degree of overlap in the role of species in an ecosystem is described as functional redundancy (Ehrlich and Ehrlich, 1981; Walker, 1992). The higher the functional redundancy in an ecosystem the more the species are "functionally expendable" (Johnson et al., 1996); i.e., they could be lost without much effect on the structure and functioning of the whole community (Gitay et al., 1996). The identification of nonredundant "key species," whose loss has dramatic consequences for the whole community, is a major but rarely achieved goal in current ecology (for a review of alternative approaches, see Ebenman and Jonsson, 2005). It has been argued that functional diversity should be closely related to trophic diversity (Petchey and Gaston, 2002; Naeem and Wright, 2003) and inspired by this we propose a new way to look at functional diversity based on characterizing the trophic niches of consumers in a community. Traditionally, the *ecological niche* of a species is defined as the n-dimensional hypervolume bounded by sets of conditions that are compatible with its persistence and success (Hutchinson, 1957; Pielou, 1972). Niche theory argues that each species of a community occupies a unique position in this n-dimensional space. High overlap along one niche dimension should then be accompanied by separation on another dimension, thus allowing coexistence (MacArthur and Pianka, 1966; Stewart et al., 2003). In fact, niche overlap is one way to look at resource use complementarity and diversity of trophic roles of consumers. The *trophic niche* describes the trophic role of consumers and can be defined as the subset of niche boundary conditions, which refers to a species' foraging strategies, such as measures of prey handling capacity or of trophic extension (Cohen, 1978; Williams and Martinez, 2000). Currently, the trophic niche is the focus of much ecosystem research (e.g., Winemiller et al., 2001; Chase and Leibold, 2003; Layman et al., 2007a), because the trophic niches of species are intimately linked to the structure and dynamics of food webs, ecosystem stability, and resilience (Williams and Martinez, 2000; Dunne et al., 2002; Brose et al., 2006b). Thus, trophic diversity may be at the core of functional diversity (Petchey and Gaston, 2002; Naeem and Wright, 2003). Although conceptually robust, it has proven difficult to find practical measures of trophic niche properties that are simple to obtain, yet provide an adequate and useful description of the trophic role of consumers (Matthews and Mazumder, 2004). Initiated by Roughgarden (1972, 1974) numerous measures have been proposed such as food item diversity (Bolnick et al., 2003), item size range (Bolnick et al., 2003), and variance in consumer stable isotope signatures (Bearhop et al., 2004). Whereas these approaches focused on single niche dimensions, we present an alternative multidimensional concept of trophic niches based on the traits prey size, prey mobility, and prey trophic position to construct a three-dimensional, resource-based proxy of the n-dimensional trophic niche of consumers.

Prey size and prey mobility are different aspects of a consumer's handling capacity while prey trophic position provides information on the trophic range of a consumer, and thus its potential impact on food web structure. We introduce a new and alternative approach for describing the trophic niche of consumers and the trophic diversity of a given system in terms of (1) the location (described by the three-element vector *trophic niche position*), (2) the width (described by *trophic flexibility*) and (3) the proximity (described by *trophic uniqueness*) of the trophic niches of the consumers and illustrate this with data from a complex marine system, the food web of Lough Hyne, a semienclosed sea lough in the southwest of Ireland.

To avoid potential misunderstanding we stress that *trophic flexibility* does not imply any switching behavior or adaptive foraging (Sargeant, 2007), but is a purely structural measurement of the width of the trophic niche. *Trophic uniqueness* is based on structural components of the community and is reciprocal to trophic redundancy. Communities with a large proportion of consumers characterized by similar trophic roles will exhibit a lower trophic uniqueness and consequently an increased trophic redundancy than a web in which consumer species are, on average, more divergent in terms of their trophic niche and role.

The distribution and characteristics of the consumer trophic niches in an ecosystem are nondynamical approaches to the study of the functional diversity and trophic structure of communities. As such the new metrics presented here could provide important insight into some of the driving forces that structure ecological communities by revealing trade-offs between and constraints on trophic traits of consumers. We hypothesize that species characterized by high trophic flexibility might be capable of buffering or compensating for species loss in contrast to species with low trophic flexibility. Thus, the new approach to the study of the functional diversity and trophic structure of communities presented here also complements previous topological and dynamical analyses of community robustness and it may be used to predict the response of food webs to species loss.

Lough Hyne, Ireland's first marine nature reserve (since 1981), is an internationally renowned marine ecological site with a very long history of biodiversity research. It offers a broad range of habitats and communities and species typical of the west of Ireland. Lough Hyne contains reefs that are very exposed to wave action on the open coast, as well as extremely sheltered reefs within the lough. Many of the communities found on the reefs are more characteristic of the exposed open coast. Lough Hyne is known for a very high species diversity and very high species richness for such a small area, it also inhabits many species near the edge of their biogeographic range. Such species may act as very good indicators for changes in biodiversity.

The Lough Hyne data set analyzed here is characterized by an extraordinary complex food web, owing to high species numbers and interactions, to a great variety in foraging strategies, and to an enormous range in body mass of species. Our new approach presented here is capable of bringing some order out of the apparent "chaos" that the trophic complexity of this real world ecosystem at first presents and some insights into its biodiversity value.

MATERIALS AND METHODS

Trophic Niche Dimensions and Parameters

Consider a food web with a total of n prey species. For each species i we compute the parameters body size S_i, trophic height T_i, and mobility M_i by the following procedures:

Body size S_i is represented by average wet body mass (g) of the species, which is either measured directly or, if no appropriate data are available, inferred from average body length (m) by means of well-established mass—length relationships for animal (Peters, 1983) and plant species (Niklas and Enquist, 2001):

$$S_i = a^* L_i^b \qquad (1)$$

where parameters a and b are specific for broad taxonomic groups. Following Brose et al. (2005) we use different relationships for marine mammals ($a = 40{,}790$; $b = 2.47$), birds ($a = 7390$; $b = 2.74$), fishes ($a = 10{,}600$; $b = 2.57$), invertebrates ($a = 80$; $b = 2.1$), and plants ($a = 27$; $b = 3.79$).

There are various ways to calculate the trophic level of a species within a food web. The prey-averaged trophic level is the TL calculation many prior studies have used, which is equal to 1 plus the mean trophic height of all the consumer's trophic resources (Williams and Martinez, 2004). Here, we use the short weighted trophic height, where the prey-averaged trophic height is weighted by the shortest chain within the network, as it is a better estimate of trophic height (Williams and Martinez, 2008).

$$T_i = \text{prey-averaged } TL_i \qquad (2)$$

Mobility M_i is categorized according to the 4-level scale "sessile or passive floater" (0), "crawler" (1), "facultative swimmer" (2), and "obligate swimmer" (3):

$$M_i \in \{0, 1, 2, 3\} \qquad (3)$$

In order to obtain identical data ranges [0, 1] and even distributions of the data, the trophic niche dimensions, S_i and T_i, are transformed into niche parameters X_i and Y_i by a sigmoid transfer function:

$$\begin{aligned} X_i &= \left(1 + e^{-\ln(S_i)^* \text{Gain}}\right)^{-1} \\ Y_i &= \left(1 + e^{-\ln(T_i)^* \text{Gain}}\right)^{-1} \end{aligned} \qquad (4)$$

where an even distribution is approached by minimizing the maximum difference between all frequencies of the corresponding distribution (class width $= 0.05$) through iterating "Gain." A log transformation would not have been able to achieve this.

Mobility M_i values were converted to Z_i by

$$Z_i = M_i / 3 \qquad (5)$$

to obtain a three-dimensional symmetric space where each axis x, y, z represents one species property and in which we can map all species of an ecosystem.

Consumer Trophic Niche Position

Trophic niche position of consumer j is defined by the mean coordinates $\overline{X}_j, \overline{Y}_j, \overline{Z}_j$ of all prey species of this consumer in the three-dimensional space:
Trophic Niche Position$_j = [\overline{X}_j, \overline{Y}_j, \overline{Z}_j]$ with

$$\overline{X}_j = \left(\sum_i^{m_j} X_k\right) \Big/ m_j \qquad (6)$$

where m_j is the number of prey species of consumer j and X_k is the sigmoid transformed body size of prey k (see above) of consumer j with $k = 1 \dots m_j$ and corresponding formulas for $\overline{Y}_j$ and $\overline{Z}_j$.

Consumer Trophic Uniqueness

Trophic uniqueness of consumer j describes the average trophic distance to its 10 nearest consumers in the three-dimensional niche space and is measured by the Euclidean distance between consumer *trophic niche positions* (see above). Euclidean distance $ED_{j,l}$ between consumer species j and l is computed from the corresponding *trophic niche positions* by

$$ED_{j,l} = \sqrt{(\overline{X}_j - \overline{X}_l)^2 + (\overline{Y}_j - \overline{Y}_l)^2 + (\overline{Z}_j - \overline{Z}_l)^2} \Big/ \sqrt{3} \qquad (7)$$

Trophic uniqueness of consumer j is then defined as the mean of the 10% lowest $ED_{j,l}$ values referring to consumer j, i.e., of the first percentile:

Trophic uniqueness$_j$ = Mean of first percentile $(ED_{j,l})$

Consumer Trophic Flexibility (= Trophic Niche Width)

Trophic flexibility of consumer j describes the average trophic distance between a consumers 10 most distant prey items in the three-dimensional niche space and is measured by the Euclidean distance between prey *trophic niche positions* (Figure 1). We compute the Euclidean distance $ED_{i,k}$ between prey species i and k by

$$ED_{i,k} = \sqrt{(X_i - X_k)^2 + (Y_i - Y_k)^2 + (Z_i - Z_k)^2} \Big/ \sqrt{3} \qquad (8)$$

Trophic flexibility of consumer j is then defined as the mean of the 10% highest values of $ED_{i,k}$, i.e., of the upper percentile:

Trophic flexibility$_j$ = Mean of upper percentile $(ED_{i,k})$.

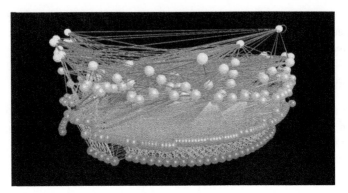

FIGURE 1 Food web of the Lough Hyne marine ecosystem.

Assignment of Ecosystem Services on the Species Level

Ecosystem services are the benefits human societies obtain from natural ecosystems. The following categories can be distinguished: provisioning, regulating, and cultural services provide direct benefits such as food, building material, medicine, genetic resources, climate regulation, clean air and water, recreational opportunities, inspiration, or spiritual values; supporting services are of indirect value to humanity but maintain all other services, as, for instance, by primary production or nutrient cycling. Here we analyze which benefits the marine ecosystem of Lough Hyne provides at the species level by assigning these services to each species in this food web. Assignment of the identified marine ecosystem services to each species of the Lough Hyne marine ecosystem communities is based on an extensive literature research. During the literature research the following keywords were used: "ecosystem service" + species name (scientific or common name) + region/ecosystem or "ecosystem service" + taxonomic group (phylum, class, order, family, etc.) + region/ecosystem. Instead of using the term "ecosystem service" the following terminology was used, too: (ecological) function, biology, ecology, feeding ecology, (ecological/economic) value, (ecological/cultural/commercially) importance/important, (ecological) significance, impact, role, conservation, benefits, (human) use/benefit, impact, role, conservation, benefits (human), use/benefit. In a second step, we searched for the species name plus the assumed ecosystem services provided by this species tourism, (commercial) fishing/landings, (commercial/recreational/sport/subsistence) fishery, aquaculture, (kelp) harvest, storm protection, bioturbation, habitat modifier, vegetative stabilization, filter feeder, medicine, drugs, antibiotic, anticancer, biotechnological/natural products, or (marine) physical/ecological engineers.

Lough Hyne Data Set

Our trophic niche approach is illustrated with a food web data set from Lough Hyne. Lough Hyne, a semienclosed marine sea lough in the southwest of

Ireland. Lough Hyne (51.50° N; 09.30° W) offers a broad range of habitats, a high species diversity, and high species richness for such a small area; it also contains many species that are near the edge of their biogeographic range. Such species act as very good indicators for changes in trophic structure and ecosystem functioning. Our Lough Hyne data set consists of 343 species, plus the three trophic entities phytodetritus, sediment, bacteria, and particulate organic matter. Diet composition of each species was inferred from a combination of field observations, stomach contents, and predation experiments; 5117 feeding links have been documented (Figure 1). For the food web construction, following the approach of Martinez (1991), a directional feeding link was assigned to any pair of species A and B whenever an investigator reports that A consumes B. Species were not divided further into larvae, juveniles, or adults, but treated as "adults": consequently, with the data used here we cannot address ontogenetic diet shifts. The average body mass of the species populations was either directly measured (>90%) or in the case of marine mammals and seabirds taken from published accounts (Riede et al., 2010, 2011).

Statistical Analysis

To explore whether (1) our niche parameters/functional traits body size, trophic level, and mobility or (2) our indices of *trophic flexibility* and *trophic uniqueness* are related, and whether consumer's trophic generality (in terms of number of prey items; e.g., Memmott et al., 2000) is related to our indices or if the number of ecosystem services provided by the different species is correlated to traits and indices noted above, we conducted simple pairwise correlations.

RESULTS

Average body mass of Lough Hyne species stretched across several orders of magnitude, from 1.43×10^{-14} g in small unicellular algae to 173,100 g in the Grey Seal *Halichoerus grypus*. The trophic height of species ranged from 1.00 in diatoms or the Dead Man's finger algae *Codium fragile* to 4.75 in the Grey Seal *H. grypus*. All mobility levels are well represented: sessile or floating species such as porifera, bryozoans, detritus, and diatoms, crawlers like asteroids, echinoids, and holothurians, facultative swimmers like some amphipods and isopods, and obligate swimmers like copepods, fishes, and seals.

Trophic Flexibility and Trophic Uniqueness of Lough Hyne Consumers

The marine Lough Hyne ecosystem contains populations with a wide range of *trophic flexibility*, from 0.005 in the pelagic herbivorous grazer

Eutemora hirundoides to 0.750 in the benthic predator, the Thornback Ray *Raja clavata*. The index of *trophic uniqueness* ranges from zero in sessile benthic suspension feeders, which mainly consume particle organic matter and bacteria, to 0.337 in the Grey Seal *H. grypus*, a dominant carnivorous species. The marine Lough Hyne ecosystem is dominated by species with a lower *trophic uniqueness* in general, and just few species have intermediate values of *trophic uniqueness*. *Trophic flexibility* is not significantly correlated to consumer size, consumer mobility, or trophic level even though some interesting patterns emerge (Figure 2). There is no systematic or significant relationship between the *trophic uniqueness* of a consumer and its body size, its trophic level, or mobility (Figure 3).

Ecosystem Service Provisioning by Species of Lough Hyne

The number of services provided per species ranged between a minimum of seven services per species and up to 13 ecosystem services provided at the species level. We could detect a positive correlation between species body weight and trophic level and number of services provided (Figure 4), but no positive correlations between the number of ecosystem services provided and the species uniqueness or flexibility (Figure 5).

DISCUSSION

There has been increasing recognition that external forcing, either anthropogenic or environmental, can profoundly impact ecosystems, from local to global scales, causing a rearrangement of their internal structure and a deviation from their original succession. The linkages between global climate perturbation and marine ecosystem responses are only just being explored. Impacts on marine systems take place against a background of complex ecological and evolutionary processes that have shaped the marine biodiversity we observe today, and that determine the form and extent of these impacts as well as the effectiveness of conservation, management, and restoration strategies. Understanding how ecological, evolutionary, and socioeconomic factors interact to determine the dynamics of biodiversity at different scales, both in space and in time, is an important challenge. The most important tasks are to identify the mechanisms by which climate change is acting on species interactions, to quantify the magnitude and direction of these effects, and to develop a number of scenarios that predict ecosystem reaction to climate change. Ignoring the direct and indirect interactions within an ecological network and between species is problematic, since it constrains our ability to forecast how species/biodiversity loss will impact future ecosystem service provisioning; i.e., ignoring species interactions has hampered well-intended human manipulations or conservation actions in natural communities. The reintroduction of rock lobsters in

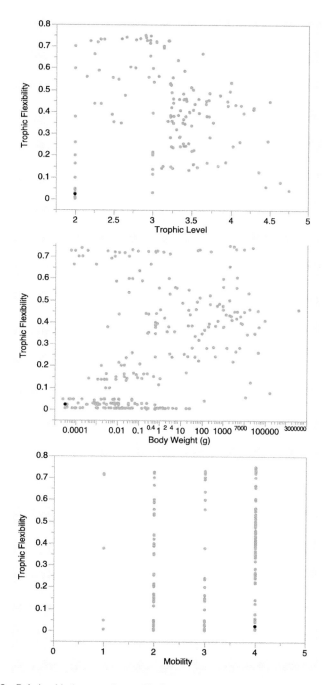

FIGURE 2 Relationship between the trophic flexibility of the marine consumers of the Lough Hyne marine ecosystem and their respective trophic level, body size, and mobility.

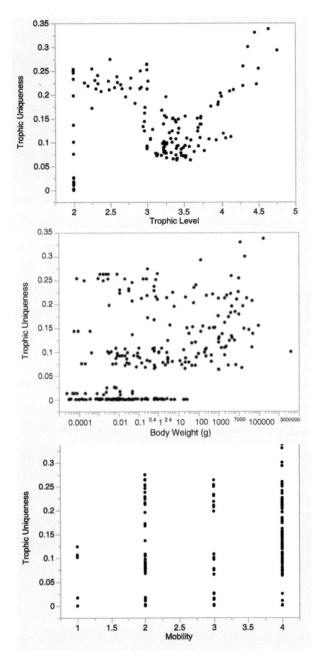

FIGURE 3 Relationship between the trophic uniqueness of the marine consumers of the Lough Hyne marine ecosystem and their respective trophic level, body size, and mobility.

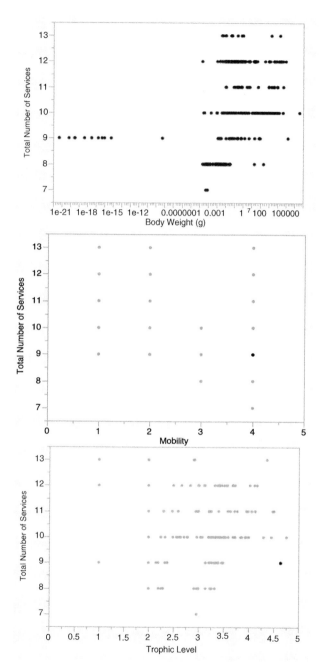

FIGURE 4 Relationship between the ecosystem service provisioning of the species of the Lough Hyne marine food web and their respective trophic level, body size, and mobility.

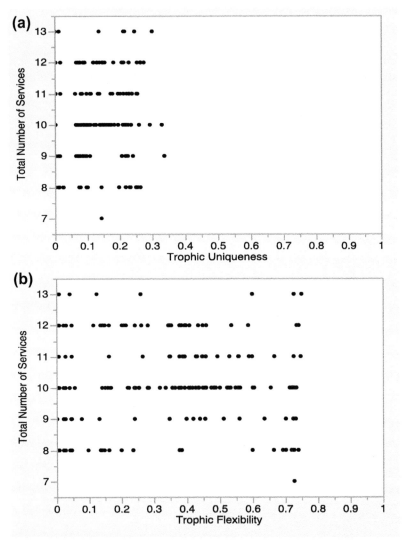

FIGURE 5 Relationship between the ecosystem service provisioning of the species of the Lough Hyne marine food web and their respective (a) *trophic uniqueness* and (b) *trophic flexibility* ((a), in terms of trophic niche width).

Marcus Island failed, as the reintroduced lobster population was diminished by its previous prey, the now overabundant predacious whelks (Doak et al., 2008).

Ecological network theory has shown that functional groups of species can be revealed based on their pattern of interactions and the majority of network structures may arise from functional traits (Eklöf et al., 2013) and certain traits (i.e., body size) possessed by extinction-prone species make the network more vulnerable to cascading extinctions (Petchey et al., 2008; Jacob et al., 2011).

Our metrics *trophic niche position, trophic uniqueness*, and *trophic flexibility* of a species do not describe simple one-dimensional characteristics of a consumer species. Instead they are based on traits of the prey consumed and measure and combine these into three metrics, related to numerous different ecological characteristics and constraints affecting a consumers foraging (Hutchinson, 1957; Chase and Leibold, 2003). The number of trophic niche dimensions in marine ecosystems is potentially indefinite and hence the number of relevant niche dimensions is a question of accuracy and precision. Even when only a modest number of likely significant parameters is considered, for example, nutritional quality of prey species, prey preference, or prey availability, the effort involved when working in an open and species-rich marine system is most likely beyond the capacity of any scientific endeavor. We use a particular subset of parameters that provides operational and meaningful descriptors of the position and characteristics of the trophic niche of consumers, while requiring a reasonable amount of measurement effort. In the following we discuss how these measures can be used to characterize distinctly different as well as trophically significant ecological properties of species as well as ecosystems.

Trophic Niche Dimensions and Parameters

Our characterization of the *trophic niche position, trophic flexibility*, and *trophic uniqueness* of consumers is based on three parameters. For each of these a number of potential caveats come readily to mind. First, the set of prey size of a consumer reflects the resource size range a consumer can handle. Ratios of consumer body size to resource body size express relative consumer sizes (Cohen et al., 1993; Brose et al., 2006a) that are positively correlated to interaction strength patterns in real ecosystems, which in turn may determine food web stability (Jonsson and Ebenman, 1998; Wootton and Emmerson, 2005; Cohen et al., 2005; Brose et al., 2006b; Otto et al., 2007). If logistical constraints prevent body-mass measurements of all species—as in ecosystems of high species diversity such as the Lough Hyne ecosystem—a simplified approach uses mass–length relationships, because body lengths are easier and faster to determine (Peters, 1983; Niklas and Enquist, 2001). Second, the trophic level of a resource species may be measured either through the trophic structure of the food web with trophic links weighted by energy flow (e.g., based on stomach content analysis) or by means of stable isotope ratios (Kline and Pauly, 1998; Williams and Martinez, 2004). Both methods should produce comparable results, when properly conducted, but a valid confirmation of this assumption is missing yet. Third, prey mobility reflects a consumer's capability of handling prey of different agilities. All four levels of resource mobility used are well represented in Lough Hyne: sessile or floating items, crawlers, facultative swimmers, and obligate swimmers. The mobility categories used here are a simplified view of what in the real world must be a continuum in prey mobility. Nevertheless, our data analyzed here and results obtained clearly reflect the generalist behavior of most consumer populations in the Lough Hyne ecosystem. Many consumers are able to capture prey species along the mobility dimension, irrespective of their own mobility. For example, benthic anemones, sessile species of very low mobility, are able to catch highly mobile prey such as copepods and amphipods or even fishes, when they get within the reach of the anemone's tentacles.

Trophic Flexibility and Trophic Uniqueness of Lough Hyne Consumers

Our measure of *trophic uniqueness* approximates differences among consumer trophic niches. Overall *trophic uniqueness* in the Lough Hyne system is relatively low. Most species have equivalent counterparts with a similar diet that may take over the trophic roles if another consumer dies out. Interestingly, our results suggest that the carnivorous Grey Seals are trophically unique, and their loss may not be entirely compensated for by other consumers. Given this example, our approach of measuring *trophic uniqueness* implies field-testable hypotheses on the relative species redundancy of a species and subsequently the consequences of species loss. The distribution of *trophic flexibility* reflects common trophic traits among consumers within marine benthic-dominated ecosystems. Most fish species and predatory invertebrates are opportunistic predators with a high trophic generality; they are quite flexible in their alimentation. This suggests that (1) the high trophic complexity of the Lough Hyne ecosystem results from most species' high *trophic flexibility* and (2) most consumers are capable of compensating other species' loss.

Distribution of Trophic Flexibility and Trophic Uniqueness

Body size is positively correlated to mobility, trophic position, and generality of consumers (Peters, 1983; Cohen et al., 2003). Generally, this suggests a positive relationship between body size and *trophic flexibility*. However, this relationship is poorly developed in the Lough Hyne system. Large animals typically feed on very small prey (seals → fish → amphipods → phytoplankton) and small benthic omnivorous species (e.g., amphipods, crabs, and gastropods) feed up and down the food chain irrespective of their size. Moreover, the Lough Hyne food web comprises many consumer species with a high *trophic flexibility* but no consumer with high *trophic uniqueness*. This can be explained by the high dietary overlap and generalist feeding nature of most of the species of the Lough Hyne ecosystem.

Ecosystem Service Provisioning by Species of Lough Hyne

Our results here generally confirm previously made assumptions that high trophic level species and larger bodied species provide in general higher numbers of services. But they can be assumed to be the least resilient species within a community (Dobson et al., 2006), which implements a high sensitivity of ecosystem services toward species loss, too. Beyond important functional groups there were some species identified being of particular importance by providing outstanding large numbers of ecosystem services or due to their functional role as keystone species across the Lough Hyne marine system.

CONCLUSIONS

The approach introduced here allows a quantitative description of consumer trophic niches, and thus provides important insights in the trophic structure of the highly complex Lough Hyne ecosystem. The three trophic niche dimensions prey size, prey mobility, and prey trophic level are sufficient for generating meaningful indices of trophic uniqueness and trophic flexibility. The Lough Hyne ecosystem displays distinct features: trophic uniqueness is low while trophic flexibility covers a wider range (Figure 2). We hypothesize that communities where most species are characterized by an intermediate to high trophic flexibility, and a simultaneously low trophic uniqueness, are more robust in many ways than communities with a comparably low trophic flexibility and a high trophic uniqueness. This is because (1) low trophic flexibility of consumers confers vulnerability to secondary extinctions as a consequence of losing prey species and (2) high trophic uniqueness (low redundancy) of consumers implies high vulnerability of community functions to the loss of consumer species (due to low capacity for compensation among species). This is consistent with prior findings (Petchey et al., 2008) showing that highly unique species are more vulnerable to extinction during a secondary extinction cascade. As a consequence, secondary extinctions tend to lead to a greater loss of trophic and probably functional diversity than expected if extinctions are random (Petchey et al., 2008).

We stress that our hypotheses are the result of the new approach introduced here, and that other case studies are needed to test our predictions. However, our approach could also provide important insights into the driving forces when comparing the trophic structure of different ecosystems and therefore studies from other systems are needed to challenge the findings presented here. Relationships among trophic uniqueness, trophic flexibility, and the number of species in the food web may provide one link among functional diversity and ecosystem function and ecosystem service provisioning. The niche concept is often used to explain the effects of changes in biodiversity on ecosystem functioning (e.g., Loreau and Hector, 2001; Layman et al., 2007a,b). Characteristics of a species determine how, when, and where it utilizes resources (the niche) and whether or not species can be considered as functionally unique or redundant (Naeem and Wright, 2003; Gaston and Chown, 2005). Loss of species will therefore lead to fewer utilized niches, stronger competition, and lower energy turnover rates, thus affecting ecosystem functioning negatively as long as there are no trophically equivalent and flexible species to fill in the vacant niches with the same set of ecosystem services provided. Here we have based our characterization of the consumer's trophic niche on three trophic traits. Future studies will indicate whether these three dimensions are the most adequate and sufficient for constructing meaningful indices of trophic uniqueness and trophic flexibility and how they relate to ecosystem service provisioning. We believe that these consumer properties will be useful for predicting why certain species are abundant, which species are rare, and how their trophic roles will change in response to environmental gradients or climate change and which services will then be subsequently at stake. We are convinced that these indices or other tractable surrogate—correlates of functional diversity are needed to evaluate the effects of environmental change, species loss, and species gain on the trophic structure and functioning as well as ecosystem service provisioning of complex ecosystems.

Analyzing possible consequences of species loss or gain for the provision of ecosystem services at the species level will allow for reducing the risks of service losses and help in designing efficient conservation plans constrained by the knowledge of which species directly or indirectly affect service delivery.

REFERENCES

Bearhop, S., Adams, C.E., Waldron, S., Fuller, R.A., MacLeod, H., 2004. Determining trophic niche width: a novel approach using stable isotope analysis. J. Anim. Ecol. 73, 1007–1012.

Bolnick, D.I., Svanbäck, R., Fordyce, J.A., Yang, L.H., Davis, J.M., Hulsey, C.D., Forister, M.L., 2003. The ecology of individuals: incidence and implications of individual specialization. Am. Nat. 161, 1–28.

Brose, U., Berlow, E.L., Martinez, N.D., 2005. Scaling up keystone effects from simple to complex ecological networks. Ecol. Lett. 8, 1317–1325.

Brose, U., Jonsson, T., Berlow, E.L., Warren, P., Banasek-Richter, C., Bersier, L.-F., Blanchard, J.L., Brey, T., Carpenter, S.R., Cattin Blandenier, M.-F., Cushing, L., Dawah, H.A., Dell, T., Edwards, F., Harper-Smith, S., Jacob, U., Ledger, M.E., Martinez, N.D., Memmott, J., Mintenbeck, K., Pinnegar, J.K., Rall, B.C., Rayner, T., Reuman, D.C., Ruess, L., Ulrich, W., Williams, R.J., Woodward, G., Cohen, J.E., 2006a. Consumer-resource body-size relationships in natural food webs. Ecology 87, 2411–2417.

Brose, U., Williams, R.J., Martinez, N.D., 2006b. Allometric scaling enhances stability in complex food webs. Ecol. Lett. 9, 1228–1236.

Chase, J.M., Leibold, M.A., 2003. Ecological Niches: Linking Classical and Contemporary Approaches, 212 pp. The University of Chicago Press, Chicago.

Cohen, J.E., 1978. Food Webs and Niche Space. Princeton University Press, Princeton.
Cohen, J.E., Pimm, S.L., Yodzis, P., Saldaña, J., 1993. Body sizes of animal predators and animal prey in food webs. J. Anim. Ecol. 62, 67–78.
Cohen, J.E., Jonsson, T., Carpenter, S.R., 2003. Ecological community description using the food web, species abundance, and body size. Proc. Natl. Acad. Sci. 100, 1781–1786.
Cohen, J.E., Jonsson, T., Muller, C.B., Godfray, H.C.J., Savage, V.M., 2005. Body sizes of hosts and parasitoids in individual feeding relationships. Proc. Natl. Acad. Sci. 102, 684–689.
Díaz, S., Cabido, M., 2001. Vive la difference: plant functional diversity matters to ecosystem processes. Trends Ecol. Evol. 16, 646–655.
Dobson, A., Lodge, D., Alder, J., Cumming, G.S., Keymer, J., McGlade, J., et al., 2006. Habitat loss, trophic collapse and the decline of ecosystem services. Ecology 87, 1915–1924.
Doak, D.F., Estes, J.A., Halpern, B.S., Jacob, U., Lindberg, D.R., Lovvorn, J., et al., 2008. Understanding and predicting ecological dynamics: are major surprises inevitable? Ecology 89, 952–961.
Dunne, J.A., Williams, R.J., Martinez, N.D., 2002. Network structure and biodiversity loss in food webs: robustness increases with connectance. Ecol. Lett. 5, 558–567.
Ebenman, B., Jonsson, T., 2005. Using community viability analysis to identify fragile systems and keystone species. Trends Ecol. Evol. 20, 568–575.
Ehrlich, P., Ehrlich, A., 1981. Extinction: The Causes and Consequences of the Disappearance of Species. Random House, New York.
Eklöf, A., Jacob, U., Kopp, J., Bosch, J., Castro-Urgal, R., Chacoff, N.P., et al., 2013. The dimensionality of ecological networks. Ecol. Lett. 16, 577–583.
Gaston, K.J., Chown, S.L., 2005. Neutrality and the niche. Funct. Ecol. 19, 1–6.
Gitay, H., Wilson, J.B., Lee, W.G., 1996. Species redundancy: a redundant concept? J. Ecol. 84, 121–124.
Hutchinson, G.E., 1957. Concluding remarks. Cold Spring Harbor Symp. Quant. Biol. 22, 415–427.
Jacob, U., Thierry, A., Brose, U., Arntz, W.E., Berg, S., Brey, T., et al., 2011. The role of body size in complex food webs. Adv. Ecol. Res. 45, 181–223.
Johnson, K.H., Vogt, K.A., Clarke, H.J., Schmitz, O.J., Vogt, D.J., 1996. Biodiversity and the production and stability of ecosystems. Trends Ecol. Evol. 11, 372–377.
Jonsson, T., Ebenman, B., 1998. Effects of predator-prey body size ratios on the stability of food chains. J. Theor. Biol. 193, 407–417.
Kline Jr., T.C., Pauly, D., 1998. Cross-validation of trophic level estimates from a mass-balance model of Prince William Sound using $^{15}N:^{14}N$ data. In: Quinn, T.J., Funk, F., Heifetz, J., Ianelli, J.N., Powers, J.E., Schweigert, J.F., Sullivan, P.J., Zhang, C.I. (Eds.), Proceedings of the International Symposium on Fishery Stock Assessment Models. Alaska Sea Grant College Program Report No. 98-01. Alaska Sea Grant, Fairbanks, pp. 693–702.
Layman, C.R., Quattrochi, J.P., Peyer, C.M., Allgeier, J.E., 2007a. Niche width collapse in a resilient top predator following ecosystem fragmentation. Ecol. Lett. 10, 937–944.
Layman, C.R., Arrington, D.A., Montana, C.G., Post, D.M., 2007b. Can stable isotope ratios provide quantitative measures of trophic diversity within food webs? Ecology 88, 42–48.
Loreau, M., Hector, A., 2001. Partitioning selection and complementarity in biodiversity experiments. Nature 412, 72–76.
Loreau, M., Naeem, S., Inchausti, P., Bengtsson, J., Grime, J.P., Hector, A., 2001. Biodiversity and ecosystem functioning: current knowledge and future challenges. Science 294, 804–808.
Loreau, M., Naeem, S., Inchausti, P., 2002. Biodiversity and Ecosystem Functioning. Synthesis and Perspectives. Oxford University Press, Oxford.

MacArthur, R.H., Pianka, E.R., 1966. On the optimal use of a patchy habitat. Am. Nat. 100, 603–609.
Martinez, N.D., 1991. Artifacts or attributes? effects of resolution on the little rock lake food web. Ecol. Monogr. 61, 367–392.
Matthews, B., Mazumder, A., 2004. A critical evaluation of intra-population variation of $\delta^{13}C$ and isotopic evidence of individual specialization. Oecologia 140, 361–371.
Memmott, J., Martinez, N.D., Cohen, J.E., 2000. Predators, parasitoids and pathogens: species richness, trophic generality and body sizes in a natural food web. J. Anim. Ecol. 69, 1–15.
Naeem, S., Wright, J.P., 2003. Disentangling biodiversity effects on ecosystem functioning: deriving solutions to a seemingly insurmountable problem. Ecol. Lett. 6, 567–579.
Niklas, K.J., Enquist, B.J., 2001. Invariant scaling relationships for interspecific plant biomass production rates and body size. Proc. Natl. Acad. Sci. USA 98, 2922–2927.
Otto, S., Rall, B.C., Brose, U., 2007. Allometric degree distributions facilitate food web stability. Nature 450, 1226–1229.
Petchey, O.L., Gaston, K.J., 2002. Functional diversity (FD), species richness and community composition. Ecol. Lett. 5, 402–411.
Petchey, O.L., Downing, A.L., Mittelbach, G.G., Persson, L., Steiner, C.F., Warren, P.H., 2004a. Species loss and the structure and functioning of multitrophic aquatic ecosystems. Oikos 104, 467–478.
Petchey, O.L., Hector, A., Gaston, K.J., 2004b. How do different measures of functional diversity perform? Ecology 85, 847–857.
Petchey, O.L., Eklöf, A., Borrvall, C., Ebenman, B., 2008. Trophically unique species are vulnerable to cascading extinction. Am. Nat. 171, 568–579.
Peters, R.H., 1983. The Ecological Implications of Body Size. Cambridge University Press, New York, NY, USA.
Pielou, E.C., 1972. Niche width and niche overlap: a method for measuring them. Ecology 53, 687–692.
Riede, J.O., Rall, B.C., Banasek-Richter, C., Navarrete, S.A., Wieters, E.A., Emmerson, M.C., Jacob, U., Brose, U., 2010. Scaling of food-web properties with diversity and complexity across ecosystems. Adv. Ecol. Res. 42, 139–170.
Riede, J.O., Brose, U., Ebenman, B., Jacob, U., Thompson, R., Townsend, C.R., Jonsson, T., 2011. Stepping in Elton's footprints: A general scaling model for body masses and trophic levels across ecosystems. Ecol. Lett. 14, 169–178.
Roughgarden, J., 1972. Evolution of niche width. Am. Nat. 106, 683–718.
Roughgarden, J., 1974. Niche width: biogeographic patterns among anolis lizard populations. Am. Nat. 108, 429–442.
Sargeant, B.L., 2007. Individual foraging specialization: niche width versus niche overlap. Oikos 116, 1431–1437.
Stewart, K.M., Bowyer, T., Kie, J.G., Dick, B.L., Ben-David, M., 2003. Niche partitioning among mule deer, elk, and cattle: do stable isotopes reflect dietary niche? Ecoscience 10, 297–302.
Tilman, D., 1999. The ecological consequences of changes in biodiversity: a search for general principles. Ecology 80, 1455–1474.
Tilman, D., 2001. In: Levin, S.A. (Ed.), Functional Diversity. Encyclopaedia of Biodiversity. Academic Press, San Diego, California, pp. 109–120.
Walker, B., 1992. Biodiversity and ecological redundancy. Conserv. Biol. 6, 18–23.
Williams, R.J., Martinez, N.D., 2000. Simple rules yield complex food webs. Nature 404, 180–183.
Williams, R.J., Martinez, N.D., 2004. Limits to trophic levels and omnivory in complex food webs: theory and data. Am. Nat. 163 (3), 458–468.

Williams, R.J., Martinez, N.D., 2008. Success and its limits among structural models of complex food webs. J. Anim. Ecol. 77, 512–519.

Winemiller, K.O., Pianka, E.R., Vitt, L.J., Joern, A., 2001. Food web laws or niche theory? six independent empirical tests. Am. Nat. 158, 193–199.

Wootton, J.T., Emmerson, M., 2005. Measurement of interaction strength in nature. Annu. Rev. Ecol. Syst. 36, 419–444.

Section III

In the Wild: Biodiversity and Ecosystem Service Conservation

Chapter 9

The Role of Marine Protected Areas in Providing Ecosystem Services

Pierre Leenhardt[1], Natalie Low[2], Nicolas Pascal[3], Fiorenza Micheli[2] and Joachim Claudet[4]

[1]*CRIOBE, CNRS-EPHE, Perpignan, France;* [2]*Hopkins Marine Station, Stanford University, Pacific Grove, CA, USA;* [3]*Ecole Pratique des Hautes Etudes, CRIOBE, CNRS-EPHE, Perpignan, France;* [4]*National Centre for Scientific Research, CRIOBE, CNRS-EPHE, Perpignan, France*

INTRODUCTION

Marine ecosystems experience constant change and adaptation processes because they are under the influences of a suite of pressures (Hooper et al., 2012). Human impacts can affect the ecosystem functioning of marine ecosystems and reduce the associated production of goods and services required for human well-being (Cardinale et al., 2012; Mora et al., 2011). For example, major concerns are rising over observed declines in the abundance of particular species as well as reductions in functional diversity and changes in food web structure due to the intensity of some anthropogenic activities (De'ath et al., 2012; Hughes et al., 2010). These changes induce strong modifications of whole ecosystems or some of their components, resulting in loss of function (Graham et al., 2013). Such ecosystem disruptions may affect the flow of ecosystem services (such as food provision) that are vital for human well-being (Carpenter et al., 2006; Chapin et al., 2000; Díaz et al., 2006). As a result, the conservation and/or restoration of marine biodiversity and its derived ecosystem goods and services are major concerns. To this end, marine protected areas (MPAs) are being established worldwide to maintain biodiversity, ecosystem functions, and the flow of ecosystem services (Gaines et al., 2010). MPAs are a specific type of management zone—they may allow some uses, including scuba diving and some types of fishing; may be strictly no-take such as marine reserves; or they may be completely no-access zones where neither extractive nor nonextractive uses are allowed (Day and Dobbs, 2013). Most MPAs include another layer of complexity by combining different levels of protection within a spatially zoned management scheme. Zones may be dedicated to strict conservation, act as buffer zones that can be used for research, education, or traditional uses, and/or allow nonconsumptive and limited-consumptive uses (Agardy et al., 2003).

Today, MPAs are commonly used around the world as management tools to promote the sustainable use of marine resources (Hargreaves-Allen et al., 2011). In this chapter, we will review the different impacts of MPAs on ecosystem functioning and service production. We will focus especially on the relationship between the effects of MPAs on ecosystem functioning and the benefits provided to people. The livelihoods and well-being of coastal communities rely on ecosystem services produced by marine ecosystems. Thus, it is assumed that MPAs secure human livelihoods and well-being by protecting marine ecosystems and ecosystem services. However, the links between ecological effects of MPAs and services have rarely been explored.

The aims of this synthesis are to (1) identify relationships between the effects of MPAs on ecosystem functioning and service provision; (2) identify knowledge gaps on which future research efforts could focus; and (3) empower marine resource managers to make more informed decisions and maximize the value derived from their natural resource base. We propose that quantification and monitoring of species' functional trait distribution and assemblages' functional diversity are promising approaches for assessing the effects of MPAs on ecosystem functioning and services.

INTRODUCTION TO MARINE PROTECTED AREAS

MPAs are globally important management tools that are expected to (1) control and manage human activities and marine uses; (2) promote the recovery of exploited marine populations; (3) conserve or restore habitats, biodiversity, and food webs; and (4) manage and enhance ecosystem services such as food production, water purification, or recreational activities (Halpern, 2014; Liquete et al., 2013). Most MPAs are implemented to mitigate some of the human-induced modifying forces on marine ecosystems, especially by reducing or removing fishing mortality (Claudet, 2012). Originally, MPAs and especially "no-take" marine reserves were conceived as pragmatic means to eliminate harvest pressure and thereby protect marine depleted and endangered species, habitats, fisheries, and ecosystems, and to provide public enjoyment of the oceans (Mora et al., 2011). Today they are also used as management tools regulating fishing, tourism, and industrial activities. Thanks to different types of zoning, each established according to specific management goals, MPAs can reduce conflict and allow coexistence of different resource uses. Establishment of different-use zones must be combined with the establishment of: easily identifiable borders to reduce possible impacts of incidental intrusions; public information about uses permitted in different zones; and the participation of local communities and diverse users who contribute to the process (Hargreaves-Allen et al., 2011). Compliance with spatial zoning regulations, such as those within an MPA, depends on whether users understand the regulations designed to ensure the orderly and sustainable use of marine resources. If compliance is good, additional management costs to ensure zoning enforcement will be reduced.

In recent years, MPA research has made several advances. First, empirical data and analyses have shown how MPA effects are driven by different factors such as MPA age, size, fish life history traits, and the level of enforcement (Claudet et al., 2008, 2010; Guidetti et al., 2008). These findings have important implications for MPA design and management. For example, if even young and small MPAs can be effective in increasing fish population density, then old, large, and isolated MPAs may show even greater positive responses (e.g., Edgar et al., 2014). Meanwhile, no positive responses should be expected from MPAs with low levels of enforcement (Guidetti et al., 2008). Second, major advances were made on the numerous indirect ecological effects of protection such as functional diversity and delivery of ecosystem services, which are also time-dependent (Fletcher et al., 2011). Third, the potential socioeconomic benefits of MPAs are now becoming clearer. Studies show, for example, that MPAs can enhance food security, empower local communities (Mascia et al., 2010), and lead to jobs and/or revenue increases in activities linked to MPAs such as fishing and tourism, as well as to the maintenance of traditional activities (McCook et al., 2010; Pascal, 2014), although negative impacts on some users have also been documented (Mascia et al., 2010). Fourth, the general agreement among scientists that MPA networks can optimize conservation and fishery benefits has led to significant advances in network design and evaluation.

Considering these recent findings, it is clear that MPAs can provide different types of benefits. They can ensure the protection and/or the restoration of marine biodiversity that provide multiple ecosystem functions and human benefits. Below we provide a definition and overview of ecosystem services, then review the expected and documented effects of MPAs on the delivery of selected ecosystem services vital for human activities (e.g., fishing and recreational activities) and well-being.

INTRODUCTION TO ECOSYSTEM SERVICES AND THE LINK TO HUMAN WELL-BEING

Ecosystem services are the benefits people derive from nature (Liquete et al., 2013). They are the cornerstones of marine resource systems and are widely used to describe human–nature interactions (Diaz et al., 2011). Thus, ecosystem services support natural ecosystems, livelihoods, and human well-being through direct and indirect processes (Liquete et al., 2013). A conceptual model that represents those different interactions is a cascade linking the

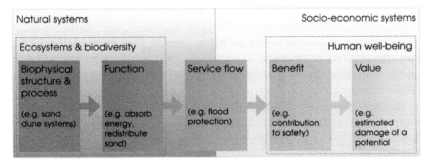

FIGURE 1 The cascade framework showing how natural provision of ecosystem services contributes to human well-being. *From Liquete et al. (2013).*

biophysical structure and processes with the benefit that people eventually derive (Figure 1). It highlights not only that ecosystems provide services but also that services do not exist in isolation from people's needs (Haines-young and Potschin, 2010).

The Millennium Ecosystem Assessment of 2005 (MA, 2005) classified ecosystem services into four categories: *provisioning services* such as food, water, timber, fiber, genetic resources, and pharmaceuticals; *regulating services* controlling climate, air and water quality, erosion, disease, pests, wastes, and natural hazards; *cultural services* providing recreational, aesthetic, and spiritual benefits; and *supporting services* such as nutrient and water cycling, soil formation, and primary production. According to the MA, approximately 60% of ES are degraded, including capture fisheries.

MPAs are key management tools established to secure the delivery of marine ecosystem services and thus contribute to human well-being (Fletcher et al., 2011). Indeed, improvement in the quality of the natural environment provided by MPAs is expected to strengthen the capacity of coastal ecosystems to produce goods and services for local people, local and nonlocal entrepreneurs, and the global community (TEEB, 2010). From a socioeconomic perspective, MPAs may be seen as public investments in marine ecosystems conservation (Laurans et al., 2013). As such, a basic question is the magnitude of MPAs' contributions to individual and societal well-being. This question has been tackled by cost–benefit analysis.

Alban et al. (2011) synthesized assessments of income and jobs related to the presence of 12 Mediterranean MPAs. A distinction was made between users obtaining commodities (commercial fishers) and recreational users (recreational fishers and scuba divers). Income generated by MPAs was generally high, particularly for commercial fishing and recreational scuba diving. The average yearly money incomes locally generated by uses in MPAs amounted to 710,000 € per MPA (between 48,000 € in Medes Islands and 1,573,000 € in Columbretes) in the case of professional fishing, 551,000 € per MPA (between 16,000 € in Tabarca and 1,099,000 € in Medes Island) in the case of scuba diving, and 88,000 € per MPA (between 35,000 € in La Graciosa and 211,000 € in Monte da Guia) in the case of recreational fishing (Roncin et al., 2008). These figures should be compared with yearly MPA management costs, which amounted on average to 588,000 € per year per MPA. However, the contribution to different economic sectors varied greatly from place to place. On a relatively remote MPA (Columbretes, Spain), the economic contribution of commercial fishing was dominant. This activity generated nearly 90% of all income provided by the ecosystem use. On an MPA closer to densely populated areas (Medes), incomes generated by commercial fishing amount to only 5% of those generated by scuba diving (Alban et al., 2011). In the Great Barrier Reef Marine Park in Australia, the estimated distribution of economic value between recreational uses and commercial fishing is approximately 4:1 (Stoeckl et al., 2011). However, despite significant improvements in recent years, this type of assessment faces substantial difficulties. First, the limited availability of economic data at a relevant scale frequently hinders a complete assessment of the influence of MPAs on the economy of the neighboring zone (Laurans et al., 2013). Moreover, total value is always underestimated because the measurement of nonmarket values, including nonuse values such as the value of marine biodiversity, is a difficult task. Even assessing the impact of an MPA using some specific market values (e.g., fishery rent) may be problematic due to limited quantitative information on underlying ecological processes (e.g., larval and juvenile spillover from MPAs to fishing grounds) (Pascal and Seidl, 2013); the use of CPUE was suggested as a way to bypass this issue. As a result, the application of cost--benefit analysis to MPAs is generally incomplete (François et al., 2012), providing an assessment of only a part of the net benefits MPAs provide. Below we review the expectations and evidence for MPAs' contributions to a selected group of ecosystem services.

MARINE PROTECTED AREA EFFECTS ON INDIVIDUAL ECOSYSTEM SERVICES

Review of the literature reveals that MPA establishment is expected to support a suite of services. Here, we provide an overview of the conceptual or empirical basis for such effects (Table 1). As a first step toward establishing a link between ecological change in MPAs and service provision, we also discuss what functional traits of species, functional groups, and/or ecological community attributes underlie MPAs' effects on services.

Marine Protected Area Effects on Provisioning Services: The Example of Fisheries

MPAs support provisioning services through their effects on fisheries and diversity (Worm et al., 2006). The first anticipated effects of establishing

TABLE 1 Examples of the Effects of MPAs on Ecosystem Services, the Mechanism Underlying Service Provision, and the Functional Traits, Functional Groups, and/or Community Attributes Driving These Effects

Ecosystem Service Category (MA, 2005)	Ecosystem Service	Mechanism by which MPAs Provide Service	Species, Community Attribute, Functional Trait, or Functional Group Underlying Effect	References
Provisioning services	Food	Increased production/stabilization of target species biomass	Large size of target species, recovery of top predators and food web complexity	Goñi et al. (2010)
	Ornamental resources	Increased production/stabilization of ornamental fish biomass	Species diversity	Williams et al. (2009)
	Raw materials	Algal and sand production	Predators controlling herbivory, bioeroders, corallivores	Karnauskas and Babcock (2014)
	Genetic resources	Protection of genetic diversity, adaptation to climate change	Response diversity, genetic diversity	Miller and Ayre (2008)
	Medicinal resources	Protection of molecular diversity	Chemically defended species, biological diversity	Schroder et al. (2004)
Regulating services	Carbon sequestration and climate regulation	Protection of plants and calcifying organisms (e.g., mangroves, sea grass, corals)	Species that have high C sequestration capacity (primary producers, calcifying species, bioconstructors)	Gonzalez-Correa et al. (2007)

Category	Service	Description	Indicators	References
Cultural services	Cultural heritage	Maintenance of traditional community-based natural resource management	Charismatic species (e.g., sharks, sea turtles, large mollusks)	Clarke and Jupiter (2010)
	Spiritual and historical heritage	Maintenance of traditional community-based natural resource management	Charismatic habitat (e.g., coral reef, kelp forests)	NA
	Recreational activities	Creation of nature-based eco-tourism opportunities (scenic beauty and emblematic species)	Charismatic species, large species, and habitat-forming species	Rios-Jara et al. (2013)
	Science and education	Creation of opportunities for research and education in placed of reduced human impacts	Biological diversity, complex food webs	Galzin et al. (2004)
Supporting services	Primary production	Protection of primary producers	Primary producers, habitat-forming species	Milazzo et al. (2002)
	Coastal protection	Protection of habitat formers (e.g., corals, sea grasses, mangroves) providing attenuation of wave intensity nature-based	Habitat-forming species	Mumby and Hardone (2010)

Babcock et al. (2010), Clarke and Jupiter (2010), Galzin et al. (2004), Goñi et al. (2010), González-Correa et al. (2007), Milazzo et al. (2002), Mumby and Harborne (2010), Rios-Jara et al. (2013), Schroder et al. (2004), and Williams et al. (2009).

no-take or limited-take regulations through MPAs can be summarized as follows. First, mortality from fishing is immediately eliminated so that targeted individuals can live longer and attain larger sizes. In the short term, the increase in fish and invertebrate densities and sizes can lead to increases in reproductive output and recruitment (Claudet et al., 2010; Micheli et al., 2012). Possible negative habitat impacts associated with the use of destructive fishing methods also cease, allowing for the recovery of biogenic habitat that in turn positively affects fish recruitment and survival (Mumby and Harborne, 2010). Thus, in the medium to long term, habitat quality is improved, the preharvesting population age and size structure is re-established, and food web complexity increases due to increased diversity and recovery of top predators, which are often major fishery targets (Babcock et al., 2010; McCook et al., 2010; Micheli and Halpern, 2005). Consequently, one of the most commonly described indirect effects of marine reserves involves a trophic cascade, which is classically defined as the indirect effects of apical species in the food web (e.g., carnivores) on basal species (e.g., primary producers), mediated by intermediate consumers (e.g., herbivores) (Babcock et al., 2010).

Fishery effects of protection can only take place if an export of fish individuals occurs over the boundaries of the MPA ("spillover"; McClanahan and Mangi, 2000), and/or if eggs and larvae are exported from the MPA outwards ("seeding"; e.g., Planes et al., 2009). In MPAs with permeable boundaries, spillover can induce increases in catch per unit of effort (CPUE) of target species in surrounding fisheries' grounds. These increases constitute a yield surplus and fishers' CPUEs tend to be higher, although often more variable due to seasonal processes underlying spillover (Goñi et al., 2006; McClanahan and Kaunda-Arara, 1996). Spillover can also induce increases in total catch, catch per unit of area, species mean size in catch, and species diversity in catch (Goñi et al., 2010). These increases in turn can lead to increases in fishing effort along the MPA boundaries. For fishers, catch of adult spillover focuses in most cases on the borders of the reserve. The fishers' behavior in response to the MPA establishment is usually the concentration of effort at the boundaries of the reservation to take advantage of export adults. This mechanism known as "fishing the line" (Roberts et al., 2001) can be interpreted as evidence of a spillover mechanism and becomes so severe in some cases that it may be affecting densities inside the reserve (Halpern et al., 2008). Figure 2 describes the CPUE decreases for lobster fishery from the border of the reserve and thus shows the concentration of effort in this area.

Although this effect has been poorly quantified (Sale et al., 2005), experience shows that the profit generated by the spillover generally has an impact limited to the local fishery and does not seem to significantly increase the densities for large fishing areas. It also has been debated whether the catch from spillover offsets harvest losses due to closure. A recent synthesis by

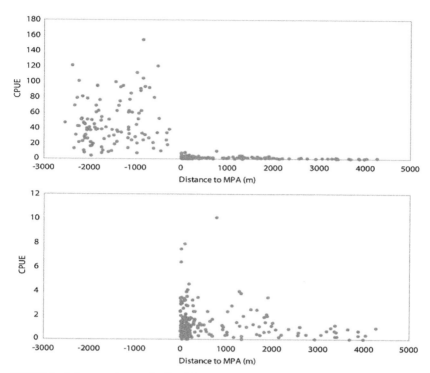

FIGURE 2 Lobster catch per unit effort (CPUE—number of lobsters caught per 600 m of net per day) versus distance from fishing set to the Columbretes Islands Marine Reserve boundary. Positive values are outward the MPA boundaries. Negative values are inward the MPA boundaries. (Upper panel—commercial and experimental data combined; lower panel—commercial fishery data on expanded y-axis scale.) *Adapted from Goñi et al. (2006, 2011).*

Halpern et al. (2010) concluded that even though the spatial extent of the contribution from the MPAs to fisheries is limited (600–1500 m from the MPA edge), in a majority of studies this contribution compensated for the loss of fishing grounds in MPAs. The average magnitude of these effects, however, should be considered with caution because (1) this study pooled very different species—e.g., with different mobility; and (2) studies on spillover focus primarily on species for which some form of spillover is expected.

Studies comparing effects of MPAs on surrounding CPUEs are scarce (Halpern et al., 2010; Harrison et al., 2012; Roberts et al., 2001). Gell et al. (2003) conducted a study selection of nine reserves in several locations with different designs and tested a significant improvement on different species' CPUEs. In their review, the authors quoted two cases of reserves in St Lucia—in Bermuda (Roberts et al., 2003) CPUE for large traps had improved by 46% after five years, and in Nabq (Egypt) the improvement was 66% for the net fishery. It also referred to reserves in Apo (Philippines) with

10-fold increases in longline CPUE after 20 years of protection, with the largest and most stable total catch in the Philippines over a 15-year span. In Tabarca, fishers benefited from a 50%—85% higher CPUE for key species, compared with before closing. Other cases in Mombasa (Kenya) (McClanahan and Mangi, 2000) have the highest catch in the region even with a major effort. However this does not increase the CPUE but only slows its decline. Also in South Africa, Tunley (2009) shows that the reserve has "only" stabilized catch. Other reserves in Chile showed that fishers are benefiting from a CPUE that is 4—10 times superior for a specific bivalve fishery, and in Columbretes (Spain) from a CPUE that is between 6 and 58 times higher for the lobster (Goñi et al., 2006).

Marine Protected Area Effects on Cultural Service: The Example of Recreational Activities

MPAs enhance the development of nonextractive activities, making recreational users perhaps the main beneficiaries of marine conservation (Christie and White, 2007). MPAs provide critical recreational services through nature-based tourism revenue (Balmford et al., 2009). The effects of MPAs stem directly from the fact that the marine environment within an MPA (particularly within a no-take zone) is granted a high level of protection against anthropogenic pressures. Protection in turn is likely to improve the quality of some attributes, such as large charismatic species and/or habitat-forming species that are valuable to visitors (Graham and Nash, 2012). For example, coral reefs are valued as cultural heritage (Hicks et al., 2009). Charismatic habitats (e.g., corals) and species (e.g., reef sharks) serve as focal points for local tourism and ecotourism, thereby enabling residents and visitors to enjoy aesthetic and spiritual values of coral reef ecosystems and seascapes. There are several species, such as the sicklefin lemon shark in French Polynesia and the dusky grouper in the Mediterranean Sea, that increase the recreation/tourism value of tropical and temperate reefs (Clua, 2011; Vandewalle et al., 2007; Guidetti and Micheli, 2011).

Even if, MPAs are expected to be powerful attractors for tourism, quantitative evidence for this benefit remains scarce (Andersson, 2007; Asafu-Adjaye and Tapsuwan, 2008; Depondt and Green, 2006; Harrison, 2007). For example, the relationship between underwater tourism and MPA impacts on some ecological attributes is not well known (Andersson, 2007). There are scientific knowledge gaps and technical difficulties in separating MPA effects on tourism from other context variables such as access and local infrastructure. High costs of studies, late participation by social sciences in MPA science, and effects too weak to be statistically significant have been proposed as reasons for the scarcity of studies of MPA social benefits (Christie et al., 2012; Cinner et al., 2009; Pollnac and Seara, 2010; Sale et al., 2005).

Marine Protected Area Effects on Supporting Services: The Example of Coastal Protection

MPAs provide protection to foundation species such as coral reefs, sea grass, kelps, and mangroves. These species produce physical structures that are natural barriers to waves, hurricanes, typhoons, and elevated sea levels, thereby providing coastal protection to people and critical coastal habitats. Thus, MPAs can contribute to maintaining the ecosystem service of coastal protection through the protection of habitat-forming species and communities (Graham and Nash, 2012). These habitats, when under good ecological conditions, limit the phenomenon of coastal erosion by absorbing high amounts of wave energy and lessening damage from severe weather events (hurricanes, tropical storms, and typhoons) (UNEP-WCMC, 2006) (Kench and Brander, 2006). Coral reefs and mangroves protect against waves by forming barriers along the coastline. Similarly, lagoon areas protected by barrier reefs are generally quiet areas that promote the multiple uses described previously. Several studies show that the reefs act similarly to wave breakers or shallow coasts—this includes a recent meta-analysis of 27 studies conducted in the Atlantic, Pacific, and Indian Oceans revealing that coral reefs provide substantial protection against natural hazards by reducing wave energy by an average of 97% (Ferrario et al., 2014). Reef crests alone dissipate most of this energy (86%). A comparison with artificial structures indicated that coral reefs can provide comparable wave attenuation benefits to artificial defenses, but at lower costs. The median costs of reef restoration projects are in fact an order of magnitude lower than the costs of building artificial breakwaters, indicating that reef conservation and restoration are cost-effective strategies for reducing risk from natural hazards. Finally, an estimated 200 million people receive risk reduction benefits from reefs, or bear hazard mitigation and adaptation costs if reefs are degraded (Ferrario et al., 2014) (it might also be important to highlight the importance of physical processes and low impact of ecological ones for coral reefs).

MARINE PROTECTED AREA EFFECTS ON LONG-TERM ECOSYSTEM FUNCTION AND THE PROVISION OF MULTIPLE SERVICES

Several studies have highlighted the positive effects of MPAs on some aspects of ecosystem function, such as functional diversity and redundancy. MPAs can have positive effects on maintaining specific functional traits, such as large body size, as well as the diversity of functional traits within communities (Micheli and Halpern, 2005; Mouillot et al., 2008). However, few studies have addressed relationships between functional diversity and composition and ecosystem services (Micheli et al., 2014; Raffaelli, 2006). Additional future work directly quantifying ecosystem function and services and investigating relationships between ecological attributes and service provisioning will be critical for understanding the role of biodiversity protection in maintaining the suite of functions and services provided by marine ecosystems (Menzel et al., 2013; Micheli et al., 2014).

In the next sections, we review work to date exploring these relationships and defining and quantifying functional traits and attributes of marine communities. We propose that broader application of functional frameworks is a key step in linking MPAs and their ecological effects to ecosystem service provision.

The Role of Biodiversity: Expectations from Functional Diversity and Redundancy

The goals of MPAs are increasingly expanding beyond the protection and restoration of a few to the restoration of ecosystem functions and services (e.g., herbivory and maintenance of corals, predatory control of invasive species, recruitment and recovery potential, coastal protection, fisheries, and opportunities for recreation and education). MPAs also aim to maintain long-term ecosystem health and sustain multiple ecosystem functions and services within the context of changing environmental conditions (e.g., UNEP-WCMC, 2008). One suggested approach for tackling this extremely complex and multifaceted sets of goals is to use biodiversity as a target for management and a proxy for the full range of functions and services within an ecosystem (Duffy, 2009; Palumbi et al., 2009). Indeed, a majority of MPAs include biodiversity protection among their goals or anticipated benefits (Pomeroy et al., 2005). Biodiversity conservation goals stem both from a recognition of the existence and option values of species and a growing recognition that biodiversity—the degree of variation in living organisms, at the genetic, population, community, and ecosystem or landscape levels—contributes to the many important ecosystem processes that underlie marine ecosystem health and ecosystem service provision. Therefore, the global trend of declining biodiversity may lead to a similar decline in ecosystem services and human well-being—both in terms of immediate losses in ecosystem services and also in the loss of an ecosystem's capacity to adapt to environmental changes and sustain the provision of services into the future (Daily, 1997; MA, 2005; Tilman et al., 2006). For example, a study of local experiments, long-term time series, and global fisheries data by Worm et al. (2006) showed that declines in genetic and species diversity in marine systems were associated with decreases in not just the productivity of fisheries, but also in their stability and recovery across different temporal scales.

Most assessments of biodiversity effects on ecosystems have focused on species or genetic diversity and have generally reported positive relationships between biodiversity and ecosystem processes from a range of ecosystems including mudflats (Emmerson et al., 2001), sea grasses (Duffy, 2006), salt marshes (Griffin and Silliman, 2011), kelp forests (Byrnes et al., 2006), and rocky shores (O'Connor and Crowe, 2005). However, there is an increasing awareness that the nature of the relationship between species diversity and ecosystem processes is highly dependent on the link between species diversity and functional diversity (Micheli and Halpern, 2005). Functional diversity is the variation in functional characteristics represented by the diversity of living organisms, and it is these characteristics that determine the range of ecological roles and species interactions that are present, and thus mediate the relationship between biodiversity, ecosystem functioning, and service provision (Cadotte, 2011; Díaz and Cabido, 2001; Loreau, 1998; McGill et al., 2006). Specifically, biodiversity is expected to promote immediate and long-term ecosystem functioning through patterns of complementarity and redundancy in the functional characteristics it encompasses (Maestre et al., 2012; Walker et al., 1999). Therefore, the protection and restoration of functional diversity is increasingly highlighted as an important principle for management of both marine and terrestrial ecosystems (Chapin et al., 2010; Foley et al., 2010).

Two key reasons underlie the expectation that functional diversity promotes long-term ecosystem health and service provisioning. First, maintaining high levels of functional diversity in an ecosystem allows for the full range of species' ecological roles and interactions to persist and thus for maintenance of multiple ecosystem functions. Both empirical and modeling studies have found that as more ecosystem functions are considered, higher levels of biodiversity are required to sustain all functions simultaneously (Gamfeldt et al., 2008; Hector and Bagchi, 2007; Hensel and Silliman, 2013; Maestre et al., 2012; Zavaleta et al., 2010). Furthermore, in some cases functional diversity, rather than species diversity, may be more important in maintaining ecosystem multifunctionality, since it is the complementarity of species' functional contributions that allows for multiple ecosystem functions to persist (Mouillot et al., 2011). Because many ecosystem services valued by people depend on multiple ecosystem functions (Palumbi et al., 2009), and different ecosystem functions and services may trade off with each other (Bennett et al., 2009; Carpenter et al., 2009), the protection of biodiversity, particularly functional diversity, can serve as a tractable proxy for an ecosystem state that sustains a balance between a range of ecosystem functions, especially when the key drivers and interactions of those functions are not yet well known (Duffy, 2009; Palumbi et al., 2009).

Second, functional diversity may act as a form of insurance for ecosystem functions and services in the face of environmental fluctuations and global environmental change (Bernhardt and Leslie, 2013; Elmqvist et al., 2003; Naeem and Li, 1997). Specifically, functional diversity is expected to promote ecosystem resilience, defined as the ecosystem's capacity to absorb disturbance, reorganize, and maintain its functioning, structure, and feedbacks such that it does not undergo an undesirable phase shift involving the loss of key ecosystem services (Folke et al., 2004). Two aspects of functional diversity underlie this expected link resilience: functional redundancy and response diversity.

Functional redundancy occurs when multiple species contribute similarly to ecosystem functions, such that redundant species may be able to functionally compensate for the decline or loss of one or more species (Naeem and Li, 1997; Walker, 1992, 1995). Therefore, loss of species would not significantly impact the functioning of the ecosystem until the last member of a functionally redundant group is lost. However, the loss of that last member could lead to the complete loss of important ecosystem feedbacks and a complete transformation or shift of the ecosystem to an alternate state (Hughes, 1994). Because of high uncertainty and variability of species' ecological roles, the extent of redundancy, and the vulnerability of functions to environmental changes and human pressures, maintaining high levels of functional diversity and redundancy in natural communities should be a key management goal.

Quantifying and Protecting Functional Diversity and Redundancy in Marine Protected Areas

Quantifying Functional Diversity

Recent reviews of strategies for sustainable management of terrestrial and marine ecosystems have specifically highlighted functional diversity and redundancy as targets for protection or restoration (Chapin et al., 2010; Foley et al., 2010). In order to successfully manage and maintain functional diversity and redundancy in ecosystems, a first step is to develop practical ways to measure and monitor these attributes in the field.

Two approaches have generally been used to quantify functional diversity in ecological communities. The most common method is to assign species to discrete functional groups based on knowledge of species' resource use and life history strategies (Micheli and Halpern, 2005; Simberloff and Dayan, 1991), or by using a hierarchical clustering analysis on a set of measured species traits (Jaksić and Medel, 1990). Functional diversity can then be measured at the level of functional groups: functional-group richness is simply the number of functional groups, while functional-group diversity is usually assessed using the Shannon—Wiener index (H′) and incorporates a measure of the relative abundance, or evenness, of the functional groups. Functional redundancy is assessed by calculating species richness or Shannon—Wiener diversity within each functional group. Functional group approaches have a long history in ecology and have provided many insights into species interactions and community structure (Dethier et al., 2003; Simberloff and Dayan, 1991; Steneck and Watling, 1982). However, this method suffers from several problems, arising from the use of discrete groupings to model functional differences that are generally continuous in nature. Most importantly, the threshold for considering functional differences as significant is an arbitrary one, and it is assumed that all pairwise differences between species from different groups are equal in magnitude (Mouchet et al., 2010; Petchey and Gaston, 2002). In some cases, these problems may compromise the usefulness of functional groups in assessing functional diversity—ecosystem functioning relationships (Wright et al., 2006). On the other hand, particularly in applications of a functional framework to diverse communities, a lack of detailed data on functional traits for all species makes categorical functional classifications, or the use of a mix of categorical and continuous trait values, the only possible approach (Micheli et al., 2014; Stuart-Smith et al., 2013).

To address some of the weaknesses identified in the functional group classification approach and increase the explanatory power of functional diversity for ecosystem function, various trait-based multivariate measures of functional diversity have been developed (Botta-Dukát, 2005; Laliberté and Legendre, 2010; Mason et al., 2005; Mouillot et al., 2013; Petchey and Gaston, 2002; Villéger et al., 2008; Walker et al., 1999). Many of these measures are calculated by first representing species within the community as points in a multivariate functional trait space, and then assessing various aspects of the distribution of species and their relative abundances within this space (Mouillot et al., 2013; Villéger et al., 2008). Unlike the functional group approach, these measures may account for various degrees of functional difference between species. They also allow for different, complementary measures of functional diversity to be assessed, such as the relative abundances of functionally redundant and functionally unique species in the community, or community-wide shifts in specific traits (Mouillot et al., 2008). However, the use of this approach has generally been limited to low-diversity assemblages or subsets of taxa within a community, such as higher taxa that have directly comparable morphological traits (e.g., terrestrial plants, insect families, and fish).

Compared with measures of species diversity, all methods of quantifying functional diversity and redundancy are more data-intensive; they require additional information about each species' functional characteristics in the form of either knowledge about species' basic ecologies or quantitatively measured trait values for each species. The latter is especially time-consuming to obtain and may not be tractable in some species-rich ecosystems such as coral reefs (Micheli et al., 2014). As a result, incorporating functional diversity into assessments of ecosystem health, MPA performance, or MPA design still presents a challenge and has not been widely implemented. Nevertheless, the few studies that have measured functional diversity in the context of MPAs have provided some useful insights into how effective MPAs have been in protecting different aspects of functional diversity.

Spatial Protection of Functional Diversity

Functional diversity has generally not been considered explicitly in the design and location of MPAs. In siting MPAs and MPA networks, areas of high taxonomic diversity (particularly species richness) have usually been targeted as a way to achieve the protection of biodiversity. Because empirical studies have demonstrated a generally positive relationship between species richness and functional richness, it is often assumed that species richness adequately proxies functional richness for the purposes of management (Foley et al., 2010).

However, the few studies that have examined spatial variation in marine functional diversity have reported spatial mismatches between MPAs and areas of high functional diversity. This incongruence corresponds to a mismatch between hotspots of taxonomic diversity and hotspots of functional diversity that occurs at multiple spatial scales. Regionally, Mouillot et al. (2011) found that existing networks of MPAs in the Mediterranean Sea were spatially congruent with fish species diversity, but failed to cover areas of high functional diversity. At a global scale, Stuart-Smith (2013) reported that areas of high reef fish functional diversity were concentrated in temperate latitudes, in contrast with well-known patterns of species richness that peak in the Tropics. They suggest that the tropical bias for MPA formation may result in failure to protect the functional aspects of biodiversity on a global scale.

Similar mismatches in functional diversity with species diversity and protection efforts have been reported in terrestrial systems (Devictor et al., 2010) and could reflect a more general need to integrate functional diversity into management. One such integrative framework has been developed to prioritize areas for conservation within a series of floodplain water bodies in France. Maire et al. (2013) assessed a combination of fish functional diversity, taxonomic diversity and the diversity of the species' natural heritage and social-economic importance and concluded that downstream areas with high lateral connectivity to the main river channel should be prioritized for protection. A similar integration of spatial patterns of marine functional diversity with other management goals could be useful in improving spatial protection of functional diversity.

Effects of Marine Protected Areas on Functional Diversity

While the limited evidence available suggests that current MPAs fail to provide adequate coverage of areas with high functional diversity, studies addressing the direct effects of MPAs on functional diversity, and in some cases redundancy, have generally found positive effects. Most direct assessments of MPA effects on functional diversity have used functional group approaches to compare functional richness and functional redundancy between MPAs and reference areas, but the emerging use of trait-based multivariate approaches has begun to provide some more detailed insights into the effects of reserves.

In the most spatially extensive assessment of MPA effects on functional diversity, Micheli and Halpern (2005) analyzed a global dataset of reef fishes from 31 different no-take reserve sites, including both temperate and tropical reefs. They reported that in comparison with reference sites, no-take reserves generally contained more functional groups (higher functional richness) and increased functional redundancy within some groups. Studies of individual MPA systems have reported similarly positive effects of MPAs on functional diversity. In the Bahamas, fish assemblages within a no-take marine reserve contained more functional groups and greater functional redundancy within each group, compared with nearby fished areas (Micheli et al., 2014). In the Mediterranean Sea, MPAs were also associated with higher functional diversity (Villamor and Becerro, 2012) and greater functional redundancy (Stelzenmüller et al., 2009).

Across most of the studied areas, greater functional diversity within MPAs was associated with higher species diversity (Micheli and Halpern, 2005; Micheli et al., 2014; Stelzenmüller et al., 2009). This pattern is often observed at least at low levels of species richness because functional richness is positively related to species richness, although the exact shape of the relationship can vary (Micheli and Halpern, 2005; Petchey and Gaston, 2002). However, opposing or uncorrelated effects of MPAs on species diversity may also be observed when different species increase and decline simultaneously, which is relatively common due to indirect effects of protection on species through competitive and predatory interactions (Micheli and Halpern, 2005). Several reserves from the Micheli and Halpern (2005) analysis were associated with positive effects on functional diversity but negative effects on species diversity, while the Spanish MPA system studied by Villamor and Becerro (2012) reported positive effects on functional diversity but no significant effect on species diversity. Collectively, these studies suggest that though species and functional diversity are generally correlated, functional diversity is more likely to respond positively to protection, and measuring species diversity alone may lead to failure to detect reserve effects on functional diversity.

Trait-based multivariate measures of functional diversity have not yet been widely used to assess the effects of protection measures in either aquatic or terrestrial ecosystems, potentially because they are a newer set of tools that also require a fairly large amount of information. In the context of management, they have generally been applied in assessing ecosystem responses to large-scale environmental and anthropogenic impacts, especially the anthropogenic modification of terrestrial and aquatic habitats (Barragán et al., 2011; Edwards et al., 2014; Helsen et al., 2013; Magnan et al., 2010; Mouchet et al., 2010; Pakeman, 2011). These trait-based approaches are expected to be particularly suitable for assessing shifts in ecological communities for two key reasons. First, trait-based measures of functional diversity are more likely to show predictable shifts with environmental change because each individual species' response to environmental drivers is ultimately determined by its functional traits (i.e., response traits). Second, these trait-based measures are based on species abundances rather than presences or absences, so they are more sensitive to changes in species assemblages, and could provide advance signals of disturbance in ecosystems ahead of the actual loss of species (Mouillot et al., 2013). Indeed, most of the studies that have applied these measures to assess ecosystem change have reported systematic losses in functional diversity and/or shifts in trait composition consistent with some degree of environmental filtering.

Because global climate change impacts essentially all marine ecosystems (Bernhardt and Leslie, 2013; Halpern et al., 2008), local-scale impacts (e.g., fishing) and management efforts (e.g., MPAs) inevitably co-occur with global-scale environmental changes such as warming and ocean acidification (Crain et al., 2008; Halpern et al., 2008). Therefore, trait-based measures of functional diversity may become increasingly useful for assessing the performance of MPAs in the context of environmental change. The potential value of this approach is illustrated by a recent study of fish functional diversity within a global warming hotspot. Bates et al. (2013) compared species richness and multivariate functional diversity measures between a Tasmanian marine reserve and nearby reference sites over 20 years. They found no significant differences in species richness or overall functional richness between the reserve and reference sites; functional richness increased in both over the study period. However, by comparing the functional trait composition of the fish assemblages, they found that the increase in functional richness within the reserve was partly driven by an increase in large-bodied, carnivorous species that are targeted by fisheries, whereas the increase in functional richness outside the reserve was driven by the colonization of species with warmer affinities. In fact, the degree of invasion of warm-water species was significantly less within the reserve, suggesting that fish communities within the MPA were more resilient to the effects of climate change. In this case, a trait-based multivariate approach was able to detect the interaction between an MPA and a large-scale climate driver, and identify the effect of the MPA on a key function: resilience to climate change. In contrast, a traditional species diversity or functional group classification approach failed to highlight this effect of protection on ecosystem function.

Trait-based functional diversity indices may also be able to provide more specific information about important reserve effects on functional diversity. For example, the integrity and functioning of ecosystems are disproportionately impacted by the contributions of functionally unique species (O'Gorman et al., 2011; Petchey et al., 2008) because functionally unique species, by definition, perform functions with low redundancy. Mouillot et al. (2008) developed and used a trait-based index, the Conservation of Biological Originality (CBO), to examine changes in the prevalence and abundance of functionally unique fish species before and after the establishment of a French MPA. They concluded that the MPA was successful in protecting the most functionally unique members of the fish community: these species were more widely distributed and more abundant after MPA establishment. Unique combinations of functional traits may be crucially important for maintaining ecosystem functioning, as demonstrated by studies of large parrotfishes in coral reefs of the Great Barrier Reef and Pacific Line Islands (Bellwood et al., 2003).

Results of the studies by Bates et al. and Mouillot et al. suggest that: (1) MPAs can have positive effects on maintaining the diversity of functional traits within communities; and (2) trait-based multivariate measures of functional diversity are a promising approach for assessing reserve effects on functional diversity. More MPA assessments using trait-based multivariate metrics will be needed to determine if these metrics generally provide better insights into ecosystem health and functioning than the less data-intensive traditional approaches based on taxonomic diversity and other community properties such as total abundance, size structure, or species composition, and if any additional information gained is worth allocating more resources for obtaining trait data.

KEY DIRECTIONS AND OPEN QUESTIONS

Our review highlights that empirical evidence for positive effects of MPAs on ecosystem service provision by coastal marine ecosystems is accumulating. However, gaps in knowledge clearly remain. Existing studies are still largely focused on a subset of services, namely provisioning services and to a lesser extent some regulating services. Studies on some regulating, supporting, and even nonmonetary provisioning services (e.g., subsistence fishing), as well as most cultural services (e.g., aesthetic and spiritual values), are still very scarce.

We argue that a possible productive way forward is to apply functional frameworks to assessing the broader effects of MPAs on services, through the links that exist between functional diversity, redundancy, and trait composition and service flows. Developing this research program will require efforts to (1) better link functional trait or functional group assignments to actual ecosystem functioning and service provision, (2) scale up analyses to whole assemblages, and (3) assess the drivers and consequences of temporal variability in functional diversity and trait composition. Such programs would allow better identification of how MPAs can protect existing and/or provide new ecosystem services, as well as identifying which ones are the drivers and correlates. An important point to identify is the extent toward MPA borders at which MPAs still have an effect. It would also enable the clearing out of those ecosystem benefits not affected by MPAs.

A key practical aspect, particularly if the main application aim is to inform management, is to enable and facilitate the acquisition of the additional data needed for functional analyses. Acquisition of morphological and behavioral data through direct collaboration between scientists, MPA managers, and fishers, and through the development of cost-effective monitoring—e.g., through low-cost video systems, publicly available databases, and involvement of diverse users (e.g., through citizen science projects)—are promising avenues for allowing the broader application and testing of functional frameworks to MPA assessments.

To successfully develop scientific frameworks and datasets needed to address the links between MPAs, ecosystem functioning, and ecosystem services, closer collaboration is needed between natural and social scientists on the one hand, and among academics, MPA managers, and users on the other.

REFERENCES

Agardy, T., Bridgewater, P., Crosby, M.P., Day, J., Dayton, P.K., Kenchington, R., Laffoley, D., McConney, P., Murray, P.A., Parks, J.E., Peau, L., 2003. Dangerous targets? Unresolved issues and ideological clashes around marine protected areas. Aquat. Conserv. Mar. Freshw. Ecosyst. 13, 353−367. http://dx.doi.org/10.1002/aqc.583.

Alban, F., Boncoeur, J., Roncin, N., 2011. Assessing the impact of marine protected areas on society's well-being: an economic perspective. In: Marine Protected Areas − A Multidisciplinary Approach. Cambridge University Press, Cambridge, UK, pp. 226−246.

Andersson, J.E.C., 2007. The recreational cost of coral bleaching—a stated and revealed preference study of international tourists. Ecol. Econ. 62, 704−715. http://dx.doi.org/10.1016/j.ecolecon.2006.09.001.

Asafu-Adjaye, J., Tapsuwan, S., 2008. A contingent valuation study of scuba diving benefits: case study in Mu Ko Similan Marine National Park, Thailand. Tour. Manag. 29, 1122−1130. http://dx.doi.org/10.1016/j.tourman.2008.02.005.

Babcock, R.C., Shears, N.T., Alcala, A.C., Barrett, N.S., Edgar, G.J., Lafferty, K.D., McClanahan, T.R., Russ, G.R., 2010. Marine reserves special feature: decadal trends in marine reserves reveal differential rates of change in direct and indirect effects. Proc. Natl. Acad. Sci. USA 107. http://dx.doi.org/10.1073/pnas.0908012107.

Balmford, A., Beresford, J., Green, J., Naidoo, R., Walpole, M., Manica, A., 2009. A global perspective on trends in nature-based tourism. PLoS Biol. 7, e1000144. http://dx.doi.org/10.1371/journal.pbio.1000144.

Barragán, F., Moreno, C.E., Escobar, F., Halffter, G., Navarrete, D., 2011. Negative impacts of human land use on dung beetle functional diversity. PLoS One 6, e17976. http://dx.doi.org/10.1371/journal.pone.0017976.

Bates, A.E., Barrett, N.S., Stuart-Smith, R.D., Holbrook, N.J., Thompson, P.A., Edgar, G.J., 2013. Resilience and signatures of tropicalization in protected reef fish communities. Nat. Climate Change 4, 62−67.

Bellwood, D.R., Hoey, A.S., Choat, J.H., 2003. Limited functional redundancy in high diversity systems: resilience and ecosystem function on coral reefs. Ecol. Lett. 6, 281−285. http://dx.doi.org/10.1046/j.1461-0248.2003.00432.x.

Bennett, E.M., Peterson, G.D., Gordon, L.J., 2009. Understanding relationships among multiple ecosystem services. Ecol. Lett. 12, 1394−1404. http://dx.doi.org/10.1111/j.1461-0248.2009.01387.x.

Bernhardt, J.R., Leslie, H.M., 2013. Resilience to climate change in coastal marine ecosystems. Ann. Rev. Mar. Sci. 5, 371−392. http://dx.doi.org/10.1146/annurev-marine-121211-172411.

Botta-Dukát, Z., 2005. Rao's quadratic entropy as a measure of functional diversity based on multiple traits. J. Veg. Sci. 16, 533−540. http://dx.doi.org/10.1111/j.1654-1103.2005.tb02393.x.

Byrnes, J., Stachowicz, J.J., Hultgren, K.M., Randall Hughes, A., Olyarnik, S.V., Thornber, C.S., 2006. Predator diversity strengthens trophic cascades in kelp forests by modifying herbivore behaviour. Ecol. Lett. 9, 61−71. http://dx.doi.org/10.1111/j.1461-0248.2005.00842.x.

Cadotte, M.W., 2011. The new diversity: management gains through insights into the functional diversity of communities. J. Appl. Ecol. 48, 1067–1069. http://dx.doi.org/10.1111/j.1365-2664.2011.02056.x.

Cardinale, B.J., Duffy, J.E., Gonzalez, A., Hooper, D.U., Perrings, C., Venail, P., Narwani, A., Mace, G.M., Tilman, D., Wardle, D.A., Kinzig, A.P., Daily, G.C., Loreau, M., Grace, J.B., Larigauderie, A., Srivastava, D.S., Naeem, S., 2012. Biodiversity loss and its impact on humanity. Nature 486, 59–67. http://dx.doi.org/10.1038/nature11148.

Carpenter, S.R., Defries, R., Dietz, T., Mooney, H.A., Polasky, S., Reid, W.V., Scholes, R.J., 2006. Millennium Ecosystem Assessment: research needs. Science 314, 257–258.

Carpenter, S.R., Mooney, H.A., Agard, J., Capistrano, D., Defries, R.S., Díaz, S., Dietz, T., Duraiappah, A.K., Oteng-Yeboah, A., Pereira, H.M., Perrings, C., Reid, W.V., Sarukhan, J., Scholes, R.J., Whyte, A., 2009. Science for managing ecosystem services: beyond the Millennium Ecosystem Assessment. Proc. Natl. Acad. Sci. USA 106, 1305–1312. http://dx.doi.org/10.1073/pnas.0808772106.

Chapin, F.S., Carpenter, S.R., Kofinas, G.P., Folke, C., Abel, N., Clark, W.C., Olsson, P., Smith, D.M.S., Walker, B., Young, O.R., Berkes, F., Biggs, R., Grove, J.M., Naylor, R.L., Pinkerton, E., Steffen, W., Swanson, F.J., 2010. Ecosystem stewardship: sustainability strategies for a rapidly changing planet. Trends Ecol. Evol. 25, 241–249. http://dx.doi.org/10.1016/j.tree.2009.10.008.

Chapin, F.S., Zavaleta, E.S., Eviner, V.T., Naylor, R.L., Vitousek, P.M., Reynolds, H.L., Hooper, D.U., Lavorel, S., Sala, O.E., Hobbie, S.E., Mack, M.C., Díaz, S., 2000. Consequences of changing biodiversity. Nature 405, 234–242. http://dx.doi.org/10.1038/35012241.

Christie, M., Fazey, I., Cooper, R., Hyde, T., Kenter, J.O., 2012. An evaluation of monetary and non-monetary techniques for assessing the importance of biodiversity and ecosystem services to people in countries with developing economies. Ecol. Econ. 83, 67–78. http://dx.doi.org/10.1016/j.ecolecon.2012.08.012.

Christie, P., White, A.T., 2007. Best practices for improved governance of coral reef marine protected areas. Coral Reefs 26, 1047–1056. http://dx.doi.org/10.1007/s00338-007-0235-9.

Cinner, J., Fuentes, M., Randriamahazo, H., 2009. Exploring social resilience in madagascar's marine protected areas. Ecol. Soc. 14, 41.

Clarke, P., Jupiter, S., 2010. Law, custom and community-based natural resource management in Kubulau District (Fiji). Environ. Conserv. 37, 98–106. http://dx.doi.org/10.1017/S0376892910000354.

Claudet, J., 2012. Marine protected areas. eLS 1–8. http://dx.doi.org/10.1002/9780470015902.a0023605.

Claudet, J., Osenberg, C.W., Benedetti-Cecchi, L., Domenici, P., García-Charton, J.-A., Pérez-Ruzafa, A., Badalamenti, F., Bayle-Sempere, J., Brito, A., Bulleri, F., Culioli, J.-M., Dimech, M., Falcón, J.M., Guala, I., Milazzo, M., Sánchez-Meca, J., Somerfield, P.J., Stobart, B., Vandeperre, F.F., Valle, C., Planes, S., Garcia-Charton, J.-A., Perez-Ruzafa, A., Falcon, J.M., Sanchez-Meca, J., 2008. Marine reserves: size and age do matter. Ecol. Lett. 11, 481–489. http://dx.doi.org/10.1111/j.1461-0248.2008.01166.x.

Claudet, J., Osenberg, C.W., Domenici, P., Badalamenti, F., Milazzo, M., Falcón, J.M., Bertocci, I., Benedetti-Cecchi, L., García-Charton, J.A., Goñi, R., Borg, J.A., Forcada, A., De Lucia, G.A., Perez-Ruzafa, A., Afonso, P., Brito, A., Guala, I., Le Diréach, L., Sanchez-Jerez, P., Somerfield, P.J., Planes, S., 2010. Marine reserves: fish life history and ecological traits matter. Ecol. Appl. 20, 830–839.

Clua, E., Buray, N., Legendre, P., 2011. Business partner or simple catch? The economic value of the sicklefin lemon shark in French Polynesia. Mar. Freshw. 764–770.

Crain, C.M., Kroeker, K., Halpern, B.S., 2008. Interactive and cumulative effects of multiple human stressors in marine systems. Ecol. Lett. 11, 1304—1315. http://dx.doi.org/10.1111/j.1461-0248.2008.01253.x.

Daily, G.C., 1997. Nature's Services: Societal Dependence on Natural Ecosystems. Press, Island.

Day, J.C., Dobbs, K., 2013. Effective governance of a large and complex cross-jurisdictional marine protected area: Australia's Great Barrier Reef. Mar. Policy. http://dx.doi.org/10.1016/j.marpol.2012.12.020. Null.

De'ath, G., Fabricius, K.E., Sweatman, H., Puotinen, M., 2012. The 27-year decline of coral cover on the Great Barrier Reef and its causes. Proc. Natl. Acad. Sci. USA 1—5. http://dx.doi.org/10.1073/pnas.1208909109.

Depondt, F., Green, E., 2006. Diving user fees and the financial sustainability of marine protected areas: opportunities and impediments. Ocean Coast. Manag. 49, 188—202. http://dx.doi.org/10.1016/j.ocecoaman.2006.02.003.

Dethier, M.N., Harbor, F., Steneck, R.S., 2003. A Functional Group Approach to the Structure of Algal-Dominated Communities Robert S. Steneck Dept. of Oceanography and Center for Marine Studies, Inst. for Environmental Studies and Friday Harbor Laboratories, the NOAA Miami Regional Library. Minor.

Devictor, V., Mouillot, D., Meynard, C., Jiguet, F., Thuiller, W., Mouquet, N., 2010. Spatial mismatch and congruence between taxonomic, phylogenetic and functional diversity: the need for integrative conservation strategies in a changing world. Ecol. Lett. 13, 1030—1040. http://dx.doi.org/10.1111/j.1461-0248.2010.01493.x.

Díaz, S., Fargione, J., Chapin, F.S., Tilman, D., 2006. Biodiversity loss threatens human well-being. PLoS Biol. 4, 1300—1305. http://dx.doi.org/10.1371/journal.pbio.0040277.

Diaz, S., Quétier, F., Cáceres, D.M., Trainor, S.F., Pérez-Harguindeguy, N., Bret-Harte, M.S., Finegan, B., Peña-Claros, M., Poorter, L., 2011. Linking functional diversity and social actor strategies in a framework for interdisciplinary analysis of nature's benefits to society. Proc. Natl. Acad. Sci. USA 108, 895—902. http://dx.doi.org/10.1073/pnas.1017993108.

Díaz, S., Cabido, M., 2001. Vive la différence: plant functional diversity matters to ecosystem processes. Trends Ecol. Evol. 16, 646—655. http://dx.doi.org/10.1016/S0169-5347(01)02283-2.

Duffy, J., 2006. Biodiversity and the functioning of seagrass ecosystems. Mar. Ecol. Prog. Ser. 311, 233—250. http://dx.doi.org/10.3354/meps311233.

Duffy, J.E., 2009. Why biodiversity is important to the functioning of real-world ecosystems. Front. Ecol. Environ. 7, 437—444. http://dx.doi.org/10.1890/070195.

Edwards, F.A., Edwards, D.P., Larsen, T.H., Hsu, W.W., Benedick, S., Chung, A., Vun Khen, C., Wilcove, D.S., Hamer, K.C., 2014. Does logging and forest conversion to oil palm agriculture alter functional diversity in a biodiversity hotspot? Anim. Conserv. 17, 163—173. http://dx.doi.org/10.1111/acv.12074.

Elmqvist, T., Folke, C., Nyström, M., Peterson, G., Bengtsson, J., Walker, B., 2003. Response diversity, ecosystem change, and resilience. Front. Ecol. Environ. 1, 488—494.

Emmerson, M.C., Solan, M., Emes, C., Paterson, D.M., Raffaelli, D., 2001. Consistent patterns and the idiosyncratic effects of biodiversity in marine ecosystems. Nature 411, 73—77. http://dx.doi.org/10.1038/35075055.

Ferrario, F., Beck, M.W., Storlazzi, C.D., Micheli, F., Shepard, C.C., Airoldi, L., 2014. The effectiveness of coral reefs for coastal hazard risk reduction and adaptation. Nat. Commun. 5, 1—9. http://dx.doi.org/10.1038/ncomms4794.

Fletcher, S., Saunders, J., Herbert, R., 2011. A review of the ecosystem services provided by broad-scale marine habitats in England's MPA network. J. Coast. Res. 378—383.

Foley, M.M., Halpern, B.S., Micheli, F., Armsby, M.H., Caldwell, M.R., Crain, C.M., Prahler, E., Rohr, N., Sivas, D., Beck, M.W., Carr, M.H., Crowder, L.B., Emmett Duffy, J., Hacker, S.D., McLeod, K.L., Palumbi, S.R., Peterson, C.H., Regan, H.M., Ruckelshaus, M.H., Sandifer, P.A., Steneck, R.S., 2010. Guiding ecological principles for marine spatial planning. Mar. Policy 34, 955–966. http://dx.doi.org/10.1016/j.marpol.2010.02.001.

Folke, C., Carpenter, S., Walker, B., Scheffer, M., Elmqvist, T., Gunderson, L., Holling, C.S., 2004. Regime shifts, resilience, and biodiversity in ecosystem management. Annu. Rev. Ecol. Evol. Syst. 35, 557–581. http://dx.doi.org/10.1146/annurev.ecolsys.35.021103.105711.

François, O., Pascal, N., Méral, P., 2012. Cost-Benefit Analysis of Coral Reefs and Mangroves: A Review of the Literature. Technical Report, T 03IF2012, IFRECOR: Initiative Française pour les Récifs Coralliens—Plan d'action 2011-2015. TIT économie, 43pp.

Gaines, S.D., Lester, S.E., Grorud-Colvert, K., Costello, C., Pollnac, R., 2010. Evolving science of marine reserves: new developments and emerging research frontiers. Proc. Natl. Acad. Sci. USA 107, 18251–18255. http://dx.doi.org/10.1073/pnas.1002098107.

Galzin, R., Crec'hriou, R., Lenfant, P., Planes, S., 2004. Marine protected areas: a laboratory for scientific research. Revue d'écologie—la Terre et la Vie 59 (1–2), 37–48.

Gamfeldt, L., Hillebrand, H., Jonsson, P.R., 2008. Multiple functions increase the importance of biodiversity for overall ecosystem functioning. Ecology 89, 1223–1231. http://dx.doi.org/10.1890/06-2091.1.

Gell, F., Roberts, C., 2003. Marine reserves for fisheries management and conservation: a win–win strategy. *El Anzuelo* Eur. Newslett. Fish. Environ. 11, 4–6.

Goñi, R., Quetglas, A., Reñones, O., 2006. Spillover of spiny lobsters *Palinurus elephas* from a marine reserve to an adjoining fishery. Mar. Ecol. Prog. Ser. 308, 207–219. http://dx.doi.org/10.3354/meps308207.

Goñi, R., Hilborn, R., Díaz, D., Mallol, S., Adlerstein, S., 2010. Net contribution of spillover from a marine reserve to fishery catches. Mar. Ecol. Prog. Ser. 400, 233–243. http://dx.doi.org/10.3354/meps08419.

Goñi, R., Badalamenti, F., Tupper, M., 2011. evidence from empirical studies. In: Claudet, Joachim (Ed.), FISHERIES—Effects of marine protected areas on local fisheries, Marine Protected Areas. Cambridge University Press, Cambridge, pp. 72–98. http://dx.doi.org/10.1017/CBO9781139049382.006.

González-Correa, J., Bayle Sempere, J., Sánchez-Jerez, P., Valle, C., 2007. Posidonia oceanica meadows are not declining globally. Analysis of population dynamics in marine protected areas of the Mediterranean Sea. Mar. Ecol. Prog. Ser. 336, 111–119. http://dx.doi.org/10.3354/meps336111.

Graham, N.A.J., Nash, K.L., 2012. The importance of structural complexity in coral reef ecosystems. Coral Reefs 32, 315–326. http://dx.doi.org/10.1007/s00338-012-0984-y.

Graham, N.A., Bellwood, D.R., Cinner, J.E., Hughes, T.P., Norström, A.V., Nyström, M., 2013. Managing resilience to reverse phase shifts in coral reefs. Front. Ecol. Environ. 11, 541–547.

Griffin, J.N., Silliman, B.R., 2011. Predator diversity stabilizes and strengthens trophic control of a keystone grazer. Biol. Lett. 7, 79–82. http://dx.doi.org/10.1098/rsbl.2010.0626.

Guidetti, P., Micheli, F., 2011. Ancient art serving marine conservation. Front. Ecol. Environ. 9, 374–375. http://dx.doi.org/10.1890/11.WB.019.

Guidetti, P., Milazzo, M., Bussotti, S., Molinari, A., Murenu, M., Pais, A., Spano, N., Balzano, R., Agardy, T., Boero, F., 2008. Italian marine reserve effectiveness: does enforcement matter? Biol. Conserv. 141, 699–709. http://dx.doi.org/10.1016/j.biocon.2007.12.013.

Haines-young, R., Potschin, M., 2010. The links between biodiversity, ecosystem services and human well-being. In: British Ecological Society (Ed.), Ecosystem Ecology: A New Synthesis. Cambridge University Press, pp. 110–139.

Halpern, B.S., 2014. Conservation: making marine protected areas work. Nature. http://dx.doi.org/10.1038/nature13053.

Halpern, B.S., Lester, S.E., McLeod, K.L., 2010. Placing marine protected areas onto the ecosystem-based management seascape. Proc. Natl. Acad. Sci. USA 107, 18312–18317. http://dx.doi.org/10.1073/pnas.0908503107.

Halpern, B.S., Walbridge, S., Selkoe, K.A., Kappel, C.V., Micheli, F., D'Agrosa, C., Bruno, J.F., Casey, K.S., Ebert, C., Fox, H.E., Fujita, R., Heinemann, D., Lenihan, H.S., Madin, E.M.P., Perry, M.T., Selig, E.R., Spalding, M., Steneck, R., Watson, R., 2008. A global map of human impact on marine ecosystems. Science 319, 948–952. http://dx.doi.org/10.1126/science.1149345.

Hargreaves-Allen, V., Mourato, S., Milner-Gulland, E.J., 2011. A global evaluation of coral reef management performance: are MPAs producing conservation and socio-economic improvements? Environ. Manag. 17, 684–700. http://dx.doi.org/10.1007/s00267-011-9616-5.

Harrison, D., 2007. Cocoa, conservation and tourism Grande Riviere, Trinidad. Ann. Tour. Res. 34, 919–942. http://dx.doi.org/10.1016/j.annals.2007.04.004.

Harrison, H.B., Williamson, D.H., Evans, R.D., Almany, G.R., Thorrold, S.R., Russ, G.R., Feldheim, K.A., van Herwerden, L., Planes, S., Srinivasan, M., Berumen, M.L., Jones, G.P., 2012. Larval export from marine reserves and the recruitment benefit for fish and fisheries. Curr. Biol. 22, 1–6. http://dx.doi.org/10.1016/j.cub.2012.04.008.

Hector, A., Bagchi, R., 2007. Biodiversity and ecosystem multifunctionality. Nature 448, 188–190. http://dx.doi.org/10.1038/nature05947.

Helsen, K., Ceulemans, T., Stevens, C.J., Honnay, O., 2013. Increasing soil nutrient loads of european semi-natural grasslands strongly alter plant functional diversity independently of species loss. Ecosystems 17, 169–181. http://dx.doi.org/10.1007/s10021-013-9714-8.

Hensel, M.J.S., Silliman, B.R., 2013. Consumer diversity across kingdoms supports multiple functions in a coastal ecosystem. Proc. Natl. Acad. Sci. USA 110, 20621–20626. http://dx.doi.org/10.1073/pnas.1312317110.

Hicks, C.C., McClanahan, T.R., Cinner, J.E., Hills, J.M., 2009. Trade-offs in values assigned to ecological goods and services associated with different coral reef management strategies. Ecol. Soc. 14, 18.

Hooper, D.U., Adair, E.C., Cardinale, B.J., Byrnes, J.E.K., Hungate, B.A., Matulich, K.L., Gonzalez, A., Duffy, J.E., Gamfeldt, L., O'Connor, M.I., 2012. A global synthesis reveals biodiversity loss as a major driver of ecosystem change. Nature 486, 105–108. http://dx.doi.org/10.1038/nature11118.

Hughes, T., 1994. Catastrophes, phas shiftd and large-scale degradation of a Caribbean. Sci. Pap. Ed. 265 (5178), 1574–1551.

Hughes, T.P., Graham, N.A.J., Jackson, J.B.C., Mumby, P.J., Steneck, R.S., 2010. Rising to the challenge of sustaining coral reef resilience. Trends Ecol. Evol. 25, 633–642. http://dx.doi.org/10.1016/j.tree.2010.07.011.

Jaksić, F.M., Medel, R.G., 1990. Objective recognition of guilds: testing for statistically significant species clusters. Oecologia 82, 87–92. http://dx.doi.org/10.1007/BF00318537.

Karnauskas, M., Babcock, E.A., 2014. An analysis of indicators for the detection of effects of marine reserve protection on fish communities. Ecol. Indic. 46, 454–465. http://dx.doi.org/10.1016/j.ecolind.2014.07.006.

Kench, P.S., Brander, R.W., 2006. Wave processes on coral reef flats: implications for reef geomorphology using Australian case studies. J. Coast. Res. 221, 209–223. http://dx.doi.org/10.2112/05A-0016.1.

Laliberté, E., Legendre, P., 2010. A distance-based framework for measuring functional diversity from multiple traits. Ecology 91, 299–305.

Laurans, Y., Pascal, N., Binet, T., Brander, L., 2013. Economic valuation of ecosystem services from coral reefs in the South Pacific: taking stock of recent experience. J. Environ. Manage. 116, 135–144. http://dx.doi.org/10.1016/j.jenvman.2012.11.031.

Liquete, C., Piroddi, C., Drakou, E.G., Gurney, L., Katsanevakis, S., Charef, A., Egoh, B., 2013. Current status and future prospects for the assessment of marine and coastal ecosystem services: a systematic review. PLoS One 8, e67737. http://dx.doi.org/10.1371/journal.pone.0067737.

Loreau, M., 1998. Biodiversity and ecosystem functioning: a mechanistic model. Proc. Natl. Acad. Sci. USA 95, 5632–5636.

MA, 2005. Ecosystems and Human Well-Being. Biodiversity Synthesis, Washington, DC.

Maestre, F.T., Quero, J.L., Gotelli, N.J., Escudero, A., Ochoa, V., Delgado-Baquerizo, M., Garcia-Gomez, M., Bowker, M.A., Soliveres, S., Escolar, C., Garcia-Palacios, P., Berdugo, M., Valencia, E., Gozalo, B., Gallardo, A., Aguilera, L., Arredondo, T., Blones, J., Boeken, B., Bran, D., Conceicao, A.A., Cabrera, O., Chaieb, M., Derak, M., Eldridge, D.J., Espinosa, C.I., Florentino, A., Gaitan, J., Gatica, M.G., Ghiloufi, W., Gomez-Gonzalez, S., Gutierrez, J.R., Hernandez, R.M., Huang, X., Huber-Sannwald, E., Jankju, M., Miriti, M., Monerris, J., Mau, R.L., Morici, E., Naseri, K., Ospina, A., Polo, V., Prina, A., Pucheta, E., Ramirez-Collantes, D.A., Romao, R., Tighe, M., Torres-Diaz, C., Val, J., Veiga, J.P., Wang, D., Zaady, E., 2012. Plant species richness and ecosystem multifunctionality in global drylands. Science 335, 214–218. http://dx.doi.org/10.1126/science.1215442.

Magnan, P., Pool, T.K., Olden, J.D., Whittier, J.B., Paukert, C.P., 2010. Environmental drivers of fish functional diversity and composition in the Lower Colorado River Basin. Can. J. Fish. Aquat. Sci. 67, 1791–1807. http://dx.doi.org/10.1139/F10-095.

Maire, A., Buisson, L., Biau, S., Canal, J., Laffaille, P., 2013. Ecological Indicators. Ecological Indicators 34, 450–459.

Mascia, M.B., Claus, C.A., Naidoo, R., 2010. Impacts of marine protected areas on fishing communities. Conserv. Biol. 24, 1424–1429. http://dx.doi.org/10.1111/j.1523-1739.2010.01523.x.

Mason, N.W.H., Mouillot, D., Lee, W.G., Wilson, J.B., 2005. Functional richness, functional evenness and functional divergence: the primary components of functional diversity Oikos 111, 112–118.

McClanahan, T.R., Kaunda-Arara, B., 1996. Fishery recovery in a coral-reef marine park and its effect on the adjacent fishery. Conserv. Biol. 10, 1187–1199. http://dx.doi.org/10.1046/j.1523-1739.1996.10041187.x.

McClanahan, T.R., Mangi, S., 2000. Spillover of exploitable fishes from a marine park and its effect on the adjacent fishery. Ecol. Appl. 10, 1792–1805. http://dx.doi.org/10.1890/1051-0761(2000)010[1792:SOEFFA]2.0.CO;2.

McCook, L.J., Ayling, T., Cappo, M., Choat, J.H., Evans, R.D., De Freitas, D.M., Heupel, M., Hughes, T.P., Jones, G.P., Mapstone, B., Marsh, H., Mills, M., Molloy, F.J., Pitcher, C.R., Pressey, R.L., Russ, G.R., Sutton, S., Sweatman, H., Tobin, R., Wachenfeld, D.R., Williamson, D.H., 2010. Marine reserves special feature: adaptive management of the Great Barrier Reef: a globally significant demonstration of the benefits of networks of marine reserves. Proc. Natl. Acad. Sci. USA. http://dx.doi.org/10.1073/pnas.0909335107.

McGill, B.J., Enquist, B.J., Weiher, E., Westoby, M., 2006. Rebuilding community ecology from functional traits. Trends Ecol. Evol. 21, 178−185. http://dx.doi.org/10.1016/j.tree.2006.02.002.

Menzel, S., Kappel, C.V., Broitman, B.R., Micheli, F., Rosenberg, A.A., 2013. Linking human activity and ecosystem condition to inform marine ecosystem based management. Aquat. Conserv. Mar. Freshw. Ecosyst. 23, 506−514. http://dx.doi.org/10.1002/aqc.2365.

Micheli, F., Halpern, B.S., 2005. Low functional redundancy in coastal marine assemblages. Ecol. Lett. 8, 391−400. http://dx.doi.org/10.1111/j.1461-0248.2005.00731.x.

Micheli, F., Saenz-Arroyo, A., Greenley, A., Vazquez, L., Espinoza Montes, J.A., Rossetto, M., De Leo, G.A., 2012. Evidence that Marine Reserves Enhance Resilience to Climatic Impacts. PLoS ONE 7 (7), e40832. http://dx.doi.org/10.1371/journal.pone.0040832.

Micheli, F., Mumby, P.J., Brumbaugh, D.R., Broad, K., Dahlgren, C.P., Harborne, A.R., Holmes, K.E., Kappel, C.V., Litvin, S.Y., Sanchirico, J.N., 2014. High vulnerability of ecosystem function and services to diversity loss in Caribbean coral reefs. Biol. Conserv. 171, 186−194. http://dx.doi.org/10.1016/j.biocon.2013.12.029.

Milazzo, M., Chemello, R., Badalamenti, F., Riggio, S., 2002. Short-term effect of human trampling on the upper infralittoral macroalgae of Ustica Island MPA (western Mediterranean, Italy). J. Mar. Biol. Assoc. UK 82, 745−748. http://dx.doi.org/10.1017/S0025315402006112.

Miller, K.J., Ayre, D.J., 2008. Protection of genetic diversity and maintenance of connectivity among reef corals within marine protected areas. Conserv. Biol. 22, 1245−1254. http://dx.doi.org/10.1111/j.1523-1739.2008.00985.x.

Mora, C., Aburto-Oropeza, O., Ayala Bocos, A., Ayotte, P.M., Banks, S., Bauman, A.G., Beger, M., Bessudo, S., Booth, D.J., Brokovich, E., Brooks, A., Chabanet, P., Cinner, J.E., Cortés, J., Cruz-Motta, J.J., Cupul Magaña, A., Demartini, E.E., Edgar, G.J., Feary, D.A., Ferse, S.C.A., Friedlander, A.M., Gaston, K.J., Gough, C., Graham, N.A.J., Green, A., Guzman, H., Hardt, M., Kulbicki, M., Letourneur, Y., López Pérez, A., Loreau, M., Loya, Y., Martinez, C., Mascareñas-Osorio, I., Morove, T., Nadon, M.-O., Nakamura, Y., Paredes, G., Polunin, N.V.C., Pratchett, M.S., Reyes Bonilla, H., Rivera, F., Sala, E., Sandin, S.A., Soler, G., Stuart-Smith, R., Tessier, E., Tittensor, D.P., Tupper, M., Usseglio, P., Vigliola, L., Wantiez, L., Williams, I., Wilson, S.K., Zapata, F.A., 2011. Global human footprint on the linkage between biodiversity and ecosystem functioning in reef fishes. PLoS Biol. 9, e1000606. http://dx.doi.org/10.1371/journal.pbio.1000606.

Mouchet, M.A., Villéger, S., Mason, N.W.H., Mouillot, D., 2010. Functional diversity measures: an overview of their redundancy and their ability to discriminate community assembly rules. Funct. Ecol. 24, 867−876. http://dx.doi.org/10.1111/j.1365-2435.2010.01695.x.

Mouillot, D., Albouy, C., Guilhaumon, F., Ben Rais Lasram, F., Coll, M., Devictor, V., Meynard, C.N., Pauly, D., Tomasini, J.A., Troussellier, M., Velez, L., Watson, R., Douzery, E.J.P., Mouquet, N., 2011. Protected and threatened components of fish biodiversity in the Mediterranean sea. Curr. Biol. 21, 1044−1050. http://dx.doi.org/10.1016/j.cub.2011.05.005.

Mouillot, D., Culioli, J.M., Pelletier, D., Tomasini, J.A., 2008. Do we protect biological originality in protected areas? A new index and an application to the Bonifacio Strait Natural Reserve. Biol. Conserv. 141, 1569−1580. http://dx.doi.org/10.1016/j.biocon.2008.04.002.

Mouillot, D., Graham, N.A.J., Villéger, S., Mason, N.W.H., Bellwood, D.R., 2013. A functional approach reveals community responses to disturbances. Trends Ecol. Evol. 28, 167−177. http://dx.doi.org/10.1016/j.tree.2012.10.004.

Mumby, P.J., Harborne, A.R., 2010. Marine reserves enhance the recovery of corals on Caribbean reefs. PLoS One 5, e8657. http://dx.doi.org/10.1371/journal.pone.0008657.

Naeem, S., Li, S., 1997. Biodiversity enhances ecosystem reliability. Nature 390, 507−509. http://dx.doi.org/10.1038/37348.

O'Connor, N.E., Crowe, T.P., 2005. Biodiversity loss and ecosystem functioning: distinguishing between number and identity of species. Ecology 86, 1783−1796. http://dx.doi.org/10.1890/04-1172.

O'Gorman, E.J., Yearsley, J.M., Crowe, T.P., Emmerson, M.C., Jacob, U., Petchey, O.L., 2011. Loss of functionally unique species may gradually undermine ecosystems. Proc. Biol. Sci. 278, 1886−1893. http://dx.doi.org/10.1098/rspb.2010.2036.

Pakeman, R.J., 2011. Functional diversity indices reveal the impacts of land use intensification on plant community assembly. J. Ecol. 99, 1143−1151. http://dx.doi.org/10.1111/j.1365-2745.2011.01853.x.

Palumbi, S.R., Sandifer, P.A., Allan, J.D., Beck, M.W., Fautin, D.G., Fogarty, M.J., Halpern, B.S., Incze, L.S., Leong, J.-A., Norse, E., Stachowicz, J.J., Wall, D.H., 2009. Managing for ocean biodiversity to sustain marine ecosystem services. Front. Ecol. Environ. 7, 204−211. http://dx.doi.org/10.1890/070135.

Pascal, N., 2014. Economic valuation of Palau Large Marine Sanctuary—Costs and Benefits. A report for the Pew Charitable Trusts. Global Ocean Legacy, Palau Office.

Pascal, N., Seidl, A., 2013. Economic Benefits of Marine Protected Areas: Case Studies in Vanuatu and Fiji, South Pacific.

Petchey, O.L., Eklöf, A., Borrvall, C., Ebenman, B., 2008. Trophically unique species are vulnerable to cascading extinction. Am. Nat. 171, 568−579. http://dx.doi.org/10.1086/587068.

Petchey, O.L., Gaston, K.J., 2002. Functional diversity (FD), species richness and community composition. Ecol. Lett. 5, 402−411. http://dx.doi.org/10.1046/j.1461-0248.2002.00339.x.

Planes, S., Jones, G.P., Thorrold, S.R., 2009. Larval dispersal connects fish populations in a network of marine protected areas. Proc. Natl. Acad. Sci. USA 106, 5693−5697. http://dx.doi.org/10.1073/pnas.0808007106.

Pollnac, R., Seara, T., 2010. Factors influencing success of marine protected areas in the Visayas, Philippines as related to increasing protected area coverage. Environ. Manag. 584−592. http://dx.doi.org/10.1007/s00267-010-9540-0.

Pomeroy, R., Watson, L., Parks, J., Cid, G., 2005. How is your MPA doing? A methodology for evaluating the management effectiveness of marine protected areas. Coast. Manag. 48, 485−502. http://dx.doi.org/10.1016/j.ocecoaman.2005.05.004.

Raffaelli, D., 2006. Biodiversity and ecosystem functioning: issues of scale and trophic complexity. Mar. Ecol. Prog. Ser. 311, 285−294.

Ríos-Jara, E., Galván-Villa, C.M., Rodríguez-Zaragoza, F.A., López-Uriarte, E., Muñoz-Fernández, V.T., 2013. The tourism carrying capacity of underwater trails in Isabel Island National Park, Mexico. Environ. Manag. 52, 335−347. http://dx.doi.org/10.1007/s00267-013-0047-3.

Roberts, C.M., Andelman, S., Branch, G., Bustamante, R.H., Carlos Castilla, J., Dugan, J., Halpern, B.S., Lafferty, K.D., Leslie, H., Lubchenco, J., McArdle, D., Possingham, H.P., Ruckelshaus, M., Warner, R.R., 2003. Ecological criteria for evaluating candidate sites for marine reserves. Ecol. Appl. 13, 199−214. http://dx.doi.org/10.1890/1051-0761(2003)013[0199:ECFECS]2.0.CO;2.

Roberts, C.M., Bohnsack, J.A., Gell, F., Hawkins, J.P., Goodridge, R., 2001. Effects of marine reserves on adjacent fisheries. Science 294, 1920−1923. http://dx.doi.org/10.1126/science.294.5548.1920.

Roncin, N., Alban, F., Charbonnel, E., Crec'hriou, R., de la Cruz Modino, R., Culioli, J.-M., Dimech, M., Goñi, R., Guala, I., Higgins, R., Lavisse, E., Direach, L., Le, Luna, B., Marcos, C., Maynou, F., Pascual, J., Person, J., Smith, P., Stobart, B., Szelianszky, E., Valle, C., Vaselli, S., Boncoeur, J., 2008. Uses of ecosystem services provided by MPAs: how much do they impact the local economy? A southern Europe perspective. J. Nat. Conserv. 16, 256–270. http://dx.doi.org/10.1016/j.jnc.2008.09.006.

Sale, P.F., Cowen, R.K., Danilowicz, B.S., Jones, G.P., Kritzer, J.P., Lindeman, K.C., Planes, S., Polunin, N.V.C., Russ, G.R., Sadovy, Y.J., Steneck, R.S., 2005. Critical science gaps impede use of no-take fishery reserves. Trends Ecol. Evol. 20, 74–80. http://dx.doi.org/10.1016/j.tree.2004.11.007.

Schroder, H.C., Grebenjuk, V.A., Binder, M., Skorokhod, A., Hassanein, H., Muller, W., 2004. Functional molecular biodiversity: assessing the immune status of two sponge populations (*Suberites domuncula*) on the molecular level. Mar. Ecol. 25, 93–108.

Simberloff, D., Dayan, T., 1991. The guild concept and the structure of ecological communities. Annu. Rev. Ecol. Syst. 22, 115–143. http://dx.doi.org/10.1146/annurev.es.22.110191.000555.

Stelzenmüller, V., Maynou, F., Martín, P., 2009. Patterns of species and functional diversity around a coastal marine reserve: a fisheries perspective. Aquat. Conserv. Mar. Freshw. Ecosyst. 19, 554–565. http://dx.doi.org/10.1002/aqc.1003.

Steneck, R.S., Watling, L., 1982. Feeding capabilities and limitation of herbivorous molluscs: a functional group approach. Mar. Biol. 68, 299–319. http://dx.doi.org/10.1007/BF00409596.

Stoeckl, N., Hicks, C.C., Mills, M., Fabricius, K., Esparon, M., Kroon, F., Kaur, K., Costanza, R., 2011. The economic value of ecosystem services in the Great Barrier Reef: our state of knowledge. Ann. New York Acad. Sci. 1219, 113–133.

Stuart-Smith, R.D., Bates, A.E., Lefcheck, J.S., Duffy, J.E., Baker, S.C., Thomson, R.J., Stuart-Smith, J.F., Hill, N.A., Kininmonth, S.J., Airoldi, L., Becerro, M.A., Campbell, S.J., Dawson, T.P., Navarrete, S.A., Soler, G.A., Strain, E.M.A., Willis, T.J., Edgar, G.J., 2013. Integrating abundance and functional traits reveals new global hotspots of fish diversity. Nature 501, 539–542. http://dx.doi.org/10.1038/nature12529.

TEEB, 2010. The Economics of Ecosystems and Biodiversity: Mainstreaming the Economics of Nature: A Synthesis of the Approach, Conclusions and Recommendations of TEEB. Nairobi.

Tilman, D., Reich, P.B., Knops, J.M.H., 2006. Biodiversity and ecosystem stability in a decade-long grassland experiment. Nature 441, 629–632. http://dx.doi.org/10.1038/nature04742.

Tunley, K., 2009. State of Management of South Africa's Marine Protected Areas. WWF South Africa Report Series—2009/Marine/001. WWF-South Africa. Newlands, Cape Town.

UNEP-WCMC, 2006. In the Front Line: Shoreline Protection and Other Ecosystem Services from Mangroves and Coral Reefs. UNEP-WCMC, Cambridge.

UNEP-WCMC, 2008. National and Regional Networks of Marine Protected Areas: A Review of Progress. UNEP-WCMC, Cambridge.

Vandewalle, M., Sykes, M.T., Harrison, P.A., Luck, G.W., Berry, P., Bugter, R., Dawson, T.P., Feld, C.K., Harrington, R., Haslett, J.R., Hering, D., Jones, K.B., Jongman, R., Lavorel, S., Martins da Silva, P., Moora, M., Paterson, J., Rounsevell, M.D.A., Sandin, L., Settele, J., Sousa, J.P., Zobel, M., 2007. Review Paper on Concepts of Dynamic Ecosystems and Their Services.

Villamor, A., Becerro, M.A., 2012. Species, trophic, and functional diversity in marine protected and non-protected areas. J. Sea Res. 73, 109–116. http://dx.doi.org/10.1016/j.seares.2012.07.002.

Villéger, S., Mason, N.W.H., Mouillot, D., 2008. New multidimensional functional diversity indices for a multifaceted framework in functional ecology. Ecology 89, 2290–2301.

Walker, B., 1992. Biodiversity and ecological redundancy. Conserv. Biol. 6, 18–23.

Walker, B., 1995. Conserving biological diversity through ecosystem resilience. Conserv. Biol. 9, 747–752.

Walker, B., Kinzig, A., Langridge, J., 1999. Plant attribute diversity, resilience, and ecosystem function: the nature and significance of dominant and minor species. Ecosystems 2 (2), 95–113.

Williams, I.D., Walsh, W.J., Claisse, J.T., Tissot, B.N., Stamoulis, K.A., 2009. Impacts of a Hawaiian marine protected area network on the abundance and fishery sustainability of the yellow tang, *Zebrasoma flavescens*. Biol. Conserv. 142, 1066–1073. http://dx.doi.org/10.1016/j.biocon.2008.12.029.

Worm, B., Barbier, E.B., Beaumont, N., Duffy, J.E., Folke, C., Halpern, B.S., Jackson, J.B.C., Lotze, H.K., Micheli, F., Palumbi, S.R., Sala, E., Selkoe, K.A., Stachowicz, J.J., Watson,, R, 2006. Impacts of biodiversity loss on ocean ecosystem services. Science 314, 787–790. http://dx.doi.org/10.1126/science.1132294.

Wright, J.P., Naeem, S., Hector, A., Lehman, C., Reich, P.B., Schmid, B., Tilman, D., 2006. Conventional functional classification schemes underestimate the relationship with ecosystem functioning. Ecol. Lett. 9, 111–120. http://dx.doi.org/10.1111/j.1461-0248.2005.00850.x.

Zavaleta, E.S., Pasari, J.R., Hulvey, K.B., Tilman, G.D., 2010. Sustaining multiple ecosystem functions in grassland communities requires higher biodiversity. Proc. Natl. Acad. Sci. USA 107, 1443–1446. http://dx.doi.org/10.1073/pnas.0906829107.

Chapter 10

Freshwater Conservation and Biomonitoring of Structure and Function: Genes to Ecosystems

Clare Gray[1,2], Iliana Bista[3], Simon Creer[3], Benoit O.L. Demars[4], Francesco Falciani[5], Don T. Monteith[6], Xiaoliang Sun[7] and Guy Woodward[2]

[1]*School of Biological and Chemical Sciences, Queen Mary University of London, London, UK;* [2]*Department of Life Sciences, Imperial College London, Ascot, Berkshire, UK;* [3]*Molecular Ecology and Fisheries Genetics Laboratory, School of Biological Sciences, Environment Centre Wales, Bangor University, Gwynedd, UK;* [4]*The James Hutton Institute, Aberdeen, Scotland, UK;* [5]*Institute of Integrative Biology, University of Liverpool, Liverpool, UK;* [6]*Centre for Ecology & Hydrology, Lancaster Environment Centre, Lancaster, UK;* [7]*Department of Molecular Systems Biology, University of Vienna, Vienna, Austria*

INTRODUCTION

Current Focus of Aquatic Biomonitoring and Conservation

Freshwater biomonitoring—i.e., the repeated, quantitative assessment of surface waters using the presence and/or abundance of groups of organisms of known environmental sensitivity—currently provides a staple tool in aquatic management and conservation and underpins wide-reaching environmental legislation including the European Union Water Framework Directive (EU WFD), Environmental Quality Standards for Surface Water in China (GB 3838-2002), and the Clean Water Act in the United States of America. Its scientific origins can be traced back to societal changes during the industrialization of the developed world and simultaneous scientific developments in epidemiology and biological taxonomy—the impacts of rising human populations on the chemical and microbiological quality of urban water supplies necessitated the development of rapid and robust methods to assess risks to public health.

The history of aquatic biomonitoring is extensively reviewed elsewhere (e.g., Metcalfe, 1989; Rosenberg and Resh, 1993; Friberg et al., 2011) and so will not be discussed in detail here, but essentially biomonitoring hinges on two basic concepts: first that aquatic organisms tend to be unevenly distributed across environmental gradients and should therefore have value as indicators of ecosystem state, and second, that biota provide a more temporally integrated indication of ecosystem quality than many abiotic measurements, such as spot-sampled water chemistry.

Three key developments over the course of the twentieth century had major impacts on routine environmental assessment by regulatory authorities (Metcalfe, 1989). First, Kolkwitz and Marsson (1902, 1909) introduced what became the "saprobien system" in which groups of organisms were directly linked with perceived discrete levels of organic contamination, and by inference oxygen availability of waters. Second, biological diversity indices became popular around the middle of the century based largely on the premise that species richness and evenness is reduced with increasing environmental disturbance. Finally, biotic indices that combined these methodologies (such as the Trent Biotic Index, Chandler's Score System, and the Biological Monitoring Working Party) were developed. Despite the widespread adoption of these indices (in particular the average score per taxon (ASPT) approach), many surface waters are more likely to be compromised by other anthropogenic stressors such as acidification, toxins, climate change, atmospheric deposition of reactive nitrogen, and habitat modification.

During the 1980s, the need to understand the causes behind surface water acidification stimulated investigation of diatoms as paleobiological assessment tools (Renberg and Hellberg, 1982; Battarbee and Charles, 1986). These ubiquitous and chemically sensitive unicellular algae preserve well in lake sediments, thus enabling paleoecologists to reconstruct the environmental history of a water body from sediment cores. Statistical approaches based on weighted averaging procedures were developed to predict (or hindcast) lake chemistry on the basis of spatially derived "training sets" describing the chemical "optima" and tolerances of individual species (e.g., Birks et al., 1990). This approach has proved highly effective in the reconstruction of lake pH and has been applied to infer historical change in other environmental parameters with more mixed success. More recently, various community-based multivariate regression approaches have been developed to interpret the environmental significance of trends in contemporarily monitored biota including diatoms and macroinvertebrates (Monteith et al., 2005; Murphy et al., 2012), and to specifically address the extent to which biological trends can be explained by changes in water quality with time (Halvorsen et al., 2003).

In recent years, more effective water treatment regimes and environmental regulations have improved surface water quality with respect to both organic pollution and water acidity in much of the developed world. The focus of biomonitoring has consequently begun to shift from basic quantification of environmental damage to consideration of how much surface water quality, with respect to these key drivers, still deviates from a desired "reference" condition relative to a "pristine" state. The bioassessment tool RIVPACS (River Invertebrate Prediction and Classification System) pioneered this field, by quantifying the differences in the macroinvertebrate assemblage between a site under investigation relative to its "expected" assemblage at unimpacted but otherwise comparable sites. This approach and its derivatives now underpin most freshwater biomonitoring schemes across Europe (e.g., Simpson et al., 2005; Murphy et al., 2013) and other parts of the world (Simpson and Norris, 2000).

Unfortunately, despite these advances, assigning appropriate reference conditions and current status is still problematic, as preindustrial (i.e., pre-1800) target conditions are very difficult to model with confidence (Battarbee et al., 2005), as there are rarely useful paleoecological data from running waters because their sediments are well mixed, and there are mismatches between paleo and contemporary data in standing waters (as the two rarely overlap in time), so ground truthing is difficult. A notable exception is from some of the longer-term biomonitoring schemes such as the United Kingdom Acid Waters Monitoring Network (Monteith et al., 2005; Battarbee et al., 2014), where after several decades of lake biomonitoring using sediment traps, we are now finally able to compare contemporary data directly with paleoecological data (Figure 1). This has raised intriguing questions about stressor impacts: for instance, in the Acid Waters Monitoring Network (AWMN) data, the lack of evidence of clear recovery among diatom communities along the acidification trajectory evident in the sediment core records (despite improvements in water chemistry) points to hystereses in these ecosystems, and to the potential ecological importance of other factors that could be setting new environmental states that may not be reversed in the foreseeable future (Battarbee et al., 2013). The growing realization that a return to a historical preimpacted state may be unrealistic is now forcing us to consider shifting environmental baselines when assessing conservation, restoration, and the determination of when an alternative state is acceptable with respect to its function, biodiversity, and the ecosystem services it provides (UK National Ecosystem Assessment, 2011; Millenium Ecosystem Assessment, 2005). While this paleoecological reference approach to aquatic monitoring is limited to lake ecosystems (as running water sediments are turned over), there is considerable potential to extend it to other biological proxies and biogeochemical indicators such as pigments and stable isotopes, and pressures other than acidification (e.g., Smol, 2009).

All these approaches focus on linking attributes of biological assemblages to a system's chemical or physical state, and they have made important contributions to environmental assessment, policy, and legislation across ecological and evolutionary timescales. The power of these methodologies can be largely attributed to the wide variation between taxa in tolerance to specific pressures, in particular the bioavailability of oxygen, hydrogen, and aluminum ions. Newly emerging environmental threats, such as the many facets of climate change, contamination from organic micropollutants and nanoparticles etc., may not be quite so readily assessed by similar direct environment–taxa

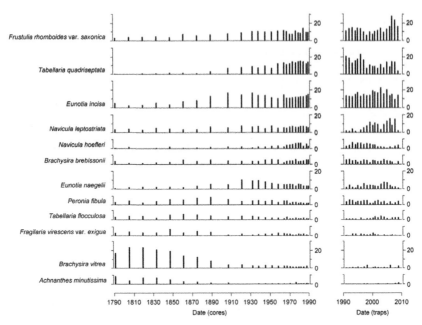

FIGURE 1 Linking a site's contemporary biomonitoring data to its historical reference condition *(redrawn from data presented in Battarbee et al. (2014))*. Percentage relative abundances of diatom species found in sediments of a UK upland lake (Round Loch of Glenhead). Species abundances in historical sediment core samples (left) shift from left to right reflecting increased water acidity during the industrial revolution. Abundances of the same species in contemporary sediment trap assemblages (right) indicate some recent reversal (decline) of some particularly acid-loving species—e.g., *Tabellaria quadriseptata*—as acidity has declined. However, other species that increased during acidification are continuing to increase in abundance while others that were common prior to acidification show little indication of recovery.

calibration-based approaches (Figure 2) (Friberg et al., 2011). In some cases, ecosystem metrics other than the relative abundance of taxa may yield clearer insights into significant environmental shifts (e.g., Layer et al., 2011). There is therefore a growing need to determine how best to assess the impact of these emerging stressors both in isolation and in combination. Also, the structural biodiversity-centric focus of these traditional methods now needs to be augmented with more explicitly functional measures, to provide complementary insights into the impacts of stressors in freshwater ecosystems (e.g., Woodward et al., 2012).

In addition to largely lacking these explicitly functional ecosystem-level metrics, another common limitation of current taxonomic-based biomonitoring schemes is that although there is an implicit evolutionary signal embedded within them (i.e., in terms of the phylogenetic relatedness of the various indicator taxa, which constrains their functional traits), there is still no explicit recognition of the role of adaptation to new stressors and the potential

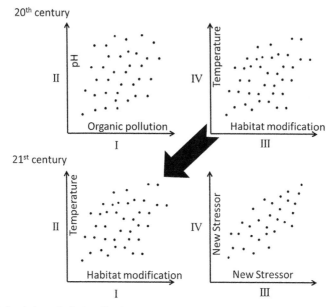

FIGURE 2 A hypothetical ordination to show the changes in the main drivers of habitat degradation in freshwaters in the developed world over time. Increasing temperature and habitat modification have become the significant drivers of change in the principal components (axes I-II) of community composition, replacing the more historical stressors of organic pollution and pH change. However, these historical stressors are still the major causes of habitat degradation in developing countries.

for evolutionary rescue from stressors within species populations, and evolutionary responses can occur surprisingly quickly in many freshwater taxa (e.g., Melián et al., 2011). This could cause mismatches between the reference and impacted conditions, if species are able to adapt to new conditions rather than acting as passive ciphers that are simply overlain on an environmental template (e.g., Bell and Gonzalez, 2011). This has resulted in a paradox of biomonitoring in which speciation is the mechanism that produces the response variables we measure but is then ignored when relating species distributions to environmental conditions. Although research is beginning to fill this gap in understanding (e.g., Thuiller et al., 2011; Vonlanthen et al., 2012) that currently exists in biomonitoring, this "inconvenient truth" is either ignored or obfuscated through attempted circumvention by removing the phylogenetic signal from the data (e.g., via trait-based approaches).

State of the Art in the Science of Biomonitoring: From Species Traits to Community Structure and Ecosystem Functioning

The earliest attempts to combine ecological and evolutionary approaches to biomonitoring included the use of additional measures of biodiversity including phylogenetic diversity (or taxonomic distinctness) and functional diversity conditioned by evolution (e.g., May 1990; Paradis et al., 2004; Webb et al., 2011), though most of the emphasis has been on the former, not the latter. A problem with focusing solely on taxonomy is that if species redundancy is high, as appears to be the case in many freshwaters (e.g., McKie et al., 2008; Perkins et al., 2010; Reiss et al., 2010, 2011), then species loss is likely to only have strong effects when entire functional guilds are lost; but it is these that we still have limited understanding of due to the longstanding reliance on more traditional measures of biodiversity (e.g., species richness). The realized species trait (or gene) profile at a local scale provides the means to link the potential effects of anthropogenic pressures on species (population) distribution and dynamics: i.e., the trait profile itself may therefore be used for diagnostic purposes (Statzner and Bêche, 2010). It is possible, however, that noncausal relationships between individual species traits and contemporary environmental conditions exist (e.g., Poff et al., 2006; Horrigan and Baird, 2008) because some traits may represent an evolutionary legacy rather than current adaptation (Gould and Lewontin, 1979). Empirical studies have confirmed the large role played by phylogeny or taxonomic distinctness in freshwater ecosystems (Willby et al., 2000; Poff et al., 2006; Demars et al., 2012) from the structural perspective, but their functional attributes remain far less well understood.

To interpret biomonitoring results (patterns in species composition), it is crucial to unravel biomonitoring's underlying mechanistic basis (processes that determine this pattern, both anthropogenically mediated or not). Species are not randomly distributed in time (e.g., Lyell and Deshayes, 1830) or space (e.g., Humboldt, 1849), and Demars and Edwards (2009) recently pointed out that even as far back as the nineteenth century, Darwin (1872) argued that environmental variables played only a subordinate role in the determination of species distribution. He offered a mechanistic explanation (pp. 318–319): immigration of individuals from a species (individuals) pool controlled by dispersal barriers and descent with modification regulated through natural selection, with competition being the most important pressure. He attributed the wide distribution of freshwater organisms to favorable means of dispersal (Darwin, 1872, pp. 323–330, 343–347, e.g., Pollux and Santamaria et al., 2005) and lessened competition (Darwin, 1872, pp. 346, e.g., Greulich and Bornette, 2003) in aquatic habitats. This debate of whether species distribution is more controlled by niche assembly (resource heterogeneity) or dispersal assembly is still ongoing (Demars and Harper, 2005; Heino, 2013). Moreover, numerous null models have reproduced biomonitoring patterns of species assembly: e.g., random (Tokeshi, 1990), niche (Tokeshi, 1993), neutral (Bell, 2001; Hubbell, 2001), metabolic scaling (Allen et al., 2002), fractal (Lennon et al., 2007), and MaxEnt (Harte, 2011).

The general consensus is that patterns in species composition and community structure emerge from the interactions of chance, dispersal, and resource heterogeneity in evolving metacommunities (Venail et al., 2008). This is supported by empirical studies using autocorrelation, spatial distances/isolation, and dispersal abilities to infer proportion of resource (niche) versus dispersal community assembly (Moilanen and Hanski, 2001; Demars and Harper, 2005; Moilanen et al., 2005, 2008; Bonada et al., 2012). Essentially, this is explicitly adding the otherwise overlooked dynamical component to biomonitoring data, which are often seen as static snapshots whereby species simply map onto the environmental template. It also starts to recognize the inherent role of dispersal and selection for particular functional traits, rather than simply focusing on the phylogenetic tree in isolation.

Every species can be characterized by not only its taxonomic identity but also its biological (response) functional traits, which may be translated into functional (effect) traits (Engelhardt, 2006; Kerkhoff and Enquist, 2006; López-Urrutia et al., 2006; Enquist et al., 2007) and eventually into ecosystem services (e.g., García-Llorente et al., 2011). Mapping traits onto the tree of life reveals a convergence (independent appearance of a trait in separate clades) or divergence (appearance of a trait in a single clade) in evolution. This is highly relevant in the context of the insurance hypothesis or portfolio effect, whereby high species (or genetic) richness maintains high and constant ecosystem (or population) productivity and services in a stochastic environment (Yachi and Loreau, 1999; Schindler et al., 2010).

The ecology of a species sets the scene in which evolution operates, while evolution may influence ecological dynamics by altering the frequency of phenotypes that are available to interact: thus, there are potentially important eco-evolutionary feedbacks, which are only now starting to be recognized (e.g., Melián et al., 2011; Moya-Larano et al., 2012). The ability of a species to adapt to a changing environment is key to how it responds to stressors: species are not simply present or absent if environmental conditions are favorable or unfavorable (Box 1). According to the old adage, there are three options—"adapt, perish, or move"—that a species is faced with in a changing environment, yet biomonitoring and conservation schemes have largely ignored the first.

An important issue here is that neither ecological nor evolutionary responses occur solely at the population level of organization: no species is an island, and its interactions with those around it will determine both species-specific and the wider community's responses to changing conditions (e.g., Rybicki and Landwehr, 2007). This explains why models derived from bioclimate envelopes and extrapolations from traditional biomonitoring techniques often fail to predict species responses in the real world, because their synecology (the ecology of communities of interacting organisms) is ignored (Woodward et al., 2010; Friberg et al., 2011). The use of trait-based approaches helps to grapple with issues related to functional biodiversity at the autecological level, but it fails to embrace the more complex, higher-level synecological functional roles that species play within multispecies systems

BOX 1 Categorizing Continuous Variables in Biomonitoring

Figure 3 maps an example of a continuous ecological variable (habitat quality) onto discrete human-made categories. This human need to categorize complexity can be seen in many aspects of ecology, not just in the biomonitoring and conservation fields. Whether it's the difficulties encountered when classifying all of life on earth into discrete species (e.g., Mayden, 1997) or the questionable practice of assigning "typologies" to a given lake or river (e.g., Friberg et al., 2011), the motivation comes from our historically poor ability to process large amounts of complex information. However, this process of classification and simplification has allowed us to make some informed generalizations and useful interpretations that otherwise would not be possible. Nevertheless, with the advent of rapidly accelerating computing power, the challenge has now shifted away from our previous inability to process complex information, to the interpretation of complex information into simple messages. With expanding analytical ability comes the need to preserve as much ecological information as possible, which will allow a deeper understanding and more informed interpretations to develop the next necessary steps forward in biomonitoring science—the shift of focus away from the simple monitoring of species composition toward the monitoring of ecosystem functions and services.

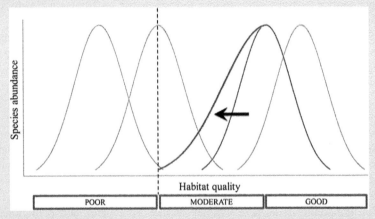

FIGURE 3 Hypothetical graph showing fluctuations in four species abundances across a habitat quality gradient, alongside the discrete criteria of habitat quality (good, moderate, poor) that these continuous variables are categorized into. The dashed line shows species loss, whereas the solid black arrow shows sublethal effects to a particular species population.

such as food webs, which may have seemingly unpredictable emergent properties (Woodward, 2009). This can be exemplified by mismatches between real-time or experimental data that track transient dynamics, versus space-for-time substitutions where the different communities across the environmental gradient may already be at equilibrium (e.g., Layer et al., 2010, 2011). Unfortunately, such data are still rare, but where they are available there is compelling evidence that the functional role of species within the food web can have important indirect and direct consequences that would be missed by relying on static data: a classic example is the seeming paradox of invertebrate abundance declining over several decades of deacidification, yet this response makes sense when the top-down effects of predators on the prey assemblage are included (Layer et al., 2011).

Figure 4 synthesizes current thinking in the role of ecology and evolution of species distribution in which taxonomic, functional, and phylogenetic diversity determine the dynamics of ecosystem functioning and services, and highlights how they can be integrated into future biomonitoring approaches.

Functional diversity provides a more direct link between species richness and ecosystem functioning, and ultimately the provision of goods and services (Naeem, 2002; Woodward, 2009). Two essential functions are primary production and decomposition, which provide the two key energy inputs into any food web, thus ultimately driving the whole system's trophic dynamics, stability, and productivity. Production and decomposition thereby provide a variety of services including the production of fish in fisheries and for recreational angling, and the processing of pollutants and waste products to produce clean water. These vital ecosystem processes, however, are not routinely measured in current biomonitoring techniques. Decomposition rates have been measured in some large-scale studies, but these too are still largely ignored in routine biomonitoring, and the responses remain complex and poorly understood (Woodward et al., 2012). Some functional measures, such as organic matter decomposition, have been the focus of attention (e.g., Young et al.,

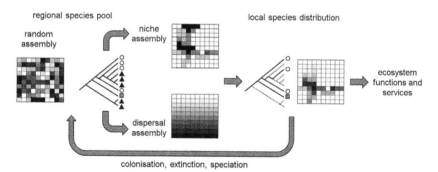

FIGURE 4 Ecology and evolution of species distribution generates diversity patterns in species (grids), species traits (symbols), and phylogeny (trees). From a hypothetical null model (e.g., random assemblage) and species pool at regional scale, species are sorted through the effects of niche assembly (heterogeneity of resources) and species dispersal into patterns of local species distribution. Over time, local extinction, colonization, and speciation alter the regional species pool and associated phylogeny and trait diversity. The dimensions of diversity—taxonomic, abundance, functional, and phylogenetic—determine the dynamics of ecosystem functions and services.

2008), and methods for standardizing this measure across ecosystems have been developed (e.g., Kampfraath et al., 2012), crucially allowing comparisons between studies, but these methods have yet to be adopted into biomonitoring schemes.

Functional indicators, and especially direct measures of ecosystem processes, should also play a larger role in quantifying ecosystem services (Millenium Ecosystem Assessment, 2005) and are being advocated increasingly for economic valuations of conservation, management, and restoration projects (Costanza et al., 1997; Everard and McInnes, 2013). Many ecosystem processes either are services in their own right (e.g., carbon sequestration and nutrient cycling) or underpin them (e.g., invertebrate production supporting fisheries), and include hydraulic retention (water transient storage), sedimentation rate, and greenhouse gas transfer. The magnitude and rate of many of these processes are sensitive to anthropogenic pressures, highlighting the scope to use functional indicators as diagnostic tools (Odum, 1969; Schindler, 1987; Sweeney et al., 2004; Mulholland et al., 2008; Yvon-Durocher et al., 2010; Demars et al., 2011).

Important insights into ecological and evolutionary responses to stressors, as well as their functional consequences, could be inferred from the large number of georeferenced and dated lists of taxa currently filling a multitude of databases in local regulatory and conservation agencies as well as natural history and conservation societies. Many databases are now being assembled that contain some or all of these elements (e.g., FishBase (Frose and Pauly, 2010) and Freshwater Life—http://www.freshwaterlife.org—supported by the Freshwater Biological Association). Scientists are collating decades of research to assemble species traits (and genes) in a phylogenetic context. Combining this with environmental data available from a wide range of government agencies and research bodies, and organizing this information into user-friendly databases (e.g., the Global Biotraits Database http://biotraits.ucla.edu/index.php)—connecting them to infer processes from patterns—offers great potential for future research (e.g., Demars and Harper, 2005; Demars and Trémolières, 2009).

The success of the next generation of biomonitoring will not come solely from assembling and interrogating these vast new databases to obtain new response variables, but also from explicitly testing ecological hypotheses and synthesizing different branches of science—e.g., eco-enzymatic stoichiometry that allows us to link the elemental composition of microbial communities to their nutrient content and biomass production (Sinsabaugh et al., 2009; Hill et al., 2012). Integrating biomonitoring schemes with experimental and modeling approaches will be crucial: combining whole ecosystem experiments with long-term monitoring can reveal spectacular responses to environmental change, although such large-scale, long-term studies are still very much in the minority. Classic examples include the work of Likens et al. (1977) at the Hubbard Brook Experimental Forest, Schindler (1990), Carpenter et al. (2001) at the

Experimental Lakes Area (ELA) in Canada (http://www.experimentallakesarea. ca), and Slavik et al. (2004) at the Kuparuk River station of the Long-Term Ecological Research (LTER) network. Other work has made use of these long-term data to develop new dynamical models to link biodiversity change to ecosystem functioning, such as Petchey et al.'s (2004) study based on the extensive time series data from the UK's Environmental Change Network. Recently, the American LTER network has been complemented by the National Ecological Observatory Network, NEON (http://www.neoninc.org/news/ lterandneon), and the STReam Experimental Observatory Network (STREON, part of NEON) is now one of the most ambitious long-term biomonitoring schemes. It combines comparative surveys across the USA with experimental design (nutrient enrichment and removal of large consumers) that extends previous LYNX programs (Mulholland et al., 2008). In the United Kingdom, the AWMN has also been very effective in providing scientific insights and in influencing policy (Hildrew, 2009; Layer et al., 2010, 2013; Friberg et al., 2011). Moreover, the value of AWMN has increased progressively over the three decades since its inception, as more subtle long-term trends such as responses to climate change can now be detected. The challenge now is to establish international networks with global coverage to tackle planet-scale issues (e.g., the Global Lakes Ecological Observatory Network) that are also integrated with regional and local monitoring. Long-term monitoring can enable us to detect early warning signals of ecosystem shifts (Scheffer et al., 2009), but it is often difficult to extract research funding for such strategic research, which often appears to fail to meet the "novelty" criteria of many research councils' remits.

Future Advances and New Perspectives—Genes to Ecosystems

Over the last 20 years huge progress has been made in understanding biodiversity—ecosystem functioning (B—EF) relationships, with an increasing emphasis on freshwater systems over the last decade in particular (Loreau et al., 2002; Woodward, 2009; Loreau, 2010; Reiss et al., 2010). While biomonitoring and conservation have tended to focus on the biodiversity end of the relationship, the functioning part of the equation as well as its relationship with biodiversity has been largely ignored in the more applied fields of freshwater ecology (but see Dangles et al., 2004; Cardinale, 2011). However, the lack of functional insights is changing, and many emerging legislative and regulatory frameworks are recognizing the need for more functional approaches (e.g., the Water Framework Directive). The main finding of B—EF research to date has been the prevalence of high levels of redundancy. Species loss may initially have little impact, but once a critical threshold is passed when entire functional groups are lost, the impacts can be extremely powerful, and sensitive to further species loss (Cardinale et al., 2006). These experiments have also revealed evidence that idiosyncratic species responses are important, harking back to earlier ideas about keystone species, where they have both strong and unique influences on a process. Despite these advances, there are still some glaring gaps in our knowledge: few studies have included more than one trophic level; most have measured just one process rather than functioning as a whole; and they have been conducted primarily in small experimental arenas over short timescales (Woodward, 2009). As such, many B—EF experiments lack the complexity of natural systems, though attempts are now being made to address these shortcomings (Reiss et al., 2010). In the context of moving from an understanding of B—EF to B—ES (biodiversity—ecosystem services) relationships, there is a huge gap to be bridged in terms of the spatiotemporal scales that are important for the latter, as ecosystem services tend to be manifested at much larger landscape scales, where source—sink, metacommunity and food web dynamics, and eco-evolutionary processes (e.g., Melián et al., 2011) are likely to be important.

The application of network-based approaches can be especially powerful here, as there is a strong food web context to where ecosystem services are located, as well as a clear trophic gradient in the scope for insurance and adaptation, which increases down the web's food chains (Figure 5). Certain stressors are associated with particular nodes in the web (e.g., biomagnification of organochlorine pesticides in apex predators; and antibiotics with the microbial loop at the base of the web), as well as different organizational levels (e.g., food web modules; functional groups; and the network as

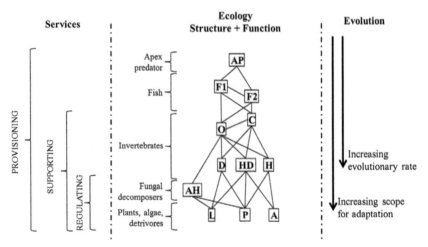

FIGURE 5 Mapping services onto the food web. When monitoring services we need to monitor the appropriate level of scale. The effects of stressors upon services would not show at all levels of the food web, although may magnify through the food web, or cause trophic cascades. AP = apex predator, F = fish, C = carnivore, O = omnivore, D = detritivore, HD = herbivore/detritivore, H = herbivore, AH = aquatic hyphomycete, L = leaf-litter, P = plant, A = algae.

a whole) acting as multiple biosensors. For instance, allometries in food web properties from the level of pairwise links, to tritrophic food chains, to the system's entire constraint space have been used recently to evaluate responses of experimental stream food webs to drought (Woodward et al., 2012; Ledger et al., 2013): these revealed that many of the more commonly used network metrics (such as connectance) were relatively robust to perturbations, whereas others were much more sensitive (e.g., allometric scaling of pairwise links and food chains). The food web provides an intuitive prism through which to view both the lower and the higher levels of organization and how they respond to stressors, as it makes the interactions between species explicit in the response variables, whereas most biomonitoring and conservation approaches focus solely on (a few) nodes and not the links between them at the system scale (Woodward et al., 2013). Considerable work has been done in freshwaters in terms of understanding how food webs respond to stressors, including acidification (e.g., Ledger and Hildrew, 2005; Layer et al., 2010, 2011), eutrophication (e.g., Rawcliffe et al., 2010), and hydrological change (e.g., Ledger et al., 2012, 2013). Such combinations of studies illustrate effectively that studying the feedbacks between the environment and the functioning of the whole system that are mediated by the food web can be extremely powerful, and may even induce regime shifts (Jones and Sayer, 2003; Scheffer and Carpenter, 2003).

Eco-evolutionary dynamics and feedbacks within the food web can be much faster than previously thought (e.g., Melián et al., 2011), and impacts on the epigenome can lead to quicker adaptation than traditional adaptation of the genome, via genetic plasticity (Johnson and Tricker, 2010). Consequently, we are starting to perceive how species evolve in the context of both the biotic and the abiotic environment, and how feedbacks and newly discovered mechanisms can accelerate evolutionary responses (Moya-Larano et al., 2012). In addition to the discovery of these ecological and evolutionary interactions, in recent years there have been rapid technological advances in next-generation sequencing (NGS, Box 2) and associated molecular techniques (Hajibabaei et al., 2011; Hajibabaei, 2012). This has allowed for significant advances in broadening the coverage of the tree of life and for adopting an eco-evolutionary approach to biomonitoring in freshwaters: emerging NGS approaches include new generations of molecular markers and the ability to characterize microbes *in situ*, allowing them to be used to monitor the functioning of ecosystems as well as determining the functions of microbes, metazoans, and macrofaunal communities directly (Purdy et al., 2010).

Novel Molecular and Microbial Approaches

An organism's molecular state results from its interaction with the environment, so measuring specific molecular machinery components can provide clues as to which stressors are present in the environment. The first generation

BOX 2 What is Next Generation Sequencing/-Omics?
The terms "next-generation" sequencing (NGS) or -omic technologies have been in use since a landmark paper (Margulies et al., 2005) detailed the use of 454 massively parallel pyrosequencings. Since then, the development of NGS platforms, accompanied by exponential increases in throughput and decreasing costs, has completely transformed the field of DNA sequencing.

For investigating functional diversity, the NGS "-omic" approaches can conveniently be broken down into discrete categories of relevance to different levels of biological organization. At the individual level, transcriptomic analyses measure differential gene expression via the analysis of expressed total RNA from specific tissues. At the community level, metagenetic and metabarcoding (Fonseca et al., 2010b; Bik et al., 2012; Taberlet et al., 2012) studies estimate environmental taxonomic richness by the en masse sequencing of environmental DNA samples (Sun et al., 2012). Shotgun metagenomic studies instead randomly sequence fragments of the total genomes present in an environmental DNA extraction (Knight et al., 2012), providing insights into both the functional and taxonomic capability of a given environment. Finally, metatranscriptomics enables researchers to investigate the actively transcribed mRNA from a community, giving an insight into the total gene expression from a local ecosystem (Filiatrault, 2011; Gilbert and Hughes, 2011).

As with microarray studies, gene expression is likely to change significantly at both short (Gilbert and Hughes, 2011) and large spatial and temporal scales, so transcriptomic analyses need to be designed around carefully and explicitly framed questions that account for environmental gene expression and short half-life of mRNA (i.e., transcript analyses are often not associated with protein composition) (Moran et al., 2013). These broad -omic categories are summarized in Figure 7.

For ecological studies, a potential disadvantage of these approaches lies in the fact that most platforms incorporate various forms of clonal amplification in the sequencing approaches, thereby introducing potential quantitative biases into datasets. New "third-generation" sequencers and technologies (Ribeiro et al., 2012; Schneider and Dekker, 2012; GridION™ and MinION™) that use single molecule sequencing approaches and therefore lack any clonal amplification step prior to sequencing could produce truly quantitative data, although these are currently tailored to analyzing shorter numbers of very long reads, and many had not reached market maturity at the time of writing.

of molecular markers (Figure 6) was developed from hypothesis-driven research and based on biochemical, histological, morphological, and physiological changes in nucleic acids and proteins measured with conventional techniques (Ryan and Hightower, 1996). The number of such biomarkers is relatively small but they include some very effective examples such as the general xenobiotic response marker CYP1A (Celander, 2011), the endocrine disruption marker vitellogenin (Celander, 2011), and the metal stress marker metallothionein (Amiard et al., 2006). However, the hypothesis-driven

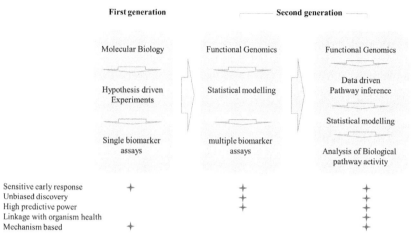

FIGURE 6 The evolution of biomarker discovery from the first generation approaches that use single genes whose expression is modulated by specific stressors, to the most recent advances that allow the discovery of multicomponent molecular signatures.

approach to biomarker discovery suffers from an important conceptual flaw, at least in this implementation: single genes whose expression is modulated in a highly specific manner are extremely rare.

In the last 10 years, new functional genomics technologies have provided a potential solution to this issue. Since these technologies allow the measurement of the expression of tens of thousands of genes, proteins, and metabolites in a single experiment, they provide the means to develop multigene signatures from the unbiased screening of genome-wide expression data (Van Aggelen et al., 2010; Figure 6).

The challenge of identifying specific molecular signatures hidden within hundreds of thousands of noisy variables has driven the development of statistical methods for the identification of molecular components that are differentially expressed in two or more sample groups (i.e., stressed versus controls). Although effective, this approach has limitations: in particular, it cannot identify synergistic effects between variables, it has a relatively low statistical power, and biological interpretation is challenging. The introduction of more complex modeling techniques that can assess the predictive power of combinations of biomarkers (Li et al., 2010) has been a significant step forward, particularly when applied to linking phenotypic responses (e.g., physiology) to molecular responses, especially in a network context. Ultimately this has allowed the identification of more effective and ecologically relevant biomarkers (Ankley et al., 2010).

Despite the potential of these approaches, the vast number of possible combinations of individual measurements drastically limits their ability to explore a large portion of the solution space and therefore makes it extremely difficult to capture biologically relevant pathways that respond specifically to particular stressors. One way to address this challenge is reverse engineering, a branch of systems biology that aims to reconstruct the underlying structure of a biological pathway from experimental data. This has been tremendously effective in biomedical research for identifying pathways predictive of clinical response, drug resistance, and novel therapeutic targets (Perkins et al., 2011). Again the biomedical-biomonitoring analogy can be used here to extend such approaches to environmental assessment. Because of the complexity of the datasets acquired using -*omics* technologies, any reverse-engineering approach must start from the identification of the high-level structure of the underlying biological networks and then progress to identifying more refined subnetworks associated with important phenotypic responses such as changes in reproductive ability following stress. Although in its infancy, this approach has already been applied by a number of groups for identifying novel stress pathways (Williams et al., 2011).

Overall, the use of these approaches allows the identification of more effective biomarkers than the ones based on differential expression and has opened up the possibility to develop specific multicomponent molecular signatures that are truly representative of a large number of stressors, with high specificity.

The use of biomarkers as a biomonitoring tool relies on inferences from molecular analyses. Returning to the more traditional approach of biomonitoring by using taxa themselves, and given that NGS technologies have finally enabled us to identify microbes in field conditions, these taxa represent ideal candidates for assessing how stressors alter community structure and ecosystem functioning. The pioneering "everything is everywhere, but the environment selects" theory proposed by Baas Becking (1934) suggests that the presence of all microorganisms is ubiquitous, but our ability to detect them via direct observation is limited by varying densities: i.e., rare microbes may be present but unobserved in ecological samples (de Wit and Bouvier, 2006). Consequently, the presence of different microbial species should be dictated by differences in environmental conditions rather than distance and biogeography (Zarraonaindia et al., 2013). If this is true, it could provide a truly global comparable framework for bioassessment and monitoring. Opposing theories exist, however, suggesting that microbial diversity is shaped by geography as well as the environment (Martiny et al., 2006; O'Malley, 2008). The key question is whether the environment enhances the presence of certain microorganisms in different locations, thus allowing us to compare components of the microbial community for the monitoring of ecosystems. High-throughput technologies with increased detection capabilities can assist here and there is huge potential for these to be exploited by ecologists for monitoring purposes (Green et al., 2008; Purdy et al., 2010; Poisot et al., 2013; Woodward et al., 2013).

Microorganisms play important functional roles in the major biogeochemical cycles at local to global scales, as well as in the recycling of nutrients and overall ecosystem functioning (Cotner and Biddanda, 2002; Nemergut et al., 2011), and many of these are also either ecosystem services in their own right or key processes that support important services (e.g., carbon sequestration). Moreover, microbial communities are themselves influenced by environmental conditions. Accordingly, bacteria have been suggested as good indicators of environmental change due to some of their attractive biomonitoring properties such as high diversity (thus broad range of environmental susceptibility), potential ubiquity, short life cycles, and minimal disturbance of the site during sampling (Lear et al., 2009; see Figure 7).

However, until recently their use was hindered by the inability to study them *in situ*, as only 5% of species are considered to be cultivable with standard techniques (Amann et al., 1995; Curtis et al., 2002), thus leading to narrowly focused approaches of single species analysis such as the targeting of specific ecotypes of pathogens, rather than whole-community detection (Hellawell, 1986; Port et al., 2012). High-throughput sequencing is already replacing historical fingerprinting approaches (Box 2) and has been used for the characterization of whole communities from a large variety of sources, from both terrestrial and aquatic systems (Roesch et al., 2007; Cole et al., 2010; Gilbert and Dupont, 2011; Foote et al., 2012; Port et al., 2012). Following sequence-based approaches, specific and identifiable microorganisms can be linked to environmental status and used as sensors for the

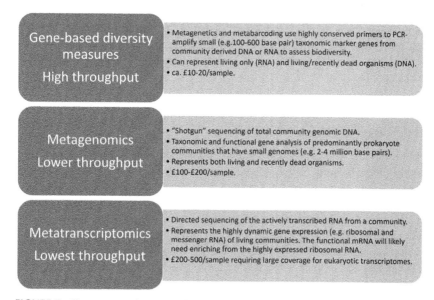

FIGURE 7 The many -omics approaches to sequencing life, from individuals to whole community techniques that can be adapted to each scenario. Methods applicable to a variety of scales are presented with their respective advantages and disadvantages.

assessment of anthropogenic threats such as eutrophication, acidification, climate change, and land use changes (Port et al., 2012; Yergeau et al., 2012; Heino, 2013). In aquatic ecosystems, whole bacterial cell analysis can also be used for the assessment of pollution effects (Lear et al., 2009) and detection of antibiotics in the water (Port et al., 2012).

Recent studies from terrestrial and marine systems (Pommier et al., 2012; Sun et al., 2012) suggest that bacterial communities are sensitive indicators of contaminant stress, and also support the theory that the presence of microorganisms is more related to environmental conditions than to dispersal or geography. However, a freshwater study by Lear et al. (2012) found that microbial communities did not differ among different environmental pressures, whereas invertebrate sampling was the more effective monitoring tool, suggesting that either the studied microbial communities were unaffected by contaminants, or the discriminatory power of the molecular fingerprinting approaches used was insufficient.

Yergeau et al. (2012) used NGS of the 16S rRNA gene to determine the effect of pollution related to oil sands mining on nearby aquatic microbial community structure. Their findings suggest that the microbial community structure was significantly altered by distance from mining sites and support the potential use of bacteria and archaea as bioindicators of pollution. Furthermore, Kisand et al. (2012) were able to compare the microbial community composition of a highly impacted area, like the port of Genoa, with that of a protected area (low anthropogenic impact) through metagenomic analysis of the microbial communities from water samples. Distinct microbial diversity and abundance counts were detected among the different sites that can be related to the differences of environmental conditions, again demonstrating the potential for the use of metagenomics for monitoring of aquatic ecosystems.

The Functional Analysis of Microbes, Metazoans, and Macrofaunal Communities

Ecologists are increasingly striving to improve predictive power not only by identifying what organisms are present, but also by asking, "What are they doing?" The majority of functional ecological studies use organismal trait information (Tilman et al., 1997; Petchey and Gaston, 2006; Hagen et al., 2012) to provide a metric for quantitative analysis, but these cannot accurately reflect all of the functional attributes of individuals and species in complex ecological communities. In theory, the -omic toolbox can be employed to address this and to understand functional diversity in ways that have not been previously possible, although synergies with traditional ecology and taxonomy are essential if we are to fully understand the connections between biodiversity and ecosystem functioning and how they respond to stressors (Loreau et al., 2001).

If we consider a hypothetical freshwater ecosystem with both benthic and aquatic habitats, these habitats can be studied first independently but then in combination by investigating both the taxonomic and the functional diversity of the entire community using the -omic toolbox (see Box 2) tailored to organismal genome size and complemented by biogeochemical and nutrient cycling analyses. Starting with the microbial fraction, taxonomy marker genes such as 16S (Caporaso et al., 2011), ITS (Nilsson et al., 2008), and 18S (Fonseca et al., 2010a; Pawlowski et al., 2012) can be used for the high-throughput assessment of bacteria, archaea, fungi, and meiobiota, respectively, from multiple samples. Phylogenetic diversity can then be used throughout all gene marker schemes as a proxy for functional diversity by employing algorithms such as UniFrac (Lozupone and Knight, 2005; Caporaso et al., 2010; Fierer et al., 2012). Metagenomic and metatranscriptomic analyses can be employed to investigate the functional capability and specific functioning of the prokaryotic size fraction characterized by organisms with small genomes (e.g., 2–4 Mb) and their relatively small transcriptomes. Metatranscriptomic analyses are likely to be robust in simple communities of eukaryotic organisms where just a few species dominate (Durkin et al., 2012), but given the current limits of sequencing power, achieving effective coverage of replicated samples of complex eukaryotic communities (Bailly et al., 2007), whose transcriptomes can be very large (e.g., 20 Mb), is still limited. Similarly, metagenomic sequencing of eukaryotic communities is unlikely to reach the appropriate depth of coverage for ecological synthesis simply because eukaryotic genomes can be very large (the human genome alone is over 3 Gb in size).

Within prokaryotic communities, a new approach (PiCrust) (Langille et al., 2013) has emerged that links marker gene 16S studies to functional diversity maps and environmental 16S reads to their closest ancestors with full genome sequences, and predicts ancestral states of functional gene ontologies. Initial analyses suggest that this outperforms low-coverage shotgun metagenomic analyses in well-characterized communities, but further testing and examples will undoubtedly provide further insight. Nevertheless, the model provides a route between high-throughput studies and full-genome capability that may also eventually feature in the eukaryotic biosphere as more genomes are sequenced.

Advances that are likely to be provided by the -omic toolbox regarding the functional diversity of eukaryotic communities (e.g., protists, fungi, meiobiota, and macrofauna) are likely to be achieved by linking genotype phenotype data with the analysis of food webs and networks (Barberan et al., 2012; Rodriguez-Lanetty et al., 2013). The Barcode of Life Project (Ratnasingham and Hebert, 2007) strives for the provision of standardized and carefully curated DNA barcode data for organisms based on official barcode markers. So far, over 200,000 species have been barcoded. Importantly, this endeavor provides a link between a standardized genotype and the taxonomy and ecology of the barcoded species. At the start of the barcoding movement, sequencing technologies were not mature enough to consider assessing multiple communities of organisms, but recently a multitude of "metabarcoding" studies (Epp et al., 2012; Taberlet et al., 2012) have shown that approaches used for microbial communities can be conveniently transferred to macrofaunal communities. If the featured species in the metabarcoding datasets have barcode reference data, these can provide very powerful links to the functional attributes of the organisms comprising the sequenced communities. The maturation of the field of metabarcoding not only provides a huge boost for our ability to assess large numbers of macrofaunal samples simultaneously (Ji et al., 2013), but also reasserts the need for generation of reference barcode libraries to provide the necessary links between -omic technologies and functional ecology. Moreover, since gene marker-based studies do not respect the boundaries between the living and the recently deceased or even ingested species, dietary and food web analyses can be conveniently performed using either individual or species-based sequencing of gut contents to investigate trophic interactions (Pompanon et al., 2012).

Overlying these possibilities is the further opportunity to deduct functional relationships using the analysis of ecological networks at multispecies levels of organization (Ings et al., 2009; Hagen et al., 2012). Following marker-based approaches and even metagenomic analyses, the resulting data are represented by a familiar taxon-by-sample frequency matrix of genotype occurrence (Ji et al., 2013) that can be related back to phenotype occurrence (i.e., species). The quantitative nature of the associations can be estimated on the basis of the mode of evolution and genomic content of the markers used (while acknowledging potential PCR bias), but the co-occurrence incidence matrices will reflect the distribution of species in space and time. Such power potentially enables us to delimit co-occurring ecological networks (in space and/or time) and how individual networks respond to external drivers. Moreover, some components of the sequence data matrices will be annotated to a high degree of accuracy (e.g., species level for barcoded metabarcoding data) and for all other groups potentially genus, order, family, etc., but at least phylum, enabling the researcher to characterize biological interactions (parasitism, predation, commensalism, mutualism, competition, etc.) and ecological processes (Faust and Raes, 2012). The additional strength of -omic high-throughput marker-based approaches is that with the now routine analysis of ca. 50 complex samples simultaneously, a high degree of replication and sample coverage can be achieved on scales that are simply not possible using traditional approaches for either microbial or macrofaunal samples. The combination of these emerging technologies and approaches promises a possible means of truly integrating ecological and evolutionary perspectives to responses to stressors across all the major domains of life in aquatic (and terrestrial) ecosystems.

CONCLUDING REMARKS

With an ever-increasing human population, the need to monitor and predict our effects on the natural world has never been more important. In the developed world the predominant stressors have changed, presenting new challenges to biomonitoring science (Figure 2), while developing nations such as India and China are facing the same stressors the Western world was exposed to in the twentieth century, but on a far greater scale (Abate, 1995; Yagishita, 1995; Aggarwal et al., 2001). An eco-evolutionary approach to biomonitoring will allow us to better understand the dynamics between the selective forces of evolution and the ecology of species. The ability of a community to adapt to change is key to its response to a particular stressor (Woodward et al., 2010; Moya-Larano et al., 2012), and this needs to be considered alongside biomonitoring results. With new technologies such as the rise of new molecular markers (e.g., Van Aggelen et al., 2010; Williams et al., 2011), the use of microbes (e.g., Lear et al., 2009), and advances in NGS techniques (Box 2), there is a great variety in approaches now available to monitor the functional response of aquatic communities to environmental stress.

A shift in the culture surrounding legislative biomonitoring, governance, and stakeholder implementation will be required before these advanced and promising approaches can be integrated into current protocols. There will likely be far fewer "traditional" taxonomists as NGS technologies take over, but many more bioinformaticians will be needed to process and analyze the NGS samples. The rate-limiting step in biomonitoring will shift from the slow and laborious process of identifying individuals through microscopy (data acquisition) to limitations in the efficiency with which large volumes of data can be processed. It is not impossible to imagine a future where remote sensing stations monitor environmental DNA or RNA and send sequence data back to the laboratory via telemetry as weather stations do now—unmanned and automated transmitting of results back to a central point. As bioinformatics solutions to data analysis and synthesis continue to develop over time as well as developing bioinformatics' huge potential to the biomonitoring world, it is likely to be simply a matter of "when" and not "if" this revolution will take place on a truly global scale.

ACKNOWLEDGMENTS

Thanks go to Rick Battarbee, University College London, and the Upland Waters Monitoring Network for allowing us to reproduce their data for Figure 1. CG was supported by Queen Mary University of London and the Freshwater Biology Association. BD was funded by the Scottish Government Rural and Environment Science and Analytical Services (RESAS). The project was partly supported by the Grand Challenges in Ecosystems and the Environment initiative at Imperial College London.

REFERENCES

Abate, T., 1995. Swedish scientists take acid-rain research to developing-nations. Bioscience 45, 738–740.

Aggarwal, S.G., Chandrawanshi, C.K., Patel, R.M., Agarwal, S., Kamavisdar, A., Mundhara, G.L., 2001. Acidification of surface water in central India. Water Air Soil Pollut. 130, 855–862.

Allen, A.P., Brown, J.H., Gillooly, J.F., 2002. Global biodiversity, biochemical kinetics, and the energetic-equivalence rule. Science 297, 1545–1548.

Amann, R.I., Ludwig, W., Schleifer, K.H., 1995. Phylogenetic identification and *in-situ* detection of individual microbial-cells without cultivation. Microbiol. Rev. 59, 143–169.

Amiard, J.-C., Amiard-Triquet, C., Barka, S., Pellerin, J., Rainbow, P., 2006. Metallothioneins in aquatic invertebrates: their role in metal detoxification and their use as biomarkers. Aquat. Toxicol. 76, 160–202.

Ankley, G.T., Bennett, R.S., Erickson, R.J., Hoff, D.J., Hornung, M.W., Johnson, R.D., Mount, D.R., Nichols, J.W., Russom, C.L., Schmieder, P.K., Serrrano, J.A., Tietge, J.E., Villeneuve, D.L., 2010. Adverse outcome pathways: a conceptual framework to support ecotoxicology research and risk assessment. Environ. Toxicol. Chem. 29, 730–741.

Baas Becking, L.G.M., 1934. Geobiologie of inleiding tot de milieukunde. WP Van Stockum & Zoon.

Bailly, J., Fraissinet-Tachet, L., Verner, M.C., Debaud, J.C., Lemaire, M., Wesolowski-Louvel, M., Marmeisse, R., 2007. Soil eukaryotic functional diversity, a metatranscriptomic approach. Isme J. 1, 632–642.

Barberan, A., Bates, S.T., Casamayor, E.O., Fierer, N., 2012. Using network analysis to explore co-occurrence patterns in soil microbial communities. Isme J. 6, 343–351.

Battarbee, R., Shilland, E., Simpson, G., Salgado, J., Goldsmith, B., Gray, W., Turner, S., 2013. Surface-water Acidity in the River Dart Area. Report for the West Country Rivers Trust.

Battarbee, R.W., Charles, D.F., 1986. Diatom-based pH reconstruction studies of acid lakes in Europe and North America: a synthesis. Water, Air, Soil Pollut. 30, 347–354.

Battarbee, R.W., Monteith, D.T., Juggins, S., Evans, C.D., Jenkins, A., Simpson, G.L., 2005. Reconstructing pre-acidification pH for an acidified Scottish loch: a comparison of palaeolimnological and modelling approaches. Environ. Pollut. 137, 135–149.

Battarbee, R.W., Simpson, G.L., Shilland, E.M., Flower, R.J., Kreiser, A., Yang, H., Clarke, G., 2014. Recovery of UK lakes from acidification: an assessment using combined palaeoecological and contemporary diatom assemblage data. Ecol. Indic. 37 (Part B), 365–380.

Bell, G., 2001. Neutral macroecology. Science 293, 2413–2418.

Bell, G., Gonzalez, A., 2011. Adaptation and evolutionary Rescue in metapopulations experiencing environmental deterioration. Science 332, 1327–1330.

Bik, H.M., Porazinska, D.L., Creer, S., Caporaso, J.G., Knight, R., Thomas, W.K., 2012. Sequencing our way towards understanding global eukaryotic biodiversity. Trends Ecol. Evol. 27, 233–243.

Birks, H.J.B., Line, J.M., Juggins, S., Stevenson, A.C., Terbraak, C.J.F., 1990. Diatoms and Ph reconstruction. Philos. Trans. R. Soc. B-Biol. Sci. 327, 263–278.

Bonada, N., Dolédec, S., Statzner, B., 2012. Spatial autocorrelation patterns of stream invertebrates: exogenous and endogenous factors. J. Biogeogr. 39, 56–68.

Caporaso, J.G., Kuczynski, J., Stombaugh, J., Bittinger, K., Bushman, F.D., Costello, E.K., Fierer, N., Pena, A.G., Goodrich, J.K., Gordon, J.I., 2010. QIIME allows analysis of high-throughput community sequencing data. Nat. Methods 7, 335–336.

Caporaso, J.G., Lauber, C.L., Walters, W.A., Berg-Lyons, D., Lozupone, C.A., Turnbaugh, P.J., Fierer, N., Knight, R., 2011. Global patterns of 16S rRNA diversity at a depth of millions of sequences per sample. Proc. Natl. Acad. Sci. 108, 4516–4522.

Cardinale, B.J., 2011. Biodiversity improves water quality through niche partitioning. Nature 472, 86–89.

Cardinale, B.J., Srivastava, D.S., Duffy, J.E., Wright, J.P., Downing, A.L., Sankaran, M., Jouseau, C., 2006. Effects of biodiversity on the functioning of trophic groups and ecosystems. Nature 443, 989–992.

Carpenter, S.R., Cole, J.J., Hodgson, J.R., Kitchell, J.F., Pace, M.L., Bade, D., Cottingham, K.L., Essington, T.E., Houser, J.N., Schindler, D.E., 2001. Trophic cascades, nutrients, and lake productivity: whole-lake experiments. Ecol. Monogr. 71, 163–186.

Celander, M.C., 2011. Cocktail effects on biomarker responses in fish. Aquat. Toxicol. 105, 72–77.

Cole, J.R., Konstandinidis, K., Farris, R.J., 2010. Microbial Diversity and Phylogeny: Extending from RRNAs to Genomes. In: Environmental Molecular Microbiology. Caister Academic Press, Norfolk, UK, pp. 1–21.

Costanza, R., d'Arge, R., De Groot, R., Farber, S., Grasso, M., Hannon, B., Limburg, K., Naeem, S., O'neill, R.V., Paruelo, J., 1997. The value of the world's ecosystem services and natural capital. Nature 387, 253–260.

Cotner, J.B., Biddanda, B.A., 2002. Small players, large role: microbial influence on biogeochemical processes in pelagic aquatic ecosystems. Ecosystems 5, 105–121.

Curtis, T.P., Sloan, W.T., Scannell, J.W., 2002. Estimating prokaryotic diversity and its limits. Proc. Natl. Acad. Sci. USA 99, 10494–10499.

Dangles, O., Gessner, M.O., Guerold, F., Chauvet, E., 2004. Impacts of stream acidification on litter breakdown: implications for assessing ecosystem functioning. J. Appl. Ecol. 41, 365–378.

Darwin, C., 1872. On the Origin of Species by Means of Natural Selection. John Murray, London.

Demars, B.O., Trémolières, M., 2009. Aquatic macrophytes as bioindicators of carbon dioxide in groundwater fed rivers. Sci. Total Environ. 407, 4752–4763.

Demars, B.O.L., Edwards, A.C., 2009. Distribution of aquatic macrophytes in contrasting river systems: a critique of compositional-based assessment of water quality. Sci. Total Environ. 407, 975–990.

Demars, B.O.L., Harper, D.M., 2005. Distribution of aquatic vascular plants in lowland rivers: separating the effects of local environmental conditions, longitudinal connectivity and river basin isolation. Freshwater Biol. 50, 418–437.

Demars, B.O.L., Kemp, J.L., Friberg, N., Usseglio-Polatera, P., Harper, D.M., 2012. Linking biotopes to invertebrates in rivers: biological traits, taxonomic composition and diversity. Ecol. Indic. 23, 301–311.

Demars, B.O.L., Manson, J.R., Olafsson, J.S., Gislason, G.M., Gudmundsdottir, R., Woodward, G., Reiss, J., Pichler, D.E., Rasmussen, J.J., Friberg, N., 2011. Temperature and the metabolic balance of streams. Freshwater Biol. 56, 1106–1121.

Durkin, C.A., Marchetti, A., Bender, S.J., Truong, T., Morales, R., Mock, T., Armbrust, E.V., 2012. Frustule-related gene transcription and the influence of diatom community composition on silica precipitation in an iron-limited environment. Limnol. Oceanogr. 57, 1619–1633.

Engelhardt, K.A., 2006. Relating effect and response traits in submersed aquatic macrophytes. Ecol. Appl. 16, 1808–1820.

Enquist, B.J., Kerkhoff, A.J., Stark, S.C., Swenson, N.G., McCarthy, M.C., Price, C.A., 2007. A general integrative model for scaling plant growth, carbon flux, and functional trait spectra. Nature 449, 218–222.

Epp, L.S., Boessenkool, S., Bellemain, E.P., Haile, J., Esposito, A., Riaz, T., Erseus, C., Gusarov, V.I., Edwards, M.E., Johnsen, A., Stenoien, H.K., Hassel, K., Kauserud, H., Yoccoz, N.G., Brathen, K., Willerslev, E., Taberlet, P., Coissac, E., Brochmann, C., 2012. New environmental metabarcodes for analysing soil DNA: potential for studying past and present ecosystems. Mol. Ecol. 21, 1821–1833.

Everard, M., McInnes, R., 2013. Systemic solutions for multi-benefit water and environmental management. Sci. Total Environ. 461, 170–179.

Faust, K., Raes, J., 2012. Microbial interactions: from networks to models. Nat. Rev. Microbiol. 10, 538–550.

Fierer, N., Leff, J.W., Adams, B.J., Nielsen, U.N., Bates, S.T., Lauber, C.L., Owens, S., Gilbert, J.A., Wall, D.H., Caporaso, J.G., 2012. Cross-biome metagenomic analyses of soil microbial communities and their functional attributes. Proc. Natl. Acad. Sci. USA 109, 21390–21395.

Filiatrault, M.J., 2011. Progress in prokaryotic transcriptomics. Curr. Opin. Microbiol. 14, 579–586.

Fonseca, V.G., Carvalho, G.R., Sung, W., Johnson, H.F., Power, D.M., Neill, S.P., Packer, M., Blaxter, M.L., Lambshead, P.J.D., Thomas, W.K., 2010a. Second-generation environmental sequencing unmasks marine metazoan biodiversity. Nat. Commun. 1, 98.

Fonseca, V.G., Carvalho, G.R., Sung, W., Johnson, H.F., Power, D.M., Neill, S.P., Packer, M., Blaxter, M.L., Lambshead, P.J.D., Thomas, W.K., Creer, S., 2010b. Second-generation environmental sequencing unmasks marine metazoan biodiversity. Nat. Commun. 1.

Foote, A.D., Thomsen, P.F., Sveegaard, S., Wahlberg, M., Kielgast, J., Kyhn, L.A., Salling, A.B., Galatius, A., Orlando, L., Gilbert, M.T.P., 2012. Investigating the potential use of environmental DNA (eDNA) for genetic monitoring of marine mammals. Plos One 7.

Friberg, N., Bonada, N., Bradley, D.C., Dunbar, M.J., Edwards, F.K., Grey, J., Hayes, R.B., Hildrew, A.G., Lamouroux, N., Trimmer, M., Woodward, G., 2011. Biomonitoring of human impacts in freshwater ecosystems: the good, the bad and the ugly. Adv. Ecol. Res. 44, 1–68.

Froese, R., Pauly, D., 2010. FishBase. International Center for Living Aquatic Resources Management.

García-Llorente, M., Martín-López, B., Díaz, S., Montes, C., 2011. Can ecosystem properties be fully translated into service values? an economic valuation of aquatic plant services. Ecol. Appl. 21, 3083–3103.

Gilbert, J.A., Dupont, C.L., 2011. Microbial metagenomics: beyond the genome. Annu. Rev. Mar. Sci. 3, 347–371.

Gilbert, J.A., Hughes, M., 2011. Gene Expression Profiling: Metatranscriptomics. High-throughput Next Generation Sequencing. Springer, pp. 195–205.

Gould, S.J., Lewontin, R.C., 1979. Spandrels of San-Marco and the Panglossian paradigm - a critique of the adaptationist program. Proc. R. Soc. Ser. B-Biol. Sci. 205, 581–598.

Green, J.L., Bohannan, B.J.M., Whitaker, R.J., 2008. Microbial biogeography: from taxonomy to traits. Science 320, 1039–1043.

Greulich, S., Bornette, G., 2003. Being evergreen in an aquatic habitat with attenuated seasonal contrasts - a major competitive advantage? Plant Ecol. 167, 9–18.

Hagen, M., Kissling, W.D., Rasmussen, C., De Aguiar, M.A.M., Brown, L.E., Carstensen, D.W., Alves-Dos-Santos, I., Dupont, Y.L., Edwards, F.K., Genini, J., Guimaraes, P.R., Jenkins, G.B., Jordano, P., Kaiser-Bunbury, C.N., Ledger, M.E., Maia, K.P., Marquitti, F.M.D., Mclaughlin, O., Morellato, L.P.C., O'Gorman, E.J., Trojelsgaard, K., Tylianakis, J.M., Vidal, M.M., Woodward, G., Olesen, J.M., 2012. Biodiversity, species interactions and ecological networks in a fragmented world. Adv. Ecol. Res. 46 (Pt 1), 189–210.

Hajibabaei, M., 2012. The golden age of DNA metasystematics. Trends Genet.: TIG 28, 535–537.

Hajibabaei, M., Shokralla, S., Zhou, X., Singer, G.A.C., Baird, D.J., 2011. Environmental barcoding: a next-generation sequencing approach for biomonitoring applications using river benthos. Plos One 6.

Halvorsen, G.A., Heegaard, E., Fjellheim, A., Raddum, G.G., 2003. Tracing recovery from acidification in the western Norwegian Nausta watershed. Ambio 32, 235–239.

Harte, J., 2011. Maximum Entropy and Ecology: A Theory of Abundance, Distribution, and Energetics. Oxford University Press.

Heino, J., 2013. The importance of metacommunity ecology for environmental assessment research in the freshwater realm. Biol. Rev. 88, 166–178.

Hellawell, J.M., 1986. Biological Indicators of Freshwater Pollution and Environmental Management.

Hildrew, A.G., 2009. Sustained research on stream communities: a model system and the comparative approach. Adv. Ecol. Res. 41, 175–312.

Hill, B.H., Elonen, C.M., Seifert, L.R., May, A.A., Tarquinio, E., 2012. Microbial enzyme stoichiometry and nutrient limitation in US streams and rivers. Ecol. Indic. 18, 540–551.

Horrigan, N., Baird, D.J., 2008. Trait patterns of aquatic insects across gradients of flow-related factors: a multivariate analysis of Canadian national data. Can. J. Fish. Aquat. Sci. 65, 670–680.

Hubbell, S.P., 2001. The Unified Neutral Theory of Biodiversity and Biogeography. Princeton University Press.

Humboldt, A., 1849. Cosmos: A Sketch of a Physical Description of the Universe. EC Otté, I. Henry G. Bohn, London.

Ings, T.C., Montoya, J.M., Bascompte, J., Bluthgen, N., Brown, L., Dormann, C.F., Edwards, F., Figueroa, D., Jacob, U., Jones, J.I., Lauridsen, R.B., Ledger, M.E., Lewis, H.M., Olesen, J.M., van Veen, F.J.F., Warren, P.H., Woodward, G., 2009. Ecological networks - beyond food webs. J. Anim. Ecol. 78, 253–269.

Ji, Y., Ashton, L., Pedley, S.M., Edwards, D.P., Tang, Y., Nakamura, A., Kitching, R., Dolman, P.M., Woodcock, P., Edwards, F.A., 2013. Reliable, verifiable and efficient monitoring of biodiversity via metabarcoding. Ecol. Lett. 16, 1245–1257.

Johnson, L.J., Tricker, P.J., 2010. Epigenomic plasticity within populations: its evolutionary significance and potential. Heredity 105, 113–121.

Jones, J.I., Sayer, C.D., 2003. Does the fish-invertebrate-periphyton cascade precipitate plant loss in shallow lakes? Ecology 84, 2155–2167.

Kampfraath, A.A., Hunting, E.R., Mulder, C., Breure, A.M., Gessner, M.O., Kraak, M.H.S., Admiraal, W., 2012. DECOTAB: a multipurpose standard substrate to assess effects of litter quality on microbial decomposition and invertebrate consumption. Freshwater Sci. 31, 1156–1162.

Kerkhoff, A.J., Enquist, B.J., 2006. Ecosystem allometry: the scaling of nutrient stocks and primary productivity across plant communities. Ecol. Lett. 9, 419–427.

Kisand, V., Valente, A., Lahm, A., Tanet, G., Lettieri, T., 2012. Phylogenetic and functional metagenomic profiling for assessing microbial biodiversity in environmental monitoring. Plos One 7.

Knight, R., Jansson, J., Field, D., Fierer, N., Desai, N., Fuhrman, J.A., Hugenholtz, P., van der Lelie, D., Meyer, F., Stevens, R., Bailey, M.J., Gordon, J.I., Kowalchuk, G.A., Gilbert, J.A., 2012. Unlocking the potential of metagenomics through replicated experimental design. Nat. Biotechnol. 30, 513–520.

Kolkwitz, R., Marsson, M., 1902. Grundsätze für die biologische Beurteilung des Wassers nach seiner Flora und Fauna. Mitt. Prüfungsanst. Wasserversorg. Abwasserbeseit. 1, 3–72.

Kolkwitz, R., Marsson, M., 1909. Ökologie der tierischen Saprobien. Beiträge zur Lehre von der biologischen Gewässerbeurteilung. Int. Rev. Gesamten Hydrobiol. Hydrogr. 2, 126–152.

Langille, M.G.I., Zaneveld, J., Caporaso, J.G., McDonald, D., Knights, D., Reyes, J.A., Clemente, J.C., Burkepile, D.E., Thurber, R.L.V., Knight, R., Beiko, R.G., Huttenhower, C., 2013. Predictive functional profiling of microbial communities using 16S rRNA marker gene sequences. Nat. Biotechnol. 31, 814–821.

Layer, K., Hildrew, A., Monteith, D., Woodward, G., 2010. Long-term variation in the littoral food web of an acidified mountain lake. Global Change Biol. 16, 3133–3143.

Layer, K., Hildrew, A.G., Jenkins, G.B., Riede, J.O., Rossiter, S.J., Townsend, C.R., Woodward, G., 2011. Long-term dynamics of a well-characterised food web: four decades of acidification and recovery in the Broadstone stream model system. Adv. Ecol. Res. 44, 69–117.

Layer, K., Hildrew, A.G., Woodward, G., 2013. Grazing and detritivory in 20 stream food webs across a broad pH gradient. Oecologia 171, 459–471.

Lear, G., Ancion, P.Y., Harding, J., Lewis, G.D., 2012. Use of bacterial communities to assess the ecological health of a recently restored stream. N. Z. J. Mar. Freshwater Res. 46, 291–301.

Lear, G., Boothroyd, I.K.G., Turner, S.J., Roberts, K., Lewis, G.D., 2009. A comparison of bacteria and benthic invertebrates as indicators of ecological health in streams. Freshwater Biol. 54, 1532–1543.

Ledger, M.E., Brown, L.E., Edwards, F., Milner, A.M., Woodward, G., 2012. Drought alters the structure and functioning of complex food webs. Nat. Clim. Change.

Ledger, M.E., Hildrew, A.G., 2005. The ecology of acidification and recovery: changes in herbivore-algal food web linkages across a stream pH gradient. Environ. Pollut. 137, 103–118.

Ledger, M.E., Milner, A., Brown, L., Edwards, F., Hudson, L., Woodward, G., 2013. Extreme climatic events alter aquatic food webs. A synthesis of evidence from a Mesocosm drought Experiment. Adv. Ecol. Res. 48, 343–395.

Lennon, J.J., Kunin, W.E., Hartley, S., Gaston, K.J., 2007. In: Storch, D., Marquet, P.A., Brown, J.H. (Eds.), Species Distribution Patterns, Diversity Scaling and Testing for Fractals in Southern African Birds. Scaling Biodiversity. Cambridge University Press, Cambridge, UK.

Li, Y., Wang, N., Perkins, E.J., Zhang, C., Gong, P., 2010. Identification and optimization of classifier genes from multi-class earthworm microarray dataset. Plos One 5, e13715.

Likens, G.E., Bormann, F.H., Pierce, R.S., Eaton, J.S., Johnson, N.M., 1977. Biogeo-chemistry of a Forested Ecosystem. Springer-Verlag, New York.

López-Urrutia, Á., San Martin, E., Harris, R.P., Irigoien, X., 2006. Scaling the metabolic balance of the oceans. Proc. Natl. Acad. Sci. 103, 8739–8744.

Loreau, M., 2010. From Populations to Ecosystems: Theoretical Foundations for a New Ecological Synthesis (MPB-46). Princeton University Press.

Loreau, M., Naeem, S., Inchausti, P., 2002. Biodiversity and Ecosystem Functioning: Synthesis and Perspectives. Oxford University Press.

Loreau, M., Naeem, S., Inchausti, P., Bengtsson, J., Grime, J.P., Hector, A., Hooper, D.U., Huston, M.A., Raffaelli, D., Schmid, B., Tilman, D., Wardle, D.A., 2001. Ecology - biodiversity and ecosystem functioning: current knowledge and future challenges. Science 294, 804–808.

Lozupone, C., Knight, R., 2005. UniFrac: a new phylogenetic method for comparing microbial communities. Appl. Environ. Microbiol. 71, 8228–8235.

Lyell, C., Deshayes, G.P., 1830. Principles of Geology: Being an Attempt to Explain the Former Changes of the Earth's Surface, by Reference to Causes Now in Operation. John Murray.

Margulies, M., Egholm, M., Altman, W.E., Attiya, S., Bader, J.S., Bemben, L.A., Berka, J., Braverman, M.S., Chen, Y.-J., Chen, Z., 2005. Genome sequencing in microfabricated high-density picolitre reactors. Nature 437, 376–380.

Martiny, J.B.H., Bohannan, B.J.M., Brown, J.H., Colwell, R.K., Fuhrman, J.A., Green, J.L., Horner-Devine, M.C., Kane, M., Krumins, J.A., Kuske, C.R., Morin, P.J., Naeem, S., Ovreas, L., Reysenbach, A.L., Smith, V.H., Staley, J.T., 2006. Microbial biogeography: putting microorganisms on the map. Nat. Rev. Microbiol. 4, 102–112.

May, R.M., 1990. Taxonomy as destiny. Nature 347, 129–130.

Mayden, R.L., 1997. A hierarchy of species concepts: the denouement in the saga of the species problem. In: Claridge, M.F., Dawah, H.A., Wilson, M.R. (Eds.), Systematics Association Special Volume Series; Species: The Units of Biodiversity, pp. 381–424.

McKie, B.G., Woodward, G., Hladyz, S., Nistorescu, M., Preda, E., Popescu, C., Giller, P.S., Malmqvist, B., 2008. Ecosystem functioning in stream assemblages from different regions: contrasting responses to variation in detritivore richness, evenness and density. J. Anim. Ecol. 77, 495–504.

Melián, C., Vilas, C., Baldo, F., González-Ortegón, E., Drake, P., Williams, R.J., 2011. Eco-evolutionary dynamics of individual-based food webs. Adv. Ecol. Res. 45, 225–268.

Metcalfe, J.L., 1989. Biological water-quality assessment of running waters based on macro-invertebrate communities - history and present status in europe. Environ. Pollut. 60, 101–139.

Millenium Ecosystem Assessment, 2005. Ecosystems and Human Well-being: Current State and Trends, vol. 1. Washington.

Moilanen, A., Franco, A.M., Early, R.I., Fox, R., Wintle, B., Thomas, C.D., 2005. Prioritizing multiple-use landscapes for conservation: methods for large multi-species planning problems. Proc. R. Soc. B: Biol. Sci. 272, 1885–1891.

Moilanen, A., Hanski, I., 2001. On the use of connectivity measures in spatial ecology. Oikos 95, 147–151.

Moilanen, A., Leathwick, J., Elith, J., 2008. A method for spatial freshwater conservation prioritization. Freshwater Biol. 53, 577–592.

Monteith, D.T., Hildrew, A.G., Flower, R.J., Raven, P.J., Beaumont, W.R.B., Collen, P., Kreiser, A.M., Shilland, E.M., Winterbottom, J.H., 2005. Biological responses to the chemical recovery of acidified fresh waters in the UK. Environ. Pollut. 137, 83–101.

Moran, M.A., Satinsky, B., Gifford, S.M., Luo, H.W., Rivers, A., Chan, L.K., Meng, J., Durham, B.P., Shen, C., Varaljay, V.A., Smith, C.B., Yager, P.L., Hopkinson, B.M., 2013. Sizing up metatranscriptomics. Isme J. 7, 237–243.

Moya-Larano, J., Verdeny-Vilalta, O., Rowntree, J., Melguizo-Ruiz, N., Montserrat, M., Laiolo, P., 2012. Climate change and eco-evolutionary dynamics in food webs. Adv. Ecol. Res. 47, 1–80.

Mulholland, P.J., Helton, A.M., Poole, G.C., Hall, R.O., Hamilton, S.K., Peterson, B.J., Tank, J.L., Ashkenas, L.R., Cooper, L.W., Dahm, C.N., 2008. Stream denitrification across biomes and its response to anthropogenic nitrate loading. Nature 452, 202–205.

Murphy, J.F., Davy-Bowker, J., McFarland, B., Ormerod, S.J., 2013. A diagnostic biotic index for assessing acidity in sensitive streams in Britain. Ecol. Indic. 24, 562–572.

Murphy, J.F., Winterbottom, J.H., Orton, S., Simpson, G.L., Shilland, E.M., Hildrew, A.G., 2012. Evidence of recovery from acidification in the macroinvertebrate assemblages of UK fresh waters: a 20-year time series. Ecol. Indic.

Naeem, S., 2002. Disentangling the impacts of diversity on ecosystem functioning in combinatorial experiments. Ecology 83, 2925–2935.

Nemergut, D.R., Costello, E.K., Hamady, M., Lozupone, C., Jiang, L., Schmidt, S.K., Fierer, N., Townsend, A.R., Cleveland, C.C., Stanish, L., Knight, R., 2011. Global patterns in the biogeography of bacterial taxa. Environ. Microbiol. 13, 135–144.

Nilsson, R.H., Kristiansson, E., Ryberg, M., Hallenberg, N., Larsson, K.-H., 2008. Intraspecific ITS variability in the kingdom fungi as expressed in the international sequence databases and its implications for molecular species identification. Evol. Bioinf. Online 4, 193.

O'Malley, M.A., 2008. Everything is everywhere but the environment selects: ubiquitous distribution and ecological determinism in microbial biogeography. Stud. Hist. Philos. Sci. Part C: Stud. Hist. Philos. Biol. Biomed. Sci. 39, 314–325.

Odum, E.P., 1969. The strategy of ecosystem development. An understanding of ecological succession provides a basis for resolving man's conflict with nature. Science 164, 262–269.

Paradis, E., Claude, J., Strimmer, K., 2004. APE: analyses of phylogenetics and evolution in R language. Bioinformatics 20, 289–290.

Pawlowski, J., Audic, S., Adl, S., Bass, D., Belbahri, L., Berney, C., Bowser, S.S., Cepicka, I., Decelle, J., Dunthorn, M., 2012. CBOL protist working group: barcoding eukaryotic richness beyond the animal, plant, and fungal kingdoms. PLoS Biol. 10, e1001419.

Perkins, D.M., McKie, B.G., Malmqvist, B., Gilmour, S.G., Reiss, J., Woodward, G., 2010. Environmental warming and biodiversity-ecosystem functioning in freshwater Microcosms: Partitioning the effects of species identity, richness and metabolism. Adv. Ecol. Res. 43, 177–209.

Perkins, E.J., Chipman, J.K., Edwards, S., Habib, T., Falciani, F., Taylor, R., Van Aggelen, G., Vulpe, C., Antczak, P., Loguinov, A., 2011. Reverse engineering adverse outcome pathways. Environ. Toxicol. Chem. 30, 22–38.

Petchey, O.L., Downing, A.L., Mittelbach, G.G., Persson, L., Steiner, C.F., Warren, P.H., Woodward, G., 2004. Species loss and the structure and functioning of multitrophic aquatic systems. Oikos 104, 467–478.

Petchey, O.L., Gaston, K.J., 2006. Functional diversity: back to basics and looking forward. Ecol. Lett. 9, 741–758.

Poff, N.L., Olden, J.D., Vieira, N.K.M., Finn, D.S., Simmons, M.P., Kondratieff, B.C., 2006. Functional trait niches of North American lotic insects: traits-based ecological applications in light of phylogenetic relationships. J. North Am. Benthol. Soc. 25, 730–755.

Poisot, T., Pequin, B., Gravel, D., 2013. High-throughput sequencing: a roadmap toward community ecology. Ecol. Evol. 3, 1125–1139.

Pollux, B.J.A., Santamaria, L., Ouborg, N.J., 2005. Differences in endozoochorous dispersal between aquatic plant species, with reference to plant population persistence in rivers. Freshwater Biol. 50, 232–242.

Pommier, T., Douzery, E.J.P., Mouillot, D., 2012. Environment drives high phylogenetic turnover among oceanic bacterial communities. Biol. Lett. 8, 562–566.

Pompanon, F., Deagle, B.E., Symondson, W.O., Brown, D.S., Jarman, S.N., Taberlet, P., 2012. Who is eating what: diet assessment using next generation sequencing. Mol. Ecol. 21, 1931–1950.

Port, J.A., Wallace, J.C., Griffith, W.C., Faustman, E.M., 2012. Metagenomic profiling of microbial composition and antibiotic resistance determinants in Puget Sound. Plos One 7.

Purdy, K.J., Hurd, P.J., Moya-Larano, J., Trimmer, M., Oakley, B.B., Woodward, G., 2010. Systems biology for ecology: from molecules to ecosystems. Adv. Ecol. Res. 43, 87–149.

Ratnasingham, S., Hebert, P.D.N., 2007. BOLD: the barcode of life data system. Mol. Ecol. Notes 7, 355–364. www.barcodinglife.org.

Rawcliffe, R., Sayer, C.D., Woodward, G., Grey, J., Davidson, T.A., Jones, J.I., 2010. Back to the future: using palaeolimnology to infer long-term changes in shallow lake food webs. Freshwater Biol. 55, 600–613.

Reiss, J., Bailey, R.A., Cassio, F., Woodward, G., Pascoal, C., 2010. Assessing the contribution of micro-organisms and macrofauna to biodiversity-ecosystem functioning relationships in freshwater microcosms. Adv. Ecol. Res. 43, 151–176.

Reiss, J., Bailey, R.A., Perkins, D.M., Pluchinotta, A., Woodward, G., 2011. Testing effects of consumer richness, evenness and body size on ecosystem functioning. J. Anim. Ecol. 80, 1145–1154.

Renberg, I., Hellberg, T., 1982. The Ph history of lakes in southwestern Sweden, as calculated from the subfossil diatom flora of the sediments. Ambio 11, 30–33.

Ribeiro, F.J., Przybylski, D., Yin, S.Y., Sharpe, T., Gnerre, S., Abouelleil, A., Berlin, A.M., Montmayeur, A., Shea, T.P., Walker, B.J., Young, S.K., Russ, C., Nusbaum, C., MacCallum, I., Jaffe, D.B., 2012. Finished bacterial genomes from shotgun sequence data. Genome Res. 22, 2270–2277.

Rodriguez-Lanetty, M., Granados-Cifuentes, C., Barberan, A., Bellantuono, A.J., Bastidas, C., 2013. Ecological inferences from a deep screening of the Complex Bacterial Consortia associated with the coral, Porites astreoides. Mol. Ecol. 22, 4349–4362.

Roesch, L.F., Fulthorpe, R.R., Riva, A., Casella, G., Hadwin, A.K.M., Kent, A.D., Daroub, S.H., Camargo, F.A.O., Farmerie, W.G., Triplett, E.W., 2007. Pyrosequencing enumerates and contrasts soil microbial diversity. Isme J. 1, 283–290.

Rosenberg, D.M., Resh, V.H., 1993. Freshwater Biomonitoring and Benthic Macroinvertebrates. Chapman and Hall, New York.

Ryan, J.A., Hightower, L.E., 1996. Stress proteins as molecular biomarkers for environmental toxicology. Exs 77, 411–424.

Rybicki, N.B., Landwehr, J.M., 2007. Long-term changes in abundance and diversity of macrophyte and waterfowl populations in an estuary with exotic macrophytes and improving water quality. Limnol. Oceanogr. 52, 1195–1207.

Scheffer, M., Bascompte, J., Brock, W.A., Brovkin, V., Carpenter, S.R., Dakos, V., Held, H., Van Nes, E.H., Rietkerk, M., Sugihara, G., 2009. Early-warning signals for critical transitions. Nature 461, 53–59.

Scheffer, M., Carpenter, S.R., 2003. Catastrophic regime shifts in ecosystems: linking theory to observation. Trends Ecol. Evol. 18, 648–656.

Schindler, D.E., Hilborn, R., Chasco, B., Boatright, C.P., Quinn, T.P., Rogers, L.A., Webster, M.S., 2010. Population diversity and the portfolio effect in an exploited species. Nature 465, 609–612.

Schindler, D.W., 1987. Detecting ecosystem responses to anthropogenic stress. Can. J. Fish. Aquat. Sci. 44, s6–s25.

Schindler, D.W., 1990. Experimental perturbations of whole lakes as tests of hypotheses concerning ecosystem structure and function. Oikos 25–41.

Schneider, G.F., Dekker, C., 2012. DNA sequencing with nanopores. Nat. Biotechnol. 30, 326–328.

Simpson, G.L., Shilland, E.M., Winterbottom, J.M., Keay, J., 2005. Defining reference conditions for acidified waters using a modern analogue approach. Environ. Pollut. 137, 119–133.

Sinsabaugh, R.L., Hill, B.H., Shah, J.J.F., 2009. Ecoenzymatic stoichiometry of microbial organic nutrient acquisition in soil and sediment. Nature 462, 795–798.

Simpson, J.C., Norris, R.H., 2000. Biological Assessment of River Quality: Development of AUSRIVAS Models and Outputs. Freshwater Biological Association, The Ferry House, Far Sawrey, Ambleside, Cumbria, LA22 0LP, UK.

Slavik, K., Peterson, B., Deegan, L., Bowden, W., Hershey, A., Hobbie, J., 2004. Long-term responses of the Kuparuk River ecosystem to phosphorus fertilization. Ecology 85, 939–954.

Smol, J.P., 2009. Pollution of Lakes and Rivers: A Paleoenvironmental Perspective. John Wiley & Sons.

Statzner, B., Bêche, L.A., 2010. Can biological invertebrate traits resolve effects of multiple stressors on running water ecosystems? Freshwater Biol. 55, 80–119.

Sun, M.Y., Dafforn, K.A., Brown, M.V., Johnston, E.L., 2012. Bacterial communities are sensitive indicators of contaminant stress. Mar. Pollut. Bull. 64, 1029–1038.

Sweeney, B.W., Bott, T.L., Jackson, J.K., Kaplan, L.A., Newbold, J.D., Standley, L.J., Hession, W.C., Horwitz, R.J., 2004. Riparian deforestation, stream narrowing, and loss of stream ecosystem services. Proc. Natl. Acad. Sci. USA 101, 14132–14137.

Taberlet, P., Coissac, E., Pompanon, F., Brochmann, C., Willerslev, E., 2012. Towards next-generation biodiversity assessment using DNA metabarcoding. Mol. Ecol. 21, 2045–2050.

Thuiller, W., Lavergne, S., Roquet, C., Boulangeat, I., Lafourcade, B., Araujo, M.B., 2011. Consequences of climate change on the tree of life in Europe. Nature 470, 531–534.

Tilman, D., Knops, J., Wedin, D., Reich, P., Ritchie, M., Siemann, E., 1997. The influence of functional diversity and composition on ecosystem processes. Science 277, 1300–1302.

Tokeshi, M., 1990. Niche apportionment or random assortment - species abundance patterns revisited. J. Anim. Ecol. 59, 1129–1146.

Tokeshi, M., 1993. Species abundance patterns and community. Adv. Ecol. Res. 24, 111.

UK National Ecosystem Assessment, 2011. The UK National Ecosystem Assessment: Synthesis of the Key Findings. Cambridge.

Van Aggelen, G., Ankley, G.T., Baldwin, W.S., Bearden, D.W., Benson, W.H., Chipman, J.K., Collette, T.W., Craft, J.A., Denslow, N.D., Embry, M.R., 2010. Integrating omic technologies into aquatic ecological risk assessment and environmental monitoring: hurdles, achievements, and future outlook. Environ. Health Perspect. 118, 1.

Venail, P., MacLean, R., Bouvier, T., Brockhurst, M., Hochberg, M., Mouquet, N., 2008. Diversity and productivity peak at intermediate dispersal rate in evolving metacommunities. Nature 452, 210–214.

Vonlanthen, P., Bittner, D., Hudson, A.G., Young, K.A., Muller, R., Lundsgaard-Hansen, B., Roy, D., Di Piazza, S., Largiader, C.R., Seehausen, O., 2012. Eutrophication causes speciation reversal in whitefish adaptive radiations. Nature 482, 357–U1500.

Webb, C., Ackerly, D., Kembel, S., 2011. Software for the Analysis of Phylogenetic Comunity Structure and Character Evolution (With Phylomatic and Ecovolve) User's Manual Version 4.2. Arnold Arboretum of Harvard University.

Willby, N.J., Abernethy, V.J., Demars, B.O.L., 2000. Attribute-based classification of European hydrophytes and its relationship to habitat utilization. Freshwater Biol. 43, 43–74.

Williams, T.D., Turan, N., Diab, A.M., Wu, H.F., Mackenzie, C., Bartie, K.L., Hrydziuszko, O., Lyons, B.P., Stentiford, G.D., Herbert, J.M., Abraham, J.K., Katsiadaki, I., Leaver, M.J., Taggart, J.B., George, S.G., Viant, M.R., Chipman, K.J., Falciani, F., 2011. Towards a system level understanding of non-model organisms sampled from the environment: a network biology approach. Plos Comput. Biol. 7.

de Wit, R., Bouvier, T., 2006. 'Everything is everywhere, but, the environment selects'; what did Baas Becking and Beijerinck really say? Environ. Microbiol. 8, 755–758.

Woodward, G., 2009. Biodiversity, ecosystem functioning and food webs in fresh waters: assembling the jigsaw puzzle. Freshwater Biol. 54, 2171–2187.

Woodward, G., Benstead, J.P., Beveridge, O.S., Blanchard, J., Brey, T., Brown, L.E., Cross, W.F., Friberg, N., Ings, T.C., Jacob, U., Jennings, S., Ledger, M.E., Milner, A.M., Montoya, J.M., O'Gorman, E.J., Olesen, J.M., Petchey, O.L., Pichler, D.E., Reuman, D.C., Thompson, M.S.A., Van Veen, F.J.F., Yvon-Durocher, G., 2010. Ecological networks in a changing climate. Adv. Ecol. Res. 42, 71–138.

Woodward, G., Brown, L.E., Edwards, F.K., Hudson, L.N., Milner, A.M., Reuman, D.C., Ledger, M.E., 2012. Climate change impacts in multispecies systems: drought alters food web size structure in a field experiment. Philos. Trans. R. Soc. B-Biol. Sci. 367, 2990–2997.

Woodward, G., Gray, C., Baird, D.J., 2013. A critique of biomonitoring in an age of globalisation and emerging environmental threats. Limnetica.

Yachi, S., Loreau, M., 1999. Biodiversity and ecosystem productivity in a fluctuating environment: the insurance hypothesis. Proc. Natl. Acad. Sci. 96, 1463–1468.

Yagishita, M., 1995. Establishing an acid deposition monitoring network in East Asia. Water, Air, Soil Pollut. 85, 273–278.

Yergeau, E., Lawrence, J.R., Sanschagrin, S., Waiser, M.J., Korber, D.R., Greer, C.W., 2012. Next-generation sequencing of microbial communities in the Athabasca river and its tributaries in relation to oil sands mining activities. Appl. Environ. Microbiol. 78, 7626–7637.

Young, R.G., Matthaei, C.D., Townsend, C.R., 2008. Organic matter breakdown and ecosystem metabolism: functional indicators for assessing river ecosystem health. J. North Am. Benthol. Soc. 27, 605–625.

Yvon-Durocher, G., Jones, J.I., Trimmer, M., Woodward, G., Montoya, J.M., 2010. Warming alters the metabolic balance of ecosystems. Philos. Trans. R. Soc. B-Biol. Sci. 365, 2117–2126.

Zarraonaindia, I., Smith, D.P., Gilbert, J.A., 2013. Beyond the genome: community-level analysis of the microbial world. Biol. Philos. 1–22.

Epilogue

The Robustness of Aquatic Biodiversity Functioning under Environmental Change: The Ythan Estuary, Scotland

David Raffaelli
Environment Department, University of York, York, UK

INTRODUCTION

Aquatic ecosystems provide a broad range of final ecosystem services, including fish, shellfish, genetic resources, climate regulation, natural hazard protection, clean water and sediments, places, seascapes, and landscapes (Beaumont et al., 2007). These in turn are underpinned by ecosystem processes or functions (intermediate services), many of which are modulated by the biodiversity present (Schmid et al., 2009; Raffaelli and White, 2013). A key question for managers and policy makers is deciding between different competing management strategies and policy options for bundles of desired goods and services that will inevitably involve changing biodiversity and hence the functioning of the system, possibly beyond its normal, safe operating limits. While there exists a wealth of literature on relevant theory and concepts such as resilience, stability, and persistence, practical application of this knowledge and understanding to real-world management is lacking. Resilience has been viewed conceptually as the return time to equilibrium, or as resistance to change following a perturbation, but both concepts are problematic at the whole-system level where many species are present: it is impractical to measure the return times or resistance of all species with any degree of confidence. Also, managers may not be interested in the marginal changes of all species when they perturb (manage) a system for a desired bundle of services, but rather whether the system is capable of accommodating those changes without closing down future options for different kinds of services. That is, whether it retains its capacity to function in a desired way by remaining within its "safe-operating" space, and hence is resilient.

Here I pose two major questions. First, are there system-level metrics which can be used to define such safe-operating limits for an ecosystem? Second, what evidence is there from natural systems that they are confined (or not) to operate within those limits in spite of changes in their structure and composition (i.e., whether the system exhibits resilience)? Specifically, I describe a network approach which originates from Systems Ecology for defining such limits and examine its application, with respect to an aquatic system that has experienced large-scale shifts in its biodiversity and functioning.

SYSTEMS APPROACHES TO DEFINING OPERATIONAL BOUNDS FOR AQUATIC ECOSYSTEMS

Systems approaches using network analysis have a long history in ecology (reviewed in Raffaelli and Frid, 2010; Jorgensen et al., 2007), but have yet to be fully applied to the ecosystem service management agenda. Indeed, Norris (2012) has claimed that "a mechanistic understanding of biodiversity change and its consequences for ES can only be addressed using systems approaches," a view echoed by Loreau (2010). A systems approach using network analysis is well suited for exploring the consequences of simultaneous changes in the different biodiversity elements that underpin ecosystem services as a result of different management policy options. Importantly, such analyses can potentially identify the safe operating space within which changes in different system elements can be allowed to vary without compromising system resilience. System-level measures, as opposed to species component measures of stability (Donohue et al., 2013), are most relevant when considering the resilience of the wider system and several such measures are available from network analysis. Thus Ulanowicz (2011) has argued that the network metric, "ascendancy," defined as the size of a network (the sum of all the trophic exchanges, termed total system throughput) scaled by the network's organization (how the flows are arranged, termed the average mutual information), has a restricted set of values for real-world ecosystems, around 30–40% (Figure 1). Too low a value of ascendancy and the system tends to disorder, having insufficient cohesiveness, whereas too high a value and the system becomes vulnerable to perturbations, both externally and internally generated through "self-organizing catastrophe" (Holling, 1986; Bak, 1996). Ulanowicz has shown that such properties can be adequately captured using more easily measured metrics, such as the topological properties of networks (Figure 2), with real networks falling within a "Window of Vitality" (Ulanowicz, 2005), so that trade-offs between different services and the network configurations (species abundances and their flows) that support them can be more easily explored in relation to a safe operating space using concepts and language more familiar to mainstream ecologists.

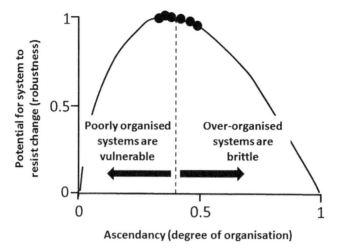

FIGURE 1 Ecological networks are only robust for limited bounds of ascendancy. *Modified from Ulanowicz (2011).*

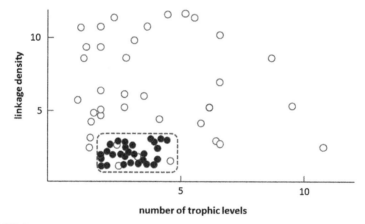

FIGURE 2 The safe operating zone (delineated by dotted lines) defined by ascendency considerations for real (solid circles) and random networks (open circles) may be captured by two simple topological properties of food webs: linkage density and number of trophic levels. *Modified from Ulanowicz (2005).*

OPERATIONAL BOUNDS UNDER DIFFERENT POLICY OPTIONS: THE YTHAN EXAMPLE

The Ythan estuary, Scotland, lies about 20 km north of the city of Aberdeen, and, because of its small size (8 km in length, only a few 100 m across at its widest point) and the presence of a permanently manned University of Aberdeen field station on its banks, is probably one of the best documented and understood in the world (Gorman and Raffaelli, 1993). The river Ythan

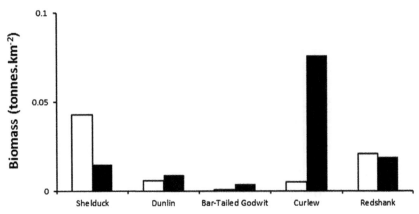

FIGURE 3 Biomass of main invertebrate species on the Ythan estuary in 1967 (open bars) and 1993 (solid bars).

itself drains just over 640 km² of productive agricultural land (>90% of the catchment is under agriculture), which, like the rest of the UK, has responded since 1965 or so to market incentives and to the Common Agricultural Policy leading to large-scale agricultural change, summarized here as a marked shift toward fertilizer-intensive crops such as wheat and oil seed rape, as well as intensive pig rearing. Over the same period there has been a two- to threefold increase in river nitrogen and increasingly heavy blooms of opportunistic green macro-algae (*Ulva* spp) in the estuary, which in turn have resulted in degradation of benthic invertebrate populations and concomitant changes in numbers of shorebirds (Raffaelli et al., 1989, 1999, and papers therein). These changes led to the estuary being designated eutrophic and declared a Nitrate Vulnerable Zone under the EU Nitrates Directive in 2000.

Due to the sustained research effort on the estuary's tropho-dynamics (Baird and Milne, 1981; Hall and Raffaelli, 1991; Raffaelli and Hall, 1992; Gorman and Raffaelli, 1993), pre-eutrophic (1967) and late eutrophic (1993) versions of the trophic network web exist. The biomasses of the main biodiversity elements for the two periods changed markedly.

The biomass of green macroalgae (mainly (*Ulva intestinalis* and *Chaetomorpha* spp)) increased in biomass by about 150%, from 308 to 762 ton·km^{-2}, forming extensive thick mats on the mudflats that resulted in deoxygenation of the underlying sediment. The mudflat invertebrate communities responded to the presence of mats in different ways (Figure 3). There was a decline in the biomass of *Corophium volutator*, the main prey species for the estuary's birds, by about 90%, due to the hostile sediment environment, but large increases in the biomasses of *Hediste diversicolor*, *Hydrobia ulvae*, and *Macoma balthica*, which were able to take advantage of the additional organic carbon and algal resources (Raffaelli et al., 1999). For the predators of these invertebrates, there were declines in two species, the Shelduck (*Tadorna*

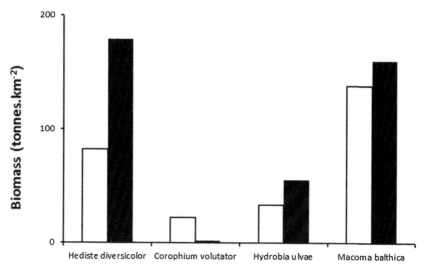

FIGURE 4 Biomass of main shorebird species on the Ythan estuary in 1967 (open bars) and 1993 (solid bars).

tadorna) and the Redshank (*Tringa totanus*) (65% and 10%, respectively), both of which have their foraging behavior dramatically affected by the presence of algal mats, while there were increases in other species such as Dunlin (*Calidris alpina*), Bar-tailed Godwit (*Limosa lapponica*), and Curlew (*Numenius arquata*) by 50%, 300%, and 1420%, respectively, presumably because of their predation on species like *Hediste Hydrobia* and *Macoma* (Figure 4).

HOW DID THE NETWORK CHANGE BETWEEN 1967 AND 1993?

The 1967 and 1993 networks were individually analyzed for a range of ecosystem-level attributes using the mass-balance software Ecopath, the full details of which are reported elsewhere (Raffaelli, 2011; Raffaelli and Friedlander, 2012). Despite, dramatic shifts in the relative abundance of species in this network between 1967 and 1993 as a result of a shift toward a production policy option for agriculture in the Ythan catchment, Ascendancy was very similar in the two periods (Table 1). While this value is lower than the 40% identified by Ulanowicz (see above), it is fairly typical for estuarine systems (Baird and Ulanowicz, 1993) and of course the important point is that the value is similar for the two periods. With respect to Ulanowicz's Window of Vitality, linkage density and number of trophic levels are very similar for the two periods (Table 1) and would locate the Ythan within the left-hand boundary of the box in Figure 2. These network values indicate that the Ythan as a system was able to accommodate the large-scale changes in

TABLE 1 Network Properties of the Ythan Estuary System in 1967 (Pre-eutrophic) and 1993 (Post-eutrophic) Periods

	Ascendancy	Linkage Density	Number of Trophic Levels
1967	24.853	2.767	1.77
1993	26.498	2.333	1.84

Linkage density and number of trophic levels were estimated after Zorach and Ulanowicz (2003), as $LD = b^{OH/2}$, where b is the logarithm to the base 10 and OH is system overhead and number of trophic levels estimated as b^A, where A is Ascendancy (see Ulanowicz, 2005, for further details).

nutrient loading, primary production, and invertebrate and bird biomasses over two periods.

DISCUSSION

It is clear that the Ythan as a system changed significantly between 1967 and 1993 with respect to the relative abundance of its major biodiversity elements. Furthermore, these changes were sufficient to see the designation of the Ythan as a Nitrogen Vulnerable Zone (the first such designation in the UK) under European environmental legislation. Yet the system still functioned to maintain itself within safe operating limits, despite a massive increase in the flows of biomass (Total System Throughput) of c. 38% (Raffaelli, 2011). The biodiversity changes were in terms of relative abundance of species and, as far as is known, no species were lost from the Ythan altogether. Thus, species richness did not change, implying that functional richness must also have remained the same. The preservation of function over the eutrophication period must also be partly due to the generalist nature of species typically present in high-latitude estuaries. For instance, while the diets of many of the predators (crustaceans, fish, and birds) on the estuary are dominated by the amphipod *Corophium*, which showed the most dramatic decline of all invertebrates (Raffaelli et al., 1999), most of those predators are capable of taking a much wider range of prey than they actually do on the Ythan (Hall and Raffaelli, 1997) allowing them to switch in response to changes in relative prey abundance. The only predator species to decline over the period were Redshank and Shelduck, both of which have their normal feeding behavior compromised by the physical presence of the algal mats, rather than declines in any particular prey (Raffaelli et al., 1999). In addition, many of the mudflat invertebrates are extremely plastic in their functional type expressed, in particular the dominant species *Hediste*. This polychaete can behave as a filter feeder, predator, or deposit feeder according to prevailing conditions. Unlike systems such as coral reefs where species may be highly specialized with respect to traits (Mora, this volume), this is probably not the case in high latitude estuaries like the Ythan, much of the functional insurance residing within a single species (e.g., ontogenic size shifts in fish and crustaceans; see also Palkovacs et al., this volume; Pawar et al., this volume) or within the same individuals, as in the case of *Hediste*.

The mass-balance modeling approach described here can capture aspects of natural capital and seems to have potential for exploring biodiversity—ecosystem processes—services relationships in a quite different way than the experimental approaches that have dominated this field to date (Cardinale et al., 2012), not only by employing a metric of biodiversity (stocks of natural capital and the flows between them) that resonate more with stakeholders, but also by being able to accommodate the landscape scales at which environmental management policies are implemented and at which ecosystem services are delivered. The system resilience measures used here indicate that even quite large-scale shifts in biodiversity do not move the system per se out of its safe space, which might give grounds for optimism for our traditionally poor management of such systems, and their ability to function. However, it should be noted that both versions of the Ythan network lie just on the left of the Ulanowicz's ascendancy curve, and at the very top left corner of his Window of Vitality. The question remains as to whether future additional stressors acting orthogonally or synergistically with eutrophication (i.e., over-harvesting or climate change) could push the Ythan out of its safe space. Furthermore, in the case of those real systems where tipping points that are essentially nonreversible (exhibit hysteresis dynamics) have been compellingly demonstrated, it is possible that they too will have appeared to remain within a safe-operating space before their catastrophic collapse. Further research is needed to establish just how far systems like the Ythan can be changed before they slip out of that safe space, and what the system's dynamics would be (linear? nonlinear?) when they enter "unsafe" space.

ACKNOWLEDGEMENTS

This manuscript was greatly improved through rigorous discussion with staff and students at the Universities of Leicester and Cardiff, and the National Oceanographic Centre of Southampton, and through reflections permitted by the author's membership of the Tansley Working Group *Funkey Traits*, funded by NERC. I am especially grateful for insightful comments by the group leader Dr Tom Oliver.

REFERENCES

Baird, D., Milne, H., 1981. Energy flow in the Ythan estuary, Aberdeenshire, Scotland. Estuarine, Coastal Mar. Sci. 13, 217–232.

Baird, D., Ulanowicz, R.W.E., 1993. A comparative study on the trophic structure, cycling and ecosystem properties of four tidal estuaries. Mar. Ecol. Prog. Ser. 99, 221–237.

Bak, P., 1996. How Nature Works: The Science of Self-organised Criticality. Copernicus Press, New York.

Beaumont, N.J., Austen, M.C., Atkins, J.P., Burdon, D., Degraer, S., Dentinho, T.P., et al., 2007. Identification, definition and quantification of goods and services provided by marine biodiversity: Implications for the ecosystem approach. Mar. Pollut. Bull. 54, 253–265.

Cardinale, B.J., Duffy, J.E., Gonzalez, A., Hooper, D.U., Perrings, C., Venail, P., Narwani, A., Mace, G.M., Tilman, D., Wardle, D.A., Kinzig, A.P., Daily, G.C., Loreau, M., Grace, J.B., Larigauderie, A., Srivastava, D., Naeem, S., 2012. Biodiversity loss and its impact on humanity. Nature 486, 59–67.

Donohue, I., Petchey, O., Montoya, J.M., Jackson, A.L., McNally, L., Viana, M., Healy, K., Lurgi, M., O'Connor, N.E., Emmerson, M.C., 2013. On the dimensionality of ecological stability. Ecol. Lett. 16, 421–429.

Gorman, M.L., Raffaelli, D., 1993. Classic sites - the Ythan estuary. The Biologist 40, 10–13.

Hall, S.J., Raffaelli, D., 1991. Static patterns in food webs: lessons from a large web. J. Anim. Ecol. 63, 823–842.

Hall, S.J., Raffaelli, D., 1997. Food web patterns: what do we really know? In: Gange, A.C., et al. (Eds.), Mutitrophic Interactions. Blackwells Scientific Publications, Oxford, pp. 395–417.

Holling, C.S., 1986. The resilience of terrestrial ecosystems: local surprise and global change. In: Clark, W.C., Munn, R.E. (Eds.), Sustainable Development of the Biosphere. Cambridge University Press, Cambridge, pp. 292–317.

Jorgensen, S.E., Fath, B.D., Bastianoni, S., Marques, J.C., Muller, F., Nielson, S.N., Patten, B.C., Tiezzi, E., Ulanowicz, R.E., 2007. A New Ecology. Systems Perspective. Elsevier, Amsterdam, 275 pp.

Loreau, M., 2010. From Populations to Ecosystems: Theoretical Foundations for a New Ecological Synthesis. Princeton University Press, Princeton.

Norris, K., 2012. Biodiversity in the context of ecosystem services: the applied need for systems approaches. Philos. Trans. R. Soc. B 367, 191–199.

Raffaelli, D., 2011. Contemporary concepts and models on biodiversity and ecosystem function. Treatise Estuarine Coastal Sci. 9, 5–21.

Raffaelli, D., Hall, S.J., 1992. Compartments and predation in an estuarine food web. J. Anim. Ecol. 61, 551–560.

Raffaelli, D., Hull, S., Milne, H., 1989. Long-term changes in nutrients, weed mats and shorebirds in an estuarine system. Cah. Biol. Mar. 30, 259–270.

Raffaelli, D., Balls, P., Way, S., Patterson, I.J., Hohmann, S., Corp, N., 1999. Major long-term changes in the ecology of the Ythan estuary, Aberdeenshire, Scotland; how important are physical factors? Aquat. Conserv: Mar. Freshwater Ecosyst. 9, 219–236.

Raffaelli, D.G., Frid, C.L.J., 2010. Ecosystem Ecology: A New Synthesis. Cambridge University Press, Cambridge.

Raffaelli, D.G., Friedlander, A.M., 2012. Biodiversity and ecosystem functioning: an ecosystem approach. In: Solan, M., Aspden, R.J., Paterson, D.M. (Eds.), Marine Biodiversity and Ecosystem Functioning. Oxford University Press, Oxford, pp. 149–163.

Raffaelli, D., White, P.C.L., 2013. Ecosystems and their services in a changing world: an ecological perspective. Adv. Ecol. Res. 48, 1–70.

Schmid, B., Balvanera, P., Cardinale, B.J., Godbold, J., Pfisterer, A.B.D., Solan, M., Srivastava, D., 2009. Consequences of species loss for ecosystem functioning: metaanalyses of data from biodiversity experiments. In: Naeem, S., Bunker, D.E., Hector, A., Loreau, M., Perrings, C. (Eds.), Biodiversity, Ecosystem Functioning, and Human Wellbeing: An Ecological and Economic Perspective. Oxford University Press, Oxford, pp. 14–29.

Ulanowicz, R.E., 2005. Ecological network analysis: an escape form the machine. In: Belgrano, A., Scharler, U.M., Dunne, J., Ulanowicz, R.E. (Eds.), Aquatic Food Webs. Oxford University Press, Oxford, pp. 201−207.

Ulanowicz, R.E., 2011. Quantitative methods for ecological network analysis and its application to coastal ecosystems. Treatise Estuarine Coastal Sci. 9, 35−57.

Zorach, A.C., Ulanowicz, R.E., 2003. Quantifying the complexity of flow networks: how many roles are there? Complexity 8, 68−76.

Index

Note: Page numbers followed by "f", "b" and "t" indicates figures, boxes and tables, respectively.

A
Abiotic factors, 3
Abundance, 164–165
Acid Waters Monitoring Network (AWMN), 243
Activation energy, 132–133
Adjacency matrix, 95
Alewife (*Alosa pseudoharengus*), 38f, 39t–40t, 41–42
Analogous analyses, 24
Anthropogenic force, 171–173
Aquatic ecosystem, 159–160
   body size role, 167–168
   macroecology variables and interactions, 161
      abundance, 164–165
      body size, 167–168
      geographical distribution, 165–167
      species richness, 162–164
Aquatic ecosystem functioning, ecological and evolutionary effects of fisheries, 177–179
   IEAs, 179
   macroecological distributions, 179
   relationships linking evolutionary and ecological responses, 178f
Aquatic functional diversity
   fisheries development, 174
   macroecological spatial scales, 175
   P. mariae, 177
   trait homogenization, 176f
   trait-based mechanism, 175–176
   traits-based focus on, 173–174
Average score per taxon (ASPT), 242
AWMN. *See* Acid Waters Monitoring Network

B
B-EF relationships. *See* Biodiversity–ecosystem functioning relationships

B–ES relationships. *See* Biodiversity–ecosystem services relationships
Betweenness centrality, 90–91
"Big data", 129
"Bio-energetic" models, 18–19
Biodiversity, 37, 222
   ecosystem capacity, 223
   on ecosystems, 222–223
   functional diversity, 223
   functional redundancy, 224
Biodiversity–ecosystem functioning relationships (B-EF relationships), 53–54, 136, 251–252
   body size, 136
   complementarity, 55
   context dependency of biodiversity effects, 140f
   drivers and responses in, 139
   food web context, 137
   importance of ancestral state of community, 55–56
   metabolic responses to warming, 138f
   methods, 56
      nutrients, 56
      resource competition models, 57
   model
      analysis, 62–63
      parameter definitions and values, 63t
   nontransgressive overyielding, 69f
      under competition for complementary resources, 65f
      under competition for essential resources, 67f
   reanalysis of empirical data, 63–64, 64t
   results, 64–70
   trait-based approach, 138–139
   transgressive overyielding
      under competition for complementary resources, 66f
      under competition for essential resources, 68f

284 Index

Biodiversity—ecosystem services relationships (B—ES relationships), 136
Biological indices, 98—99
Biomarkers, 253—256
Biomechanics, 6—11
  from individual metabolism to interactions, 11—12
    components of species interactions, 11f
    empirical support, 16—18
    metabolic theory for species interactions, 12—16
Body size, 3—6
  3D consumption rates, 20
  effect in aquatic ecosystems, 10
  effect on consumer and resource fitness, 23f
  fisheries, 167—168
Boltzmann—Arrhenius equation, 9—10

C

Caribbean coral reefs, 117—118
Catch per unit of effort (CPUE), 218
CBO. See Conservation of Biological Originality
Centrality metrics, 90—92
Clean Water Act, 241
Climate change, 21, 127—130
  See also Global warming
  gauging ecological and evolutionary responses, 145b
  synergies between multiple stressors, 147—149
  traits and functional diversity, 141
Closeness centrality, 90—91
Coastal protection, 221
Community-level properties, 90
  See also Metacommunity-level properties
  centrality indices and calculation with R program, 92, 92t
  centrality metrics, 90—92
  estimation of alternative indices of centrality, 91f
Complementarity processes, 53—54
Connectance, 93
Connectivity correlation. See Degree correlation
Connectivity distribution. See Degree distribution
Connectivity of node, 90—91
Conservation of Biological Originality (CBO), 228—229

Consumer
  distribution, 202
  identity, 37
  trophic flexibility, 193
  trophic niche position, 193
  trophic uniqueness, 191, 193
Consumer—resource dynamics, 3—5
  "bio-energetic" models, 18—19
  eco-evolutionary, 22
    analogous analyses, 24
    effect of body size and size mismatch, 23f
    ESS, 22—24
    evolutionary invasion analysis, 24
  ecological, 19—20
    pelagic ecosystems, 20
    temperatures ectotherms, 21
  interaction dimensionality effect, 20f
  mechanistic theoretical framework, 18
  pairs to community and ecosystem dynamics, 24—26
  physiological mismatch effects, 21—22
  Rosenzweig—MacArthur type equations, 19
Coral reefs, 115
  data quality, 116
    Caribbean, 117
    lack of historical baselines, 116
    meta-analyses, 117
    resilience, 116—117
  drivers of change, 118—119
  functionally redundant, 119—121
  pattern of change, 117—118
CPUE. See Catch per unit of effort
Cross-system subsidies in freshwaters, 144—146
"Cryptic" effects, 42—43
Cultural service, 214, 229
  MPA effects, 220

D

Daphnia, 38f, 39t—40t, 44
  divergence due to predators and toxic prey in, 43
  intraspecific effect, 45—46
Decomposition process, 141
Degree centrality, 90—91
Degree correlation, 94
Degree distribution, 93—94
Diameter of metacommunity, 93
Dispersal matrix, 94—95
Distance matrix, 94—95

Dominance, 164–165
Dynamic macroecological patterns, 171–173

E

Eco-evolutionary dynamics, 147
   and feedbacks, 247, 253
Ecological network theory, 200–201
Ecosystem functioning, 211–212
   See also Marine protected areas (MPAs)
Ecosystem services, 213–214
   Millennium Ecosystem Assessment of 2005, 214
   MPA, 214–215
Eigenvector centrality, 90–91
ELA. *See* Experimental Lakes Area
Empirical macroscopic constraints, 100–102
ESS. *See* Evolutionarily stable strategy
Essential resources, 60
   See also Partially substitutable resources
   conditions for stable coexistence of two consumers, 60
   ESS uptake rates, 62
   functional response, 60
EU WFD. *See* European Union Water Framework Directive
European Union Water Framework Directive (EU WFD), 241
Evolutionarily stable strategy (ESS), 22–24, 56–57
Evolutionary invasion analysis, 24
Experimental Lakes Area (ELA), 250–251
Exploitation, 162–163, 173

F

Fish predators, 42
Fisheries, 157–158, 215–218
   See also Macroecology
   CPUE, 218, 219f
   global, 160
   MPAs on CPUEs, 219–220
   oceanography, 170
   spillover, 218
"Fishing the line" mechanism, 218
Freshwater, 41, 44
   food webs, 141–145
   source–sink dynamics in, 144–146
Freshwater biomonitoring, 241
   aquatic biomonitoring and conservation developments, 242
      ecosystem-level metrics, 244–245
      hypothetical ordination, 245f
      linking site contemporary biomonitoring data, 244f
      RIVPACS, 242–243
   categorizing continuous variables, 248b
   future advances and new perspectives, 251–252
      eco-evolutionary dynamics and feedbacks, 253
      mapping services onto food web, 252f
      network-based approaches, 252–253
   macrofaunal communities, 258–260
   metazoans, functional analysis, 258–260
   microbes, functional analysis, 258–260
   novel molecular and microbial approaches, 253–255
      16S rRNA gene, 258
      biomarker discovery evolution, 255f
      high-throughput sequencing, 257–258
      microorganism, 256–257
      molecular signatures, 255
      NGS technologies, 256
      omics approaches, 257f
   from species traits to community and ecosystem, 245–246
      ecology and species distribution evolution, 249f
      ecology of species, 247
      functional diversity, 249–250
      functional indicators, 250
Freshwaters ecosystem, 127
   See also Marine ecosystem
   B–EF relationships, 136–139
   climate change, 127–130
      synergies between multiple stressors and modulation, 147–149
      traits and functional diversity, 141
   cross-system subsidies, 144–146
   currency of, 143–144
   eco-evolutionary dynamics, 147
   food web, 133f
   freshwater food webs, 141–143
   metabolism, 130–133
   MTE, 133–136
   network approaches, 129
   scaling up in attempts to glimpse, 128f
   source–sink dynamics, 144–146
   temperature, 130–133
Functional diversity, 88, 138–139, 249–250, 258–259
   in climate change, 141
   MPAs on, 226
      global climate change, 228

Functional diversity (*Continued*)
  with higher species diversity, 227
  positive effects on, 229
  trait-based multivariate measures, 227–228
  quantification, 224–225
  spatial protection of, 225–226
Functional feeding groups, 141–142
Functional genomics, 255
Functional redundancy, 224–225

G

Gasterosteus aculeatus. *See* Threespine stickleback
Geographical distribution, 165–167
Gibbs distribution, 100–102
Global Lakes Ecological Observatory Network (GLEON), 250–251
Global warming
  See also Climate change
  climate change components, 131f
  emerging research themes in literature, 132f
  food web, 133f
  master variables in biological responses to, 130–133
Graph theory, 76–77, 89–90

H

High-predation streams, 42
High-throughput sequencing, 257–258
Hotter-is-better pattern, 10–11
Human disturbance, 115–116, 118–119
Human well-being, 213–215
Hyperstability of catch rates, 166

I

IEAs. *See* Integrated Ecosystem Assessments
In-degree, 95
Individual metabolic rate, 3, 6–11
Integrated approach, 130
Integrated Ecosystem Assessments (IEAs), 179
International Union for Conservation of Nature (IUCN), 167–168
Intraspecific consumer biodiversity, 37, 39t–40t
  case studies, 38–45
  ecological dynamics, 37–38
  meta-analysis, 45–46
  overall ecological effects, 46f
  replacing or removing consumer species, 47f
  rma function, 46
  rule of thumb, 46–47
  species effects, 46–48
  study organisms, 38f
Island biogeography model of MW, 85–86
IUCN. *See* International Union for Conservation of Nature

J

Jaccard index, 98–99

K

k-neighborhood metric, 91–92

L

Lagrange multipliers, 100–103
Landlocked alewife populations, 41–42
Latitudinal diversity gradient, 163
Levins model, 82
Linkage density, 93
Long-Term Ecological Research (LTER), 250–251
Long-term ecosystem function, MPAs on biodiversity, 222–224
  functional diversity, 226–229
    quantification, 224–225
    spatial protection of, 225–226
  and multiple services provision, 221–222
Lough Hyne consumers
  ecosystem service provisioning, 203
  trophic flexibility, 195–196, 202
  trophic uniqueness, 195–196, 202
Lough Hyne data set, 194–195
Low-predation streams, 42
LTER. *See* Long-Term Ecological Research

M

MacArthur and Wilson (MW), 85–86
Macroecology, 157
  aquatic ecosystem, 159–160
  ecological and evolutionary effects of fisheries, 177–179
  global annual capture fisheries, 159f
  global fisheries, 160
  macroecological variables, 158f
  in terrestrial system, 157–158

traits-based focus on aquatic functional
diversity, 173—177
variables and interactions within aquatic
ecosystems, 161
abundance, 164—165
body size, 167—168
geographical distribution, 165—167
species richness, 162—164
Macrofaunal communities, 258—260
Malthusian parameter (r_{max}), 8—9
Marine ecosystem, 211—212
See also Freshwaters ecosystem
consumer
trophic flexibility, 193
trophic niche position, 193
trophic uniqueness, 193
ecological network theory, 200—201
ecosystem service provisioning, 196
by Lough Hyne species, 203
ecosystem services assignment on species
level, 194
functional redundancy, 189—191
human activity, 189
impacts on, 196—200
Lough Hyne data set, 194—195
relationships
ecosystem service and trophic level,
body size, and mobility, 199f
trophic flexibility and trophic level, body
size, and mobility, 197f
trophic uniqueness and trophic level,
body size, and mobility, 198f
statistical analysis, 195
trophic flexibility
distribution, 202
Lough Hyne consumers, 195—196, 202
trophic niche dimensions and parameters,
192—193, 201—202
trophic uniqueness
distribution, 202
Lough Hyne consumers, 195—196, 202
Marine protected areas (MPAs), 211—213
on cultural service, 220
on ecosystem services, 215, 216t—217t
ecosystem services, 213—215
key directions and open questions,
229—230
link to human well-being, 213—215
on long-term ecosystem function and
multiple services provision, 221—229
on provisioning services, 215—220
research, 213

on supporting services, 221
Mass effect, 82, 83f, 103
Levins model, 82
rescue effect and propagule rain, 82—84
richness—isolation patterns, 84
Maximum entropy formalisms (MaxEnt
formalisms), 76—77, 99—100
empirical macroscopic constraints,
100—102
Lagrange multipliers, 100—104
R software package, 104
SS paradigm, 103
Mesocosm studies, 44
Meta-analyses, 117
Metabolic rate
evolution and thermal physiology, 10—11
to fitness, 8—9
effect of body size in aquatic
ecosystems, 10
Boltzmann—Arrhenius equation, 9—10
size-and-temperature dependence, 6
decreases in, 8
PTR, 8
size-scaling component, 6—8
effect of temperature on individual
physiology, 7f
Metabolic theory for species interactions, 12
attack coefficient, 14
biomechanical and metabolic
dependencies, 14
consumer and resource individuals, 12—13
detection region, 13
mismatches between species in thermal
responses of traits, 16f
resource mass-specific handling time,
15—16
Metabolic theory of ecology (MTE), 5,
130—131, 133—134
ecological models, 135—136
temperature—size rules, 134
trait-based approaches, 135
Metacommunity networks, 76, 89—90
community-level properties, 90
centrality indices and calculation with R
program, 92, 92t
centrality metrics, 90—92
estimation of alternative indices of
centrality, 91f
mass effect, 82, 83f
Levins model, 82
rescue effect and propagule rain, 82—84
richness—isolation patterns, 84

Metacommunity networks (*Continued*)
  metacommunity-level properties, 92–94
  methodologies for estimation, 96
    local communities, 98–99
    maximization of coherence with community attribute, 99
    MST, 96–98, 97f
    percolation network for temporary pond system, 98f
    potential metacommunity network, 97f
  neutral mechanisms, 85
    island biogeography model of MW, 85–86
    modeling and empirical parameterization, 87
    neutral and niche theories, 86–87
    predictions, 87f
    SAD, 86
  paradigms, 77, 78f
  patch dynamics, 79
    combination of demographic parameters, 81–82
    extinction rates, 81
    predictions of metapopulation models, 80f–81f
    superior competitor species, 79–81
  species sorting, 84
    gradient of dispersal rates, 85
    local communities, 84–85
  theory data, 88
    degree of isolation of local community, 89
    spatial autocorrelation in community structure, 88–89
  weighted, 94–95
    adjacency matrix, 95
    degree centrality, 95
    immigration rates, 96
    patch isolation, 95–96
Metacommunity-level properties, 92–93
  See also Community-level properties
  characteristic structural properties of network, 93
  degree distribution, 93–94
  modularity, 94
  topological metric, 94
Metazoans, functional analysis, 258–260
Microbes, functional analysis, 258–260
Minimum spanning tree (MST), 96–98, 97f
Modularity, 94
MPAs. *See* Marine protected areas
MST. *See* Minimum spanning tree

MTE. *See* Metabolic theory of ecology
MW. *See* MacArthur and Wilson

N

National Ecological Observatory Network (NEON), 250–251
Natural force, 171–173
NEON. *See* National Ecological Observatory Network
Neutral dynamics, 103
Neutral mechanisms, 85
  island biogeography model of MW, 85–86
  modeling and empirical parameterization, 87
  neutral and niche theories, 86–87
  predictions, 87f
  SAD, 86
Next-generation sequencing (NGS), 129, 253, 254b
Niche theories, 86–87, 189–191
Nontransgressive overyielding, 55–56

O

Omics technologies, 254b
Out-degree, 95

P

Pale chub (*Zacco platypus*), 38f, 39t–40t
  within-population variation in feeding behavior in, 45
Partially substitutable resources, 57
  See also Essential resources
  chemical nutrients, 57
  ESS uptake rates, 60
  functional responses of consumers, 59
  resource compositional ratios, 59
Patch dynamics, 79, 103
  combination of demographic parameters, 81–82
  extinction rates, 81
  predictions of metapopulation models, 80f–81f
  superior competitor species, 79–81
Patch isolation, 95–96
Phase shifts, 117
Photorespiration, 9–10
Physical and biological associations with patterns, 170
Physiological temperature range (PTR), 6, 8
Poecilia reticulata. *See* Trinidadian guppy
Population dynamics, 3

Prey mobility, 189–191, 201–202
Prey size, 189–191, 201–202
Prey trophic position, 189–191
Prochilodus mariae (*P. mariae*), 177
Provisioning services, MPAs on, 215–220
PTR. *See* Physiological temperature range

R

Recreational activities, 220
Resilience, 115–117
  solutions to ensure, 121
Resource competition models, 57, 58f
  essential resources, 60
    conditions for stable coexistence of two consumers, 60
    ESS uptake rates, 62
    functional response, 60
  functional responses, 57
  partially substitutable resources, 57
    chemical nutrients, 57
    ESS uptake rates, 60
    functional responses of consumers, 59
    resource compositional ratios, 59
River Invertebrate Prediction and Classification System (RIVPACS), 242–243
Rosenzweig–MacArthur type equations, 19
Rubisco enzyme-catalyzed carboxylation, 9–10
Rule of thumb, 46–47
Runge–Kutta integration method, 62

S

SAD. *See* Species abundance distribution
Sampling effect. *See* Selection effect
Saprobien system, 242
SAR. *See* Species area relationships
Selection effect, 53–54
Shannon diversity measure, 100–102
Source–sink dynamics in freshwaters, 144–146
Spatial autocorrelation in community structure, 88–89
Spatial scale, 157, 160, 168–170
Species abundance distribution (SAD), 86, 164
Species area relationships (SAR), 162–163
Species richness, 162–164
Species sorting, 84
  gradient of dispersal rates, 85
  local communities, 84–85

Species sorting paradigm (SS paradigm), 103
Spillover, 218
SS paradigm. *See* Species sorting paradigm
Stable states, 117
Stoichiometry, 56
STReam Experimental Observatory Network (STREON), 250–251
Structural relationships, 170–171
Subgraph centrality, 91–92

T

Temperature, 3, 5
Thermal physiology, 10–11
Threespine stickleback (*Gasterosteus aculeatus*), 38f, 39t–40t
  foraging habitat divergence, 44
Topological metric, 94
Traits, 3–6, 4f
  variation, 158–159, 174–176
Transgressive overyielding, 55–56
Trinidadian guppy (*Poecilia reticulata*), 38f, 39t–40t
  life history divergence, 42–43
Trophic cascade, 41, 43, 45
Trophic flexibility, 191
  distribution, 202
  Lough Hyne consumers, 195–196, 202
Trophic interactions, 5
Trophic niche, 189–191
  consumer
    trophic flexibility, 193
    trophic niche position, 193
  dimensions and parameters, 192–193, 201–202
  ecosystem service assignment, 194
Trophic niche width. *See* Consumer trophic flexibility
Trophic uniqueness, 191
  consumer, 193
  distribution, 202
  Lough Hyne consumers, 195–196

W

Weighted metacommunity networks, 94–95
  adjacency matrix, 95
  degree centrality, 95
  immigration rates, 96
  patch isolation, 95–96

Z

Zacco platypus. *See* Pale chub